Advances in Polyhydroxyalkanoate (PHA) Production

Special Issue Editor
Martin Koller

MDPI • Basel • Beijing • Wuhan • Barcelona • Belgrade

MDPI

Special Issue Editor

Martin Koller
University of Graz
Austria

Editorial Office
MDPI AG
St. Alban-Anlage 66
Basel, Switzerland

This edition is a reprint of the Special Issue published online in the open access journal *Bioengineering* (ISSN 2306-5354) from 2016–2017 (available at: http://www.mdpi.com/journal/bioengineering/special_issues/PHA).

For citation purposes, cite each article independently as indicated on the article page online and as indicated below:

Author 1; Author 2. Article title. *Journal Name* **Year**, *Article number*, page range.

First Edition 2017

ISBN 978-3-03842-637-0 (Pbk)
ISBN 978-3-03842-636-3 (PDF)

Table of Contents

About the Special Issue Editor

Martin Koller, DI Dr. techn. completed his doctoral thesis under the supervision of Professor Gerhart Braunegg, one of the most eminent PHA pioneers, on PHA production from dairy surplus streams; these activities were embedded in the EU-FP5 project WHEYPOL. Afterwards, he became an experienced researcher in the PHA field, working on the development of continuous and discontinuous fermentation processes and novel downstream processes for sustainable PHA recovery. His research focus is the cost-efficiency of PHA production from surplus materials by eubacteria and haloarchaea. He has numerous articles in high ranked scientific journals, has authored several chapters in scientific books, has given numerous lectures at international conferences, and is an active member of the Editorial Boards of distinguished scientific journals. From 2010 to 2012, he acted as coordinator of the EU-FP7 granted project ANIMPOL, which investigated the conversion of waste streams from the animal processing industry towards structurally diversified PHAs and marketable follow-up products.

Preface to "Advances in Polyhydroxyalkanoate (PHA) Production"

At the present time, it is generally undisputed that alternatives for innumerable fossil-resource based products such as plastics are needed in our society. Plastics, a group of non-natural polymeric materials, are presently produced at growing quantities, currently exceeding 300 Mt per year. Especially in strongly emerging economies, such as in China, Brazil, India, and many other countries, plastic production and consumption is increasing almost exponentially. Such "traditional" plastics are applied in countless fields, inter alia the packaging sector, the automotive industry, the biomedical field, electronics, etc. In spite of their undisputed benefit to our every-day life, the present-day plastic industry faces some crucial problems, such as the ongoing depletion of fossil feedstocks, growing piles of recalcitrant plastic waste, and elevated CO_2 and toxin concentration in the atmosphere originating from plastic incineration.

To overcome the detrimental effects of the "Plastic Age" in which mankind lives today, it will be essential to finally switch from petrol-based plastics to bio-inspired alternatives. Most of all, these alternatives are expected to be based on renewable carbon feed stocks, and, after their life span, shall undergo biodegradation and composting. Among these candidates, polyhydroxyalkanoates (PHAs) are considered the most auspicious. As typical microbial materials, PHAs are produced by numerous prokaryotes as intracellular high-molecular inclusion bodies acting as carbon- and energy-storage compounds; under unfavorable environmental conditions such as nutrient starvation or exogenous stress factors, these secondary metabolites make the cells harboring them fitter for survival of such periods of famine and stress. From the technological point of view, PHAs, bio-based, bio-synthesized, bio-compatible, and bio-degradable in their nature, have the potential to replace traditional plastics in several sectors of the plastic market, mainly in the packaging field, but also in various niches such as the biomedical field. These multifaceted aspects of PHAs were my personal motivation for about twelve years of research in this field, focusing on both quality aspects of PHAs and enhancement of their economic competitiveness.

Currently, economic threats hamper the broad market penetration of PHAs: production of these materials is still significantly more expensive if compared to the well-established, large-scale manufacturing of petrol-based plastics. In particular, the allocation of feedstocks needed for PHA production drastically contributes to the still high production price of these biopolyesters. However, to make PHAs competitive, they must compete with traditional plastics both in terms of material performance and in economic aspects. Quality improvement of PHA-based materials is currently achieved by the selection of the microbial production strain, and, most of all, by adapting the microbial feeding strategy during the bioprocess allowing the generation of differently composed PHA homo-, co-, ter-, and quad-polyesters. Further, PHAs can be processed to design products of tailored properties by designing PHA-based (nano)composites with diverse compatible organic or inorganic materials, by blending PHAs with other biological or artificial polymeric materials, or by post-synthetic modification. Such composites and blends often constitute smart materials with tunable characteristics such as density, gas barrier properties, or degradability. It has to be emphasized that the entire PHA production chain, which starts with the screening and isolation of new powerful microbial productions strains, the selection of carbon-rich feedstocks, the optimization of the biotechnological fermentation strategy in suitable bioreactor facilities, details of the process engineering, and, last but not least, downstream processing for product recovery, needs to be in accordance with the criteria of sustainability. The previously often-cited myth of biopolymers being per se more sustainable than established petrochemical plastics nowadays has more or less vanished from scientific considerations, as manifested by recent studies. Without taking into account the entire life cycle of biopolymers, it is inadmissible to assume that they inherently outperform their petrochemical opponents in terms of the environmental impact. A polymer is not qualified as a "green plastic" purely on the basis of it being compostable! Vise versa, a non-degradable polymer based on renewable resources, such as "bio-poly(ethylene)" does not feature any environmental advantage over fossil-based poly(ethylene); on the contrary, the production of several so-called "green plastics" such as "bio-poly(ethylene)" or, in some cases also poly(lactic acid), requires feedstocks that are relevant for human nutrition, which defeats the object of sustainability.

Such economic, sustainability, and quality aspects are addressed in the 15 contributions to this Bioengineering Special Issue entitled "Advances in Polyhydroxyalkanoate (PHA) Production". Six of the articles describe the assessment of inexpensive carbon feedstocks to be used as carbon sources for PHA production. These second- and third-generation feedstocks constitute surplus streams from different industrial branches, such as crude glycerol as a by-product of biofuel production, lignocellulose waste from the food and wood industry, and waste from petrochemical plastic production. Importantly, all these feedstocks do not compete with human nutrition, thus their utilization in technological processes contributes to the preservation of food resources and to food security. Further, the articles show how such alternative feedstocks have to be pre-treated in order to minimize potential inhibitory effects on production strains, thus how to upgrade substance cocktails to biotechnological nutrient sources. Moreover, special emphasis is placed on the screening and assessment of novel PHA-accumulating microorganisms, often stemming from extreme habitats. This encompasses both eubacterial strains and extremophile haloarchaeal species. New intriguing insights are also provided into the metabolic processes of intracellular PHA mobilization during cell division, into metabolic flux analysis of PHA production by mixed microbial cultures, and into novel molecular diagnostic techniques to conveniently trace new PHA production strains from diverse environments. Further, PHA production is presented as an integral part of future (bio)refinery systems, as shown in the case of autotrophic PHA production by cyanobacteria using CO_2 from effluent gases of a coal power plant as a carbon source, or for a case study coupling PHA biosynthesis with the operation of a wastewater treatment plant. Additional contributions address the fine-tuning of PHA composition at the monomeric level during biosynthesis in order to facilitate the processing of the polyesters to vendible items, and give new insights into the processing of novel PHA copolyesters with other biocompatible materials to blends suitable for the generation of scaffolds for tissue engineering.

The contributions to this book were prepared over a period of almost one year, and it was a great pleasure for me to see such diversely focused research groups participating in this mutual publication project. Authors from 15 different countries from four continents made contributions, representing synergistic competences in diverse areas related to PHA research, some already established in this scientific field for years or even decades, and some of currently attracting increasing attention in the scientific community for their recent research and development accomplishments. I am very confident that this book will inspire the future endeavors of researchers worldwide from the diverse scientific fields linked to PHA, such as microbiology, genetics, chemical engineering, fermentation technology and material science. Go forth and make PHA the future "green plastics" of choice!

Martin Koller
Special Issue Editor

bioengineering

MDPI

Editorial

Advances in Polyhydroxyalkanoate (PHA) Production

Martin Koller [1,2]

[1]   Institute of Chemistry, University of Graz, NAWI Graz, Heinrichstrasse 28/III, 8010 Graz, Austria;
      martin.koller@uni-graz.at; Tel.: +43-316-380-5463
[2]   ARENA—Association for Resource Efficient and Sustainable Technologies, Inffeldgasse 21b,
      8010 Graz, Austria

Academic Editor: Liang Luo
Received: 5 September 2017; Accepted: 31 October 2017; Published: 2 November 2017

Abstract: This editorial paper provides a synopsis of the contributions to the *Bioengineering* special issue "Advances in Polyhydroxyalkanoate (PHA) Production". It illustrates the embedding of the issue's individual research articles in the current global research and development landscape related to polyhydroxyalkanoates (PHA). The article shows how these articles are interrelated to each other, reflecting the entire PHA process chain including strain selection, metabolic and genetic considerations, feedstock evaluation, fermentation regimes, process engineering, and polymer processing towards high-value marketable products.

Keywords: bacteria; copolyester; feedstocks; fermentation; haloarchaea; metabolism; mixed microbial cultures; polyhydroxyalkanoate; strain selection; process engineering; processing; pure culture; sustainability; waste streams

1. Introduction

Nowadays, it is generally undisputed that we need alternatives for various fossil-resource based products such as plastics, which make our daily life comfortable. Plastics, a group of polymeric materials not produced by Mother Nature, are currently produced at increasing quantities, now in a magnitude of about 300 Mt per year [1]. Such well-established plastics are used in innumerable fields of application, such as packaging materials, parts in the automotive industry, biomedical devices, electronic parts, and many more. Despite their high impact in facilitating our everyday life, current plastic production faces essential shortcomings, such as the ongoing depletion of fossil resources, growing piles of waste consisting of non-degradable full-carbon backbone plastics, and elevated CO_2 and toxin levels in the atmosphere caused by plastic incineration [2,3].

To overcome the abovementioned evils, the last decades were devoted to find a way out of the fatal "Plastic Age" we live in today. Switching from petrol-based plastics to bio-alternatives with plastic-like properties, which are based on renewable resources, and which can be subjected towards biodegradation and composting, is regarded as one of these exit strategies [3]. Polyhydroxyalkanoates (PHA), microbial storage materials produced by numerous prokaryotes, are generally considered auspicious candidates to replace traditional plastics in several market sectors, such as the packaging field, or in biomedical applications [4–7]. To make PHA competitive, they must compete with petrol-based plastics both in terms of quality and economic aspects. Quality improvement of PHA-based materials is currently achieved by the microbial feeding strategy during the bioprocess [8,9], by the generation of (nano)composites with diverse compatible materials [10,11], or by blending with suitable other polymers [11–13]. Importantly, the entire PHA production chain, encompassing the isolation of new robust productions strains, feedstock selection, fermentation technology, process engineering, bioreactor design, and, last but not least, downstream processing, needs to meet the criteria of sustainability [14]. The previously often-cited myth of biopolymers being per se more sustainable than established petrochemical plastics nowadays has finally been abandoned, as recently comprehensively reviewed; without conceiving the entire life cycle of biopolymers, it is impossible to conclude if they inherently outperform their petrochemical counterparts in terms environmental benefit [15].

Such economic, sustainability, and quality aspects are addressed in the 14 contributions to this special issue, "Advances in Polyhydroxyalkanoate (PHA) Production". This issue evolved during a period of almost one year, and it was an outstanding pleasure for me to see so differently focused research groups participating in this mutual publication project. Authorships from 15 different countries from four continents were selected, having synergistic competences in diverse areas related to PHA research, some of them well-known in this scientific field already for years or even decades, and some of them currently attracting increasing attention in the scientific community for their recent research and development (R&D) achievements.

In principle, these contributions are dedicated to four major impact directions of PHA research:

First, six articles deal with the assessment of inexpensive [16–20] or exotic [21] feedstocks to be used as carbon sources for PHA production. Importantly, these feedstocks constitute carbonaceous (agro)industrial waste streams, such as waste glycerol from biofuel production [16,17], lignocellulose waste from the food industry [18,20] and forestry [19], and even petrochemical plastic waste [21]. These works aim to find alternatives to commonly used feedstocks of value for human nutrition, in order to avoid the current "plate vs. plastic" dispute. Further, the articles show how such alternative feedstocks have to be pre-treated in order to minimize potential inhibitory effects on production strains.

Second, new insights into metabolic processes during intracellular PHA mobilization [22], metabolic flux analysis of PHA production by mixed microbial cultures (MMCs) [23], and novel molecular diagnostic techniques to trace new PHA production strains from diverse environments in a convenient manner [24] are presented.

Third, PHA production is introduced as an integral part of future (bio)refinery systems, as shown in the case of autotrophic PHA production by the effluent gases of a power plant [25], and the coupling of PHA biosynthesis with a wastewater treatment plant (WWTP) [26].

The fourth group of contributions addresses the fine-tuning of PHA composition on the monomeric level to facilitate its processing [27], and describes the processing of new PHA blends with other biocompatible materials to generate scaffolds for tissue engineering [28].

As the roof above all these articles, a comprehensive review makes us familiar with the current state of enhancing the sustainability, economics, and product quality of PHA [29]. The subsequent paragraphs intend to provide a short overview of the individual chapters of this special issue.

2. Individual Contributions

Kourmentza and colleagues provide a comprehensive review on current challenges and opportunities in PHA production. This article covers all hot spots during the multi-facetted PHA production lines. From the microbiological point of view, the application of both pure (monoseptic) cultures and MMCs is addressed. In the case of MMCs, the coupling of PHA to WWTPs is strongly encouraged by the authors. Special emphasis is also dedicated to raw material selection, process design, and the downstream processing for PHA recovery from microbial biomass. Regarding raw materials, the authors suggest abundant lignocelluloses as the future materials of choice to run a sustainable PHA production facility, and discuss recent advances in using toxic substrates like aromatic compounds, which would provide for bioremediation coupled to PHA biosynthesis. Moreover, halophile microbes are presented as stable production strains; their application should contribute to running PHA production processes at reduced sterility requirements. Finally, the outlook of this review refers to synthetic biology as a tool to achieve competitive PHA production by facilitating downstream processing, and to boost PHA productivity [29].

Takahashi and colleagues screened marine bacteria in order to assess their potential to thrive and accumulate PHA on the inexpensive substrate combination crude glycerol from the biodiesel industry and seawater. Out of 150 isolates, the authors report the identification of two auspicious new marine strains with high potential of PHA production on this substrate combination, which, in the future, should be subjected to detailed investigation and optimization in order to assess their applicability for industrial-scale PHA production [16].

In addition, Bhattacharya and associates used crude glycerol, in this case stemming from *Jatropha curcas*-based biodiesel production, as a carbon substrate for PHA production. Here, a Gram-positive production strain of marine origin, *Bacillus licheniformis* PL26, was used [17]. Such Gram-positive strains display the advantage of generating endotoxin-free PHA especially suitable for in vivo application in the biomedical field [30]. In addition to the intracellular storage product PHA, the authors also investigated the excretion of the extracellular product poly(ε-lysine), a material of significance inter alia for food preservation, by this organism. Regarding PHA production, the authors revealed that this organism accumulates the copolyester poly(3-hydroxybutyrate-*co*-3-hydroxyvalerate) (PHBHV) from waste glycerol as the sole carbon source without the need to add 3-hydroxyvalerate (3HV)-related precursor substrates [17].

Salgaonkar and Bragança investigated *Halogeometricum borinquense*, a new haloarchael PHA producer, as a PHA production strain on hydrolyzed sugarcane bagasse. Using this abundant lignocellulose substrate, *Hgm. borinquense* exhibited a superior performance in terms of PHA productivity and intracellular PHA fraction when compared to other scientifically rather undescribed haloarchaea, which were studied in parallel to this work. Furthermore, this strain was shown to produce a PHBHV copolyester from hydrolyzed bagasse without the need for precursor compounds [18], as also detected in the previous contribution [17]. Generally, this work contributes to the current quest for extremophile PHA producers, which are frequently described as the "rising stars" in the consortium of industrially production strains [31].

A similar substrate was used by Kucera and colleagues, who cultivated two *Burkholderia* strains (*B. cepacia* and *B. sacchari*) on the hydrolysate of spruce sawdust, a lignocellulosic wood waste. Sawdust hydrolysis was carried out both by applying strong acids to hydrolyze the hemicellulose fraction, and enzymatically to hydrolyze the cellulose fraction. This approach generates considerable amounts of fermentable sugars, which are converted by the two applied organisms towards biomass and PHA. Because this hydrolysate also contains growth-limiting components like polyphenols, furfural, acetic acid, or levulinic acid, the authors present a new, convenient upstream processing strategy to remove growth-inhibiting compounds from the hydrolysate by using inexpensive lignite (brown coal) instead of overliming or the commonly used, more expensive charcoal. As a further benefit, the authors suggest the value-added use of lignite after detoxification as an energy carrier [19].

Hokamura et al. used soybean wastewater from a Japanese *miso* production process for the accumulation of an intracellular PHA blend by a recombinant *Pseudomonas* sp. 61-3 harboring two different PHA synthase enzymes. For substrate preparation, soybean wastewater was spray-dried and used as a feedstock with and without subsequent hydrolysis. The intracellular blend consisted of poly(3-hydroxybutyrate) (PHB) homopolyester, and a randomly distributed copolyester consisting of 3-hydroxybutyrate (3HB) and longer 3-hydroxyalkanoates with four to 12 carbon atoms. Using hydrolyzed spray-dried soybean wastewater as the sole carbon source, the highest achieved concentration of this PHA blend amounted to 1 g/L, which corresponds to a PHA fraction of 35% of the cell dry mass. In this case, the blend contained about 80% 3HB and 20% longer building blocks, and displayed a flexible material with considerable toughness comparable to the characteristics displayed by poly(ethylene) (PE) [20].

Johnston and colleagues directly connected the utilization of petrol-based plastic waste with the production of PHA biopolymers. These authors used non-oxidized PE wax as an unconventional, exotic substrate for PHA production by the well-known eubacterial production strain *Cupriavidus necator* H16. In order to make this hydrophobic substrate accessible for the bacteria, the authors presented a viable sonication technique to produce an emulsion, which could be used as cultivation medium to thrive the bacteria. Non-oxidized PE wax displays the advantage of being conveniently produced from waste PE by simple thermal cracking and subsequent purification of the resulting gaseous stream; moreover, it has no other industrial use. Most of all, the authors underlined the independence of this substrate from food and feed applications, making the process ethically clear. In addition, it turned out that a PHBHV copolyester with abbot 11% 3HV in PHA is produced by *C. necator* when supplied with this substrate [15], which is similar to the findings for other new production strains in this special issue [17,18,20].

In contrast to the application of pure, monoseptic microbial cultures described in the above contributions, Montana-Herrera and co-workers used MMCs to study the monomeric composition of MMC-PHA during microbial growth and concomitant PHA accumulation. Different substrate feeding strategies using volatile fatty acids (VFAs) were investigated, showing a dynamic trend in biomass formation and monomeric PHA composition dependent on the substrate feed. Metabolic flux analysis was used to gain deeper insights into the goings-on in the MMC during the cultivation; revealing the correlation of reducing equivalents' generation to the intracellular carbon flux, thus to the PHA composition on the monomeric level, which can be considered as a significant outcome of this study [24].

Karmann and colleagues focused their contribution on the investigation of population dynamics during medium chain length (*mcl*)-PHA production by *Pseudomonas putida* KT2440 at the level of individual cells under different environmental conditions. This work provides a completely new understanding of the mobilization of PHA during cell separation [22]. In contrast to the previously assumed paradigm, which suggested a balanced distribution of PHA granules to new daughter cells generated by binary division, the presented work teaches us that the distribution of PHA granules, often referred to as "carbonosomes" [32], under carbon-limited conditions occurs in an asymmetric manner; the culture segregates into a PHA-rich and a PHA-poor subculture, thus displaying a "bistable behavior" [22].

Morgana de Silva Montenegro and colleagues studied the microbial diversity of PHA-producing species by new molecular diagnosis techniques. These authors applied PHA synthase (*phaC*) by using suitable primers based on multiple alignments of PHA synthases from a total of 218 species with deciphered genomes for detecting new potential PHA producers; PHA synthases of type I and IV were used as positive controls to trace new organisms with PHA accumulation capacities. The authors describe the successful application of this new diagnostic technique to identify nine new marine PHA producers out of 16 marine isolates; when screening 37 additional isolates from different environments, about 30% among them were identified as potential PHA producers [23].

Pittmann and Steinmetz studied the production potential for PHA as a by-product of municipal WWTPs. Here, differently composed WWTP sludge lots were investigated as substrates under different operational conditions regarding pH-value, retention time, and withdrawal/refilling for optimized VFA production; short retention time and low withdrawal/refilling rate turned out to be the most beneficial for high VFA formation. In a second stage, generated VFA were used for high and stable production of PHA of constant monomeric composition in a feast/famine regime under fluctuating conditions. The authors suggest that this process, when using the entire quantity of sludge for all municipal WWTP in the European Union, could theoretically provide the feedstock for the production of about 20% of the current global PHA production [26].

Another concept for a PHA-based biorefinery was presented by Troschl and colleagues, who studied the solar-driven autotrophic pilot-scale cultivation of cyanobacteria for PHA production over extended periods in a 200-L horizontal tubular photobioreactor. As carbon source, CO_2 from the gaseous effluents of an Austrian coal power plant was used. The authors describe the challenges they were confronted with during process development, and highlighted several issues considered as especially crucial for developing a stable cyanobacterial PHA production process, namely strain selection, CO_2-availability, and process design and automation. Regarding strain selection, the authors suggest the use of rather small, unicellular cyanobacterial species like *Synechocystis* sp., which should be more resistant against shear stress when compared to the well-known filamentous cyanobacterial PHA-producer *Arthrospira* sp. [25].

Coming to defined applications of PHA, the authorship of Puppi et al. presented novel strategies to design new tissue engineering scaffolds by blending the copolyester poly(3-hydroxybutyrate-*co*-3-hydroxyhexanoate) (PHBHHx) with poly(ε-caprolactone) (PCL) as another biocompatible polymer. These blends were processed by the new method of "computer-aided wet-spinning", a novel hybrid-additive manufacturing technique suitable for processing PHA in organic solution. The processing of PHA in solution instead of the thermal treatment normally used for PHA processing avoids adverse effects on PHA molecular mass. This technique successfully provided customer-made scaffolds with pre-defined architecture regarding macro- and micro-porosity. The high biocompatibility of these scaffolds was demonstrated by showing the successful adhesion and proliferation of pre-osteoblast cells on them, underlining their suitability for biomedical applications [28].

Poly(3-hydroxybutyrate-*co*-4-hydroxybutyrate) copolyesters with tailored 4-hydroxybutyrate (4HB) fraction, suitable for convenient processing, were produced in our laboratories in Graz, Austria, by Miranda de Sousa Dias et al. This was achieved by co-feeding the direct sucrose converter *B. sacchari* with sucrose from the Brazilian sugarcane industry and fine-tuned amounts of the 4HB-precursor γ-butyrolactone. The copolyesters were generated in fed-batch bioreactor setups at high productivity, and were subjected to detailed material characterization to evaluate their physicochemical properties and molecular mass distribution. As major outcome, it was proposed that the strain could act as one of the major industrial-scale PHA copolyester producers based on sucrose. As the conditio sine qua non for economic feasibility, the integration of the PHA production facilities into existing sugar production lines is necessary [27].

3. Conclusions

Global research efforts are currently devoted to the individual aspects needed to be addressed in order to facilitate the quick success of PHA-based materials on the polymer market. I hope that reading the *Bioengineering* special issue at hand will motivate and inspire researchers all over the world (and undergraduates interested in getting their feet on the ground of biopolymers!) to dedicate intensified efforts to further improve PHA production in terms of economics, product quality, and sustainability.

Acknowledgments: Special credits go to all contributors to this special issue for their outstanding articles, and to all referees active in critically analyzing and improving them. At this point, special thanks go to Professor Gerhart Braunegg, one of the global pioneers in PHA research, for his scientific achievements and his year-long guidance of my PHA-related research.

Conflicts of Interest: The authors declare no conflict of interest.

References

1.   Možejko-Ciesielska, J.; Kiewisz, R. Bacterial polyhydroxyalkanoates: Still fabulous? *Microbiol. Res.* **2016**, *192*, 271–282. [CrossRef] [PubMed]
2.   Braunegg, G.; Bona, R.; Koller, M. Sustainable polymer production. *Polym. Plast. Technol. Eng.* **2004**, *43*, 1779–1793. [CrossRef]
3.   Iwata, T. Biodegradable and bio-based polymers: Future prospects of eco-friendly plastics. *Angew. Chem. Int. Ed.* **2015**, *54*, 3210–3215. [CrossRef] [PubMed]
4.   Keshavarz, T.; Roy, I. Polyhydroxyalkanoates: Bioplastics with a green agenda. *Curr. Opin. Microbiol.* **2010**, *13*, 321–326. [CrossRef] [PubMed]
5.   Tan, G.Y.A.; Chen, C.L.; Li, L.; Ge, L.; Wang, L.; Razaad, I.M.N.; Li, Y.; Zhao, L.; Mo, Y.; Wang, J.Y. Start a research on biopolymer polyhydroxyalkanoate (PHA): A review. *Polymers* **2014**, *6*, 706–754. [CrossRef]
6.   Zinn, M.; Witholt, B.; Egli, T. Occurrence, synthesis and medical application of bacterial polyhydroxyalkanoate. *Adv. Drug Deliv. Rev.* **2001**, *53*, 5–21. [CrossRef]
7.   Chen, G.Q. A microbial polyhydroxyalkanoates (PHA) based bio-and materials industry. *Chem. Soc. Rev.* **2009**, *38*, 2434–2446. [CrossRef] [PubMed]

8. Aziz, N.A.; Huong, K.H.; Sipaut, C.S.; Amirul, A.A. A fed-batch strategy to produce high poly (3-hydroxybutyrate-*co*-3-hydroxyvalerate-*co*-4-hydroxybutyrate) terpolymer yield with enhanced mechanical properties in bioreactor. *Bioproc. Biosyst. Eng.* **2017**. [CrossRef] [PubMed]

9. Koller, M.; de Sousa Dias, M.; Rodríguez-Contreras, A.; Kunaver, M.; Žagar, E.; Kržan, A.; Braunegg, G. Liquefied wood as inexpensive precursor-feedstock for bio-mediated incorporation of (*R*)-3-hydroxyvalerate into polyhydroxyalkanoates. *Materials* **2015**, *8*, 6543–6557. [CrossRef] [PubMed]

10. Khosravi-Darani, K.; Bucci, D.Z. Application of poly(hydroxyalkanoate) in food packaging: Improvements by nanotechnology. *Chem. Biochem. Eng. Q.* **2015**, *29*, 275–285. [CrossRef]

11. Martínez-Sanz, M.; Villano, M.; Oliveira, C.; Albuquerque, M.G.; Majone, M.; Reis, M.; Lopez-Rubio, A.; Lagaron, J.M. Characterization of polyhydroxyalkanoates synthesized from microbial mixed cultures and of their nanobiocomposites with bacterial cellulose nanowhiskers. *New Biotechnol.* **2014**, *31*, 364–376. [CrossRef] [PubMed]

12. Pérez Amaro, L.; Chen, H.; Barghini, A.; Corti, A.; Chiellini, E. High performance compostable biocomposites based on bacterial polyesters suitable for injection molding and blow extrusion. *Chem. Biochem. Eng. Q.* **2015**, *29*, 261–274. [CrossRef]

13. Jost, V.; Kopitzky, R. Blending of polyhydroxybutyrate-co-valerate with polylactic acid for packaging applications-reflections on miscibility and effects on the mechanical and barrier properties. *Chem. Biochem. Eng. Q.* **2015**, *29*, 221–246. [CrossRef]

14. Koller, M.; Maršálek, L.; Miranda de Sousa Dias, M.; Braunegg, G. Producing microbial polyhydroxyalkanoate (PHA) biopolyesters in a sustainable manner. *New Biotechnol.* **2017**, *37*, 24–38. [CrossRef] [PubMed]

15. Narodoslawsky, M.; Shazad, K.; Kollmann, R.; Schnitzer, H. LCA of PHA production—Identifying the ecological potential of bio-plastic. *Chem. Biochem. Eng. Q.* **2015**, *29*, 299–305. [CrossRef]

16. Takahashi, R.Y.U.; Castilho, N.A.S.; Silva, M.A.C.D.; Miotto, M.C.; Lima, A.O.D.S. Prospecting for marine bacteria for polyhydroxyalkanoate production on low-cost substrates. *Bioengineering* **2017**, *4*, 60. [CrossRef] [PubMed]

17. Bhattacharya, S.; Dubey, S.; Singh, P.; Shrivastava, A.; Mishra, S. Biodegradable polymeric substances produced by a marine bacterium from a surplus stream of the biodiesel industry. *Bioengineering* **2016**, *3*, 34. [CrossRef] [PubMed]

18. Salgaonkar, B.B.; Bragança, J.M. Utilization of sugarcane bagasse by *Halogeometricum borinquense* strain E3 for biosynthesis of poly(3-hydroxybutyrate-*co*-3-hydroxyvalerate). *Bioengineering* **2017**, *4*, 50. [CrossRef] [PubMed]

19. Kucera, D.; Benesova, P.; Ladicky, P.; Pekar, M.; Sedlacek, P.; Obruca, S. Production of polyhydroxyalkanoates using hydrolyzates of spruce sawdust: Comparison of hydrolyzates detoxification by application of overliming, active carbon, and lignite. *Bioengineering* **2017**, *4*, 53. [CrossRef] [PubMed]

20. Hokamura, A.; Yunoue, Y.; Goto, S.; Matsusaki, H. Biosynthesis of polyhydroxyalkanoate from steamed soybean wastewater by a recombinant strain of *Pseudomonas* sp. 61-3. *Bioengineering* **2017**, *4*, 68. [CrossRef] [PubMed]

21. Johnston, B.; Jiang, G.; Hill, D.; Adamus, G.; Kwiecień, I.; Zięba, M.; Sikorska, W.; Green, M.; Kowalczuk, M.; Radecka, I. The molecular level characterization of biodegradable polymers originated from polyethylene using non-oxygenated polyethylene wax as a carbon source for polyhydroxyalkanoate production. *Bioengineering* **2017**, *4*, 73. [CrossRef] [PubMed]

22. Karmann, S.; Panke, S.; Zinn, M. The bistable behaviour of *Pseudomonas putida* KT2440 during PHA Depolymerization under Carbon Limitation. *Bioengineering* **2017**, *4*, 58. [CrossRef] [PubMed]

23. Montano-Herrera, L.; Laycock, B.; Werker, A.; Pratt, S. The evolution of polymer composition during PHA accumulation: The significance of reducing equivalents. *Bioengineering* **2017**, *4*, 20. [CrossRef] [PubMed]

24. Montenegro, E.M.D.S.; Delabary, G.S.; Silva, M.A.C.D.; Andreote, F.D.; Lima, A.O.D.S. Molecular diagnostic for prospecting polyhydroxyalkanoate-producing bacteria. *Bioengineering* **2017**, *4*, 52. [CrossRef] [PubMed]

25. Troschl, C.; Meixner, K.; Drosg, B. Cyanobacterial PHA production—Review of recent advances and a summary of three years' working experience running a pilot plant. *Bioengineering* **2017**, *4*, 26. [CrossRef] [PubMed]

26. Pittmann, T.; Steinmetz, H. Polyhydroxyalkanoate production on waste water treatment plants: Process scheme, operating conditions and potential analysis for German and European municipal waste water treatment plants. *Bioengineering* **2017**, *4*, 54. [CrossRef] [PubMed]

27. Miranda de Sousa Dias, M.; Koller, M.; Puppi, D.; Morelli, A.; Chiellini, F.; Braunegg, G. Fed-batch synthesis of Poly(3-hydroxybutyrate) and poly(3-hydroxybutyrate-*co*-4-hydroxybutyrate) from sucrose and 4-hydroxybutyrate precursors by *Burkholderia sacchari* strain DSM 17165. *Bioengineering* **2017**, *4*, 36. [CrossRef] [PubMed]

28. Puppi, D.; Morelli, A.; Chiellini, F. Additive manufacturing of poly(3-hydroxybutyrate-*co*-3-hydroxyhexanoate)/poly(ε-caprolactone) blend scaffolds for tissue engineering. *Bioengineering* **2017**, *4*, 49. [CrossRef] [PubMed]

29. Kourmentza, C.; Plácido, J.; Venetsaneas, N.; Burniol-Figols, A.; Varrone, C.; Gavala, H.N.; Reis, M.A. Recent advances and challenges towards sustainable polyhydroxyalkanoate (PHA) production. *Bioengineering* **2017**, *4*, 55. [CrossRef] [PubMed]

30. Valappil, S.P.; Boccaccini, A.R.; Bucke, C.; Roy, I. Polyhydroxyalkanoates in Gram-positive bacteria: Insights from the genera *Bacillus* and *Streptomyces*. *Antonie Van Leeuwenhoek* **2007**, *91*, 1–17. [CrossRef] [PubMed]

31. Yin, J.; Chen, J.C.; Wu, Q.; Chen, G.Q. Halophiles, coming stars for industrial biotechnology. *Biotechnol. Adv.* **2015**, *33*, 1433–1442. [CrossRef] [PubMed]

32. Jendrossek, D.; Pfeiffer, D. New insights in the formation of polyhydroxyalkanoate granules (carbonosomes) and novel functions of poly(3-hydroxybutyrate). *Environ. Microbiol.* **2014**, *16*, 2357–2373. [CrossRef] [PubMed]

Review

Recent Advances and Challenges towards Sustainable Polyhydroxyalkanoate (PHA) Production

Constantina Kourmentza [1,*], Jersson Plácido [2], Nikolaos Venetsaneas [3,4], Anna Burniol-Figols [5], Cristiano Varrone [5], Hariklia N. Gavala [5] and Maria A. M. Reis [1]

[1] UCIBIO-REQUIMTE, Department of Chemistry, Faculdade de Ciências e Tecnologia/Universidade Nova de Lisboa, 2829-516 Caparica, Portugal; amr@fct.unl.pt
[2] Centre for Cytochrome P450 Biodiversity, Institute of Life Science, Swansea University Medical School, Singleton Park, Swansea SA2 8PP, UK; j.e.placidoescobar@swansea.ac.uk
[3] Faculty of Engineering and the Environment, University of Southampton, Highfield, Southampton SO17 1BJ, UK; n.venetsaneas@aston.ac.uk
[4] European Bioenergy Research Institute (EBRI), Aston University, Aston Triangle, Birmingham B4 7ET, UK
[5] Department of Chemical and Biochemical Engineering, Center for Bioprocess Engineering, Søltofts Plads, Technical University of Denmark, Building 229, 2800 Kgs. Lyngby, Denmark; afig@kt.dtu.dk (A.B.-F.); cvar@kt.dtu.dk (C.V.); hnga@kt.dtu.dk (H.N.G.)
* Correspondence: ckourmentza@gmail.com or c.kourmentza@fct.unl.pt; Tel.: +351-917-900-929

Academic Editor: Martin Koller
Received: 30 April 2017; Accepted: 9 June 2017; Published: 11 June 2017

Abstract: Sustainable biofuels, biomaterials, and fine chemicals production is a critical matter that research teams around the globe are focusing on nowadays. Polyhydroxyalkanoates represent one of the biomaterials of the future due to their physicochemical properties, biodegradability, and biocompatibility. Designing efficient and economic bioprocesses, combined with the respective social and environmental benefits, has brought together scientists from different backgrounds highlighting the multidisciplinary character of such a venture. In the current review, challenges and opportunities regarding polyhydroxyalkanoate production are presented and discussed, covering key steps of their overall production process by applying pure and mixed culture biotechnology, from raw bioprocess development to downstream processing.

Keywords: polyhydroxyalkanoates; biopolymers; renewable feedstock; mixed microbial consortia; enrichment strategy; pure cultures; synthetic biology; downstream processing

1. Introduction

Polyhydroxyalkanoates (PHAs) are a class of renewable, biodegradable, and bio-based polymers, in the form of polyesters. Together with polylactic acid (PLA) and polybutylene succinate (PBS), they are considered the green polymers of the future since they are expected to gradually substitute conventional plastics with similar physicochemical, thermal, and mechanical properties such as polypropylene (PP) and low-density polyethylene (LDPE) [1,2]. While PLA and PBS are produced upon polymerization of lactic and succinic acid respectively, PHA polymerization is performed naturally by bacteria.

A wide variety of bacteria are able to accumulate PHAs in the form of intracellular granules, as carbon and energy reserves. PHA accumulation is usually promoted when an essential nutrient for growth is present in limited amount in the cultivation medium, whereas carbon is in excess. Although, several bacteria are able to produce PHAs during growth and do not require growth-limiting conditions. This carbon storage is used by bacteria as an alternate source of fatty acids, metabolized under stress conditions, and is the key mechanism for their survival [3]. Up to 150 different PHA structures have been identified so far [4]. In general, PHAs are classified into two groups according to the carbon atoms that comprise their monomeric unit. Short-chain-length PHAs (scl-PHAs) consist of 3–5 carbon atoms, whereas medium-chain-length PHAs (mcl-PHAs) consist of 6–14 carbon atoms. PHB, the most well-known scl-PHA member, is characterized as a stiff and brittle material and is difficult to be processed due to its crystalline nature. The incorporation of 3-hydroxyvalerate (HV) units in PHB, results in the production of the copolymer poly(3-hydroxybutyrate-co-3-hydroxyvalerate), or else PHBV. PHBV is a material that becomes tougher, more flexible, and broader to thermal processing when its molar fraction in the copolymer increases [5]. scl-PHAs are mostly used for the production of disposable items and food packaging materials. On the other hand, mcl-PHAs are characterized as elastomers and they are suitable for high value-added application, such as surgical sutures, implants, biodegradable matrices for drug delivery, etc. [4].

Taking into account the recalcitrance of conventional plastics in the environment, replacement of synthetic plastics with PHAs would have huge benefits for the society and the environment [6]. Wide commercialization and industrialization of PHAs is still struggling due to their high production cost, resulting in higher prices compared to conventional polymers. While the price of polymers such as PP and PE is around US$0.60–0.87/lb, PHA biopolymer cost is estimated to be 3–4 times higher, ranging between US$2.25–2.75/lb [7,8]. Although several companies have initiated and industrialized the production of PHAs, as presented in Table 1, there are still major issues that need to be addressed in an effort to reduce the overall production cost. The main reasons for their high cost is the high price of high purity substrates, such as glucose, production in discontinuous batch and fed-batch cultivation modes, and large amount of solvents and/or labor regarding their downstream processing. The increasing availability of raw renewable materials and increasing demand and use of biodegradable polymers for bio-medical, packaging, and food applications along with favorable green procurement policies are expected to benefit PHA market growth. According to a recent report, published in 2017, the global PHA market is expected to reach US$93.5 million by 2021, from an estimated US$73.6 million within 2016, characterized by a compound annual growth rate (CAGR) of 4.88% [9].

Table 1. Pilot and industrial scale PHA manufacturers currently active worldwide.

Name of Company	Product (Trademark)	Substrate	Biocatalyst	Production Capacity
Biomatera, Canada	PHA resins (Biomatera)	Renewable raw materials	Non-pathogenic, non-transgenic bacteria isolated from soil	
Biomer, Germany	PHB pellets (Biomer®)	Sugar (sucrose)		
Bio-On Srl., Italy	PHB, PHBV spheres (minerv®-PHA)	Sugar beets	*Cupriavidus necator*	10,000 t/a
BluePHA, China	Customized PHBVHHx, PHV, P3HP3HB, P3HP4HB, P3HP, P4HB synthesis		Development of microbial strains via synthetic biology	
Danimer Scientific, USA	mcl-PHA (Nodax® PHA)	Cold pressed canola oil		
Kaneka Corporation, Japan	PHB-PHHx (AONILEX®)	Plant oils		3500 t/a
Newlight Technologies LLC, USA	PHA resins (AirCarbon™)	Oxygen from air and carbon from captured methane emissions	Newlight's 9X biocatalyst	
PHB Industrial S.A., Brazil	PHB, PHBV (BIOCYCLE®)	Saccharose	*Alcaligenes* sp.	3000 t/a
PolyFerm, Canada	mcl-PHA (VersaMer™ PHA)	Sugars, vegetable oils	Naturally selected microorganisms	
Shenzhen Ecomann Biotechnology Co. Ltd., China	PHA pellets, resins, microbeads (AmBio®)	Sugar or glucose		5000 t/a
SIRIM Bioplastics Pilot Plant, Malaysia	Various types of PHA	Palm oil mill effluent (POME), crude palm kernel oil		2000 t/a
TianAn Biologic Materials Co. Ltd., China	PHB, PHBV (ENMAT™)	Dextrose deriving from corn of cassava grown in China	*Ralstonia eutropha*	10,000 t/a, 50,000 t/a by 2020
Tianjin GreenBio Material Co., China	P (3, 4HB) films, pellets/foam pellets (Sogreen®)	Sugar		10,000 t/a

PHB, P3HB: poly(3-hydroxybutyrate); PHBV: poly(3-hydroxybutyrate-co-3-hydroxyvalerate); PHBVHHx: poly(3-hydroxybutyrate-co-3-hydroxyvalerate-co-3-hydroxyhexanoate); PHV: poly-3-hydroxyvalerate; P3HP3HB: poly(3-hydroxypropionate-co-3-hydroxybutyrate); P3HP4HB: poly(3-hydroxypropionate-co-4-hydroxybutyrate); P3HP: poly(3-hydroxypropionate); P4HB: poly(4-hydroxybutyrate); mcl-PHA: medium-chain length PHA; P(3,4HB): poly(3-hydroxybutyrate-co-4-hydroxybutyrate).

In the following sections, the advantages and drawbacks of PHA production employing both pure and mixed culture biotechnology are presented and discussed, as well as several approaches regarding their downstream processing in order to identify bottlenecks and opportunities to leverage PHA production.

2. PHA Production by Pure Bacterial Cultures

Pure culture biotechnology is implemented on an industrial scale, since a wide variety of food, pharmaceutical, and cosmetic agents derive as metabolic compounds from certain bacterial strains. Within the last few decades, research has been focused on finding ways to decrease the high production cost of PHAs. One of the main contributors to their high cost is the use of high purity substrates, which can account for 45% of the total production cost [10]. Instead, renewable feedstocks are being explored and researchers have been developing bioprocesses for the valorization of waste streams and by-products. In addition, current legislation and policies promote biodegradable waste management solutions other than disposal in landfills. Since every type of waste stream or by-product has different composition, selecting the appropriate biocatalyst is of great importance. In cases where the raw material is rich in carbon and nutrients, a growth-associated PHA producer would be selected, such as *Alcaligenes latus* or *Paracoccus denitrificans*. Conversely, in cases where the feedstock lacks an essential nutrient for growth such as nitrogen, phosphorous, etc., PHA accumulation using non-growth-associated bacteria would be ideal, i.e., *Cupriavidus necator*.

Apart from well-known species involved in industrial PHA production such as *Alcaligenes latus*, *Cupriavidus necator*, and *Pseudomonas putida*, bacteria need to combine several features in order be selected and regarded as promising PHA producers. Such features include their performance utilizing renewable feedstocks and/or environmental pollutants, seawater instead of fresh water, possibility of PHA production under open, non-sterile conditions, and their potential to develop contamination-free continuous bioprocesses. The use of agricultural byproducts and forest residues as an abundant and renewable source of lignocellulosic material for PHA production, is mainly considered after its physicochemical or biological hydrolysis. However, a few microorganisms possess the ability to saccharify cellulose and simultaneously produce PHAs. Moreover, PHA producers isolated from contaminated sites may be regarded for combined PHA production and bioremediation of toxic pollutants and post-consumer plastics. In addition, microorganisms isolated from hypersaline environments are considered the most promising ones, since they combine several benefits with a huge potential for reducing PHA production cost, namely in the downstream step. Last but not least, within recent years synthetic biology tools are continuously being developed in order to provide solutions to industrial challenges such as maximizing cellular capacity to 'make more space' for PHA accumulation, manipulating PHA composition to design polymers for high value-added applications and enhancing PHA efficiency.

2.1. Lignocellulose Degraders

2.1.1. Saccharophagus Degradans

Saccharophagus degradans, formerly known as *Microbulbifer degradans*, refers to a species of marine bacteria capable of degrading complex marine polysaccharides, such as cellulose of algal origin and higher plant material [11–13]. So far, the only strain reported is *S. degradans* 2-40, that was isolated from decaying marsh cord grass *Spartina alterniflora*, in the Chesapeake Bay watershed [14]. It is a Gram-negative, aerobic, rod-shaped, and motile γ-proteobacterium and is able to use a variety of different complex polysaccharides as its sole carbon and energy source, including agar, alginate, cellulose, chitin, β-glucan, laminarin, pectin, pullulan, starch, and xylan [11,13,15–17].

The key enzymes involved in PHA biosynthesis—β-ketothiolase, acetoacetyl-CoA reductase, and PHA synthase—have been identified in the genome of *S. degradans* 2-40 [12,18]. Preliminary studies have been performed in order to evaluate the feasibility of *S. degradans* to produce PHAs from

D-glucose and D-cellobiose as the sole carbon source in minimal media comprised of sea salts [19]. In addition, the authors evaluated the capability of the strain to degrade lignocellulosic material in the form of tequila bagasse (*A. tequilana*). According to the results obtained, it was shown that *S. degradans* successfully degraded and utilized cellulose as the primary carbon source to grow and produce PHB. However, PHB yields were not reported, so as to evaluate the efficiency of the process, but it became evident that prior hydrolysis of the lignocellulosic material is not required. This is considered positive since it can contribute to up-stream processing cost reduction and thus encourage further research employing the certain strain. In another study, Gonzalez-Garcia et al. [20], investigated PHA production using glucose as the sole carbon source and a culture medium designed according to bacterial biomass and seawater composition. Experiments were performed using a two-step batch strategy, where in the first step bacterial growth was performed under balanced conditions for 24 h, whereas in the second step cells were aseptically transferred to a fresh nitrogen deficient medium and incubated for 48 h. Under these conditions PHA content reached up to $17.2 \pm 2.7\%$ of the cell dry weight (CDW).

PHB biosynthesis from raw starch in fed-batch mode was also investigated and the results were compared to the ones obtained using glucose as the carbon source under the same conditions [21]. When starch was used PHA yield, content, and productivity reached up to 0.14 ± 0.02 g/g, $17.5 \pm 2.7\%$ of CDW and 0.06 ± 0.01 g/L h, respectively. In the case where glucose was fed the respective values were higher but still low compared to other PHA producers. However, only a few microorganisms have been reported to directly utilize raw starch for PHA production [21,22]. During the experiments, the authors observed the simultaneous production of organic acids and exopolymers and this was the main reason for the low PHB accumulation. Higher PHA efficiency could be achieved by optimizing cultivation parameters to drive carbon flux towards PHA biosynthesis and also by applying genetic engineering to knock out genes responsible for the production of side products such as exopolymers.

PHA production in aquarium salt medium supplemented with 1% of different types of cellulosic substrates such as α-cellulose, avicel PH101, sigmacell 101, carboxymethyl cellulose (CMC), and cellobiose have also been studied [23]. In flask experiments, PHB production was 11.8, 14.6, 13.7, and 12.8% of the DCW respectively. Fed-batch cultivation strategy resulted in increased PHB contents reaching up to 52.8% and 19.2% of the DCW using glucose and avicel respectively, as carbon sources.

Recently, another approach towards PHA biosynthesis from *S. degradans* was proposed. During their experiments, Sawant et al. [24] observed that *Bacillus cereus* (KF801505) was growing together with *S. degradans* 2-40 as a contaminant and had the ability of producing high amounts of PHAs [25]. In addition, the viability and agar degradation potential of *S. degradans* increased with the presence of *B. cereus*. Taking those into account, they further investigated the ability of co-cultures of *S. degradans* and *B. cereus* to produce PHAs using 2% w/v agarose and xylan without any prior treatment. PHA contents obtained from agarose and xylan were 19.7% and 34.5% of the DCW respectively when co-cultures were used compared to 18.1% and 22.7% achieved by pure cultures of *S. degradans*. This study reported for the first time the production of PHAs using agarose. Moreover, according to the results, the highest PHA content from xylan was obtained using a natural isolate. So far, only recombinant *Escherichia coli* has been reported to produce 1.1% PHA from xylan, which increased to 30.3% and 40.4% upon supplementation of arabinose and xylose, respectively [26].

These unique features of *S. degradans* open up the possibility to use it as a source of carbohydrases in order to saccharify lignocellulosic materials. Thus, coupled hydrolysis and fermentation is a promising alternative for the production of PHAs using carbon sources that may derive from biomass residues of different origin (Table 2). However, saccharification and coupled PHA production need to be studied in detail in order to understand their potential and find ways to increase the rates of their processes.

2.1.2. Caldimonas Taiwanensis

Caldimonas taiwanensis is a bacterial strain isolated from hot spring water in southern Taiwan in 2004 [27]. Researchers had been searching for thermophilic amylase-producing bacteria since those enzymes are of high industrial importance for the food and pharmaceutical sector. In addition, since starch hydrolysis is known to be faster at relatively high temperatures, thermophilic amylases are usually preferred [28]. Upon morphological and physiological characterization, it was shown that this Gram-negative, aerobic, rod shaped bacteria can form PHB granules.

A few years later, Sheu et al. [22] investigated PHA production from a wide variety of carbon sources. At first cultivation of *C. taiwanensis* on a three-fold diluted Luria-Bertani medium supplemented with sodium gluconate, fructose, maltose, and glycerol as the sole carbon sources under optimal nitrogen limiting conditions, C/N = 30 was performed. PHB contents reached up to 70, 62, 60, and 52% of the CDW at 55 °C in shake flask experiments. In the next step, fatty acids were tested as sole carbon sources for growth and PHA production. It was observed that the strain did not grow at a temperature between 45 °C or 55 °C while no PHA was formed. Nevertheless, when mixtures of gluconate and valerate were provided bacterial growth was feasible and PHA cellular content reached up to 51% of its CDW. The presence of valerate induced the presence of HV units in the polymer resulting in the production of a PHBV copolymer constituting of 10–95 mol % HV depending on the relative valerate concentration in the mixture. Moreover, mixtures of commercially available starches and valerate were evaluated for PHA production at 50 °C. The carbon source mixture consisted of 1.5% starch and 0.05% valerate. Starch types examined were cassava, corn, potato, sweet potato, and wheat starch. After 32 h of cultivation PHBV copolymer was produced in all cases, composed of approximately 10 mol % HV. PHA contents of 67, 65, 55, 52, and 42% of its CDW were achieved respectively.

Despite the fact that biotechnological process using thermophilic bacteria need to be performed at high temperatures, they reduce the risk of contamination. Another advantage is the fact that thermophiles grow faster compared to mesophiles, therefore less time is needed to achieve maximum PHA accumulation [22]. Moreover, employing *C. taiwanensis* for PHA production using starch-based raw materials is extremely beneficial, in economic terms, since no prior saccharification is required. On the other hand, as mentioned before, *C. taiwanensis* cannot grow on fatty acids but when a mixture of valerate and gluconate/or starch is supplied bacterial growth and PHA accumulation occurs in the form of PHBV copolymer. However, the concentration of fatty acids may result in toxicity for bacterial cells, that up to a point can be overcome by the fast growth of cells. In addition, since amylose, amylopectin, and nitrogen contents vary between types of starch, prior characterization needs to be performed. The feasibility of enzymatic degradation of amylose and amylopectin is considered a key factor as it regulates the amount of sugars present in the medium. Last but not least, nitrogen content should be also controlled as high amounts favor biomass growth instead of PHA accumulation.

2.2. Bioremediation Technologies Allowing PHA Production

One of the major causes of environmental pollution is the presence of volatile aromatic hydrocarbons such as benzene, toluene, ethylbenzene, and xylene (BTEX) that are found in crude oil and petroleum products. In addition, huge amounts of starting materials for the production of petrochemical based plastics, such as styrene, are released annually [29]. Moreover, chemical additives in plastics, which are accumulated in the environment due to their recalcitrance, can leach out and are detectable in aquatic environments, dust and air because of their high volatility [30]. Also, textile dyes and effluents are one of the worst polluters of our precious water bodies and soils. All the above are posing mutagenic, carcinogenic, allergic, and cytotoxic threats to all life forms [31].

Since PHAs are known to have a functional role in bacterial survival under stress conditions, toxic environments characterized by poor nutrient availability are proven to be important sources of PHA producers [32]. Several attempts have been made within the last decade to explore contaminated sites as a resource of microorganisms that are expected to advance biotechnological production of PHAs.

Employment of such bacteria combines bioremediation with the production of a high value-added material. So far, bacterial strains that belong to the genus of *Sphingobacterium*, *Bacillus*, *Pseudomonas*, and *Rhodococcus* have been isolated and studied regarding their PHA production potential degrading environmental pollutants, as summarized in Table 2 [3,29,33–39].

Pseudomonas species are characterized by their ability to utilize and degrade a variety of carbon sources due to their wide catabolic versatility and genetic diversity. For these reasons, they are a natural choice regarding techniques of in situ and ex situ bioremediation [40]. Several *Pseudomonas* strains have been isolated from hydrocarbon-contaminated soils and together with other *Pseudomonas* sp. have been examined regarding their ability to produce PHA from hydrocarbons. In their study, Nikodinovic et al. [41], investigated PHA accumulation in several *Pseudomonas* strains from single BTEX aromatic substrates and mixed aromatic substrates as well as mixtures of BTEX with styrene. It was reported that when *P. putida* F1 was supplied with 350 µL of toluene, benzene, or ethylbenzene it accumulated PHA up to 22, 14, and 15% of its CDW respectively, while no growth was observed when *p*-xylene and styrene were supplied as the sole carbon source. In the case of *P. putida* mt-2 no growth was obtained with benzene, ethylbenzene, or styrene but when toluene and *p*-xylene were used its PHA content was 22% and 26% respectively. *P. putida* CA-3 efficiently degraded styrene but could not metabolize any of the other hydrocarbons investigated. However, a defined mixed culture of *P. putida* F1, mt-2, and CA-3 was successfully used for PHA production from BTEX and styrene mixtures, where the highest biomass concentration was achieved and PHA content reached up to 24% of the CDW. In another study, strains from petroleum-contaminated soil samples were screened on their ability to degrade toluene and synthesize mcl-PHA [42]. Among them *P. fluva* TY16 was selected to be further investigated. It was shown that the highest PHA content of 68.5% was achieved when decanoic acid was used as the carbon source. In the case of benzene, toluene, and ethylbenzene PHA contents reached up to 19.1, 58.9, and 28.6% respectively, using a continuous feeding strategy. *Pseudomonas* sp. TN301 was isolated from a river sediment sample from a site in a close proximity to a petrochemical industry [43]. Both monoaromatic and polyaromatic hydrocarbons were examined as PHA precursors and cellular mcl-PHA contents varied between 1.2% and 23% of its CDW, while this study was the first one on the ability of a bacterial strain to convert polyaromatic hydrocarbon compounds to mcl-PHA. Moreover, *Pseudomonas* strains isolated from contaminated soil and oily sludge samples from Iranian southwestern refineries accumulated 20–23% of their CDW to mcl-PHA using 2% v/v crude oil as the sole carbon source [32].

As mentioned before, styrene—used for the synthesis of polystyrene—is a major and toxic environmental pollutant. Ward et al. [29], has reported that *P. putida* CA-3 was capable of converting styrene, its metabolic intermediate phenylacetic acid and glucose into mcl-PHA under nitrogen limited conditions, characterized by conversion yields of 0.11, 0.17, and 0.22 g/g, respectively. However, higher cell density and PHA production, characterized by a conversion yield of 0.28 g/g, were observed when cells were supplied with nitrogen at a feeding rate of 1.5 mg/L/h [37]. Moreover, in a recent study the key challenges of improving transfer and increasing supply of styrene, without inhibiting bacterial growth, were addressed [35]. It was shown that by changing the feed from gaseous to liquid styrene, through the air sparger, release of styrene was reduced 50-fold, biomass concentration was five times higher, while PHA production was four-fold compared to previous experiments, with a PHA content reaching up to 32% in terms of CDW and a conversion yield of 0.17 g/g.

A two-step chemo-biotechnological approach has been proposed for the management of post-consumer polystyrene, involving its pyrolysis to styrene oil and subsequent conversion of the styrene oil to PHA by *P. putida* CA-3 [44]. According to their results, after 48 h 1.6 g of mcl-PHA were obtained from 16 g of oil with a cellular content of 57%. Following the same approach, the solid fraction of pyrolyzed polyethylene terephthalate (PET) was used as feedstock for PHA production by bacteria isolated from soil exposed to PET granules at a PET processing plant [45]. The isolated strains were identified and designated as *P. putida* GO16, *P. putida* GO19, and *P. frederiksbergensis* GO23 and they were able to accumulate mcl-PHA up to 27, 23, and 24% of their CDW respectively, using 1.1 g/L sodium terephthalate as the sole carbon source under conditions of nitrogen limitation. Recently, conversion of polyethylene (PE) pyrolysis wax to mcl-PHAs was investigated employing *P. aeruginosa* PAO-1 [46]. Addition of rhamnolipid biosurfactants in the growth medium had a positive impact on bacterial growth and PHA accumulation. Substitution of ammonium chloride with ammonium nitrate led to faster growth and earlier PHA accumulation that reached up to 25% of its CDW.

A series of studies has been focused on the degradation of textile dyes for PHA production using *Sphingobacterium, Bacillus,* and *Pseudomonas* species. When the dye Direct Blue GLL (DBGLL) was used, *Sphingobacterium* sp. ATM completely decolorized 0.3 g/L in 24 h, while simultaneous polyhydroxyhexadecanoic acid (PHD) occurred reaching up to 64% of its CDW [38]. The potential of *B. odyssey* SUK3 and *P. desmolyticum* NCIM 2112 was also investigated. It was shown that both strains were able to decolorize 0.05 g/L DBGLL by 82% and 86% and produce PHD up to 61 and 52% of their CDW, respectively. In another study, 82% decolorization of 0.8 g/L of the textile dye Orange 3R was feasible, employing *Sphingobacterium* sp. ATM which resulted in the production of 3.48 g/L of PHD and a cellular PHA content of 65% after 48 h [39]. In addition, full decolorization of 0.5 g/L of the textile dye Direct Red 5B (DR5B) was accomplished when the medium was supplemented with glycerol, glucose, starch, molasses, frying oil, and cheese whey. In those cases PHD accumulation was 52, 56, 55, 64, 46, and 10% of its CDW respectively [47].

2.3. Halophiles

Halophiles are microorganisms that require salt for their growth and are categorized, according to their halotorerance, in two groups: moderate (up to 20% salt) and extreme (20–30% salt) halophiles [48]. Their name comes from the Greek word for 'salt-loving' [49] and can be found in the three domains of life: Archaea, Bacteria, and Eukarya. They thrive in marine and hypersaline environments around the globe such as the saline lakes, salt marshes, and salterns [50,51]. With the use of halophiles the risk for contamination is reduced and/or eliminated since non-halophilic microorganisms cannot grow in media containing high salt concentrations. This is of great importance since their use combines the advantages of low energy requirements under unsterile conditions, minimal fresh water consumption, due to its substitution with seawater for medium preparation, and the possibility of operating contamination free continuous fermentation processes that are much more efficient. In addition, downstream processing cost can be reduced by treating the bacterial cells with salt-deficient water in order to cause hypo-osmotic shock [52]. The above, together with the valorization of low cost substrates, bring halophilic bacteria a step closer to being used as biocatalysts for industrial PHA production.

PHA accumulation by halophilic archaea was first observed in 1972 by Kirk and Ginzburg [53]. So far, the most well-known and best PHA halophilic archaeon producer is *Haloferax mediterranei*, which was first isolated from seawater evaporation ponds near Alicante in Spain [54]. Several studies have shown their ability to accumulate high PHA contents utilizing low cost feedstocks. Among them, vinasse, a byproduct of ethanol production from sugarcane molasses, has been utilized [55]. After pre-treatment, via adsorption on activated carbon, 25% and 50% v/v of pre-treated vinasse led to cellular PHA contents of 70% and 66%, respectively. Maximum PHBV (86% HB–14% HV) concentration reached up to 19.7 g/L, characterized by a volumetric productivity of 0.21 g/L h and a conversion yield of 0.87 g/g, for the case where 25% pre-treated vinasse was used. In another study, stillage derived from a rice-based ethanol industry, was investigated [56]. PHA accumulation was 71% of its CDW that led to 16.4 g/L of PHBV (85% HB-15% HV) with a yield coefficient of 0.35 g/g and a volumetric productivity of 0.17 g/L h. Moreover, cheese whey hydrolysate—obtained upon acid pre-treatment—has been used for the production of PHBV with low HV fraction, 1.5 mol % [57]. Batch cultivation of *H. mediterranei* led to the production of 7.54 g/L of biomass, with a PHA content of 53%, and a volumetric productivity of 0.17 g/L h. Olive mill wastewater (OMW), which is a highly polluting waste, was also utilized recently as the sole carbon source for PHA production [58]. Using a medium containing 15% of OMW up to 43% of PHBV/CDW was produced consisting of 6 mol % HV in a one-stage cultivation step.

Halophilic bacteria belong to γ-Proteobacteria and they can grow on a wide range of pH, temperature, and salinity concentrations up to 30% (w/v NaCl) and possess the ability to accumulate PHA [52]. *Halomonas* TD01 has been isolated from Aydingkol Lake in China. This strain was investigated regarding its PHA production potential under unsterile and continuous conditions [59]. Glucose salt medium was used and initial fed-batch cultivation resulted in the production of 80 g/L of biomass with a PHA content of 80%, in the form of PHB, after 56 h. A continuous and unsterile cultivation process was developed that lasted for 14 days, and that allowed cells to grow to an average of 40 g/L containing 60% PHB in the first reactor. Cells were forwarded by continuous pumping from the first to the second reactor that contained nitrogen-deficient glucose salt medium. In the second reactor PHB levels ranged from 65 to 70% of its CDW and a conversion yield of 0.5 g/g was achieved. This was the first attempt for continuous PHA production under non-sterile conditions from a halophilic bacterium. In addition, Yue et al. [60] explored the potential of *Halomonas campaniensis* LS21, isolated from the Dabancheng salt lake in China, to produce PHA in a seawater-based open and continuous process. The strain utilized a mixture of substrates mainly consisting of cellulose, starch, animal fats and proteins. Instead of fresh water fermentation was performed using artificial seawater composed of 26.7 g/L NaCl, among others under a pH around 10 and 37 °C. PHB accumulation reached up to 26% during 65 days of continuous fermentation without any contamination. Through this study the benefits of long-lasting, seawater-based, and continuous processes for PHA production under unsterile conditions were demonstrated.

Bacillus megaterium has recently drawn attention since several studies have isolated such stains from salterns. *Bacillus megaterium* H16, isolated from the solar salters of Ribandar Goa in India, was shown to accumulate up to 39% PHA in the presence (5% w/v) or absence of NaCl using glucose [61]. In another study, a mangrove isolate that was found to belong to *Bacillus* sp. could tolerate salinity up to 9% w/v [62]. The certain strain was able to utilize a wide variety of carbon sources such as monosaccharides, organic acids, acid pre-treated liquor, and lignocellulosic biomass reaching cellular PHA contents of up to 73% of its CDW. Furthermore, *Bacillus megaterium uyuni* S29, isolated from Uyuni salt lake in Bolivia, was examined in terms of its salinity tolerance and impact on biomass and PHB production [63]. It was observed that the strain could grow at 10% w/v NaCl while PHB production was observed even at high salinity levels of 25% w/v. Optimum results for biomass and PHB production were achieved in medium containing 4.5% w/v NaCl and were 5.4 and 2.2 g/L characterized by a yield coefficient of 0.13 g/g and a volumetric PHB productivity of 0.10 g/L h.

The results obtained from the studies described above are very promising and demonstrate the remarkable potential of halophiles for biotechnological production of PHAs. Although, processes performed under high salinity concentrations present disadvantages such as the corrosion of stainless steel fermenters and piping systems [51,52]. However, since no sterilization is required when halophiles are used, other types of low cost materials, such as plastics and ceramics, may be used to design and construct fermentation and piping systems in order to overcome corrosion issues. In addition, the number of halophilic bacteria that their genome is being sequenced is constantly increasing throughout the years. Subsequently, in the near future, molecular biology techniques will result in metabolically engineered strains with better performances regarding their industrial application [48,51].

Table 2. Characteristic parameters describing PHA production from different types of bacteria.

Strain	Carbon Source	PHA	Cultivation Mode	DCW (g L^{-1})	PHA (g L^{-1})	PHA (%)	$Y_{P/S}$	Ref.
	Lignocellulose Degraders							
	Glucose	PHB				17.2		[20]
	Glucose	PHB	Fed-batch	12.7	2.7	21.4	0.17	[21]
	Starch		Fed-batch	11.7	2.0	17.5	0.14	
	Glucose		Flask	2.1	0.46	22.4		
	Cellobiose		Flask	2.0	0.42	20.8		
	α-Cellulose		Flask	1.2	0.14	11.8		
	Avicel		Flask	1.0	0.15	14.6		
S. degradans	Sigmacell	PHB	Flask	1.0	0.14	13.7		[23]
	CMC		Flask	1.1	0.14	12.7		
	Glucose		Batch	1.6	0.40	25.3		
	Glucose		Fed-batch	4.2	2.20	52.8		
	Avicel		Fed-batch	2.1	0.40	19.2		
	Agarose		One-step batch		0.24	18.1		
	Xylan	PHB	One-step batch		0.20	22.7		[24]
	Agarose		Two-step batch		0.31	18.4		
	Xylan		Two-step batch		0.24	15.3		
Co-culture of	Agarose		One-step batch		0.29	19.7		
S. degradans and	Xylan	PHB	One-step batch		0.27	34.5		[24]
B. cerues	Agarose		Two-step batch		0.23	15.3		
	Xylan		Two-step batch		0.33	30.2		
	Propionate + Glc [a]	PHBV (88–12) [f]		2.0	1.04	52		
	Valerate + Glc [a]	PHBV (49–51)		1.0	0.51	51		
	Hexanoate + Glc [a]	PHBHHx [c]		2.7	1.67	62		
	Hexanoate + Glc [a] + AA [b]	PHBHHx [d]		1.2	0.56	47		
	Heptanoate + Glc [a]	PHBV (65–35)		1.7	0.56	33		
C. taiwanensis	Heptanoate + Glc [a] + AA	PHBV (15–85)	Flask	0.3	0.05	17		[22]
	Octanoate + Glc [a]	PHB		0.4	0.05	13		
	Cassava starch + Val [e]	PHBV (87–13)		2.8	1.88	67		
	Corn starch + Val [e]	PHBV (80–10)		3.3	2.14	65		
	Potato + Val [e]	PHBV (80–10)		2.6	1.43	55		
	Sweet potato + Val [e]	PHBV (80–10)		1.6	0.83	52		
	Wheat starch + Val [e]	PHBV (80–10)		4.1	1.72	42		
	Polyhydroxyalkanoates and Bioremediation							
	Benzene			0.34	0.05	14		
P. putida F1	Toluene	mcl-PHA	Flask	0.72	0.16	22		[41]
	Ethylbenzene			0.67	0.10	15		
P. putida mt-2	Toluene	mcl-PHA	Flask	0.37	0.08	22		[41]
	p-Xylene			0.53	0.14	26		
P. putida CA-3	Styrene	mcl-PHA	Flask	0.79	0.26	33		[41]
	Benzene			2.54		19	0.03	
P. fluva TY16	Toluene	mcl-PHA	Continuous feeding	3.87		59	0.11	[42]
	Ethylbenzene			2.80		29	0.04	
P. putida CA-3	Styrene pyrolysis oil	mcl-PHA	Flask	2.80	1.60	57	0.10	[44]
Sphingobacterium sp. ATM					3.48	65		
B. odysseyi SUK3	Orange 3R dye	PHA	Flask		2.10	61		[38]
P. desmolyticim NCIM 2112					1.12	52		
	Halophiles							
	25% pre-treated vinasse	PHBV (86–14)	Flask		19.7	70	0.87	[55]
H. mediterranei	Stillage	PHBV (85–15)			16.4	71	0.35	[56]
DSM 1411	Hydrolyzed cheese whey	PHBV (98.5–1.5)	Batch	7.54		54	0.78	[57]
	15% v/v olive mill wastewater	PHBV (94-6)	Flask		0.2	43		[58]
Halomonas TD01	Glucose salt medium	PHA	Continuous two-fermentor			65	0.51	[59]

Table 2. *Cont.*

Strain	Carbon Source	PHA	Cultivation Mode	DCW (g L^{-1})	PHA (g L^{-1})	PHA (%)	$Y_{P/S}$	Ref.
Halomonas campaniensis LS21	Mixed substrates (mostly comprised of kitchen waste)	PHB	Continuous pH-stat			26		[60]
B. megaterium H16	Glucose salt medium	PHB	Flask			39		[61]
B. megaterium uyuni S29	Glucose salt medium	PHB	Flask	5.42	2.22	41	0.13	[63]

[a] Mixtures consisting of 0.1% fatty acid and 1.5% gluconate; [b] 2mM acrylic acid; [c] 99.5% HB, 0.5% HHx; [d] 98.5% HB, 1.5% HHx; [e] Mixtures consisting of 1.5% starch type + 0.05% Valerate; [f] PHBV (%HB–%HV).

2.4. Synthetic Biology of PHA Producers

Synthetic biology tools may aid in developing competitive bioprocesses by engineering biocatalysts with the potential of being employed at industrial scale, producing large amounts of PHA at low prices [64]. Industrial biotechnology requires non-pathogenic, fast growing bacteria that do not produce toxins and their genome is easily manipulated. Utilization of cellulose and fast growth under a wider range of temperature and pH are considered a plus [65]. The effort of minimizing PHA production cost focuses mainly on engineering strains that show higher PHA production efficiency from raw waste material, require less energy consumption during PHA production, simplify downstream processing, and produce tailored functional polymers for high value-added applications.

In order to achieve high PHA volumetric productivities high cell densities need to be obtained of up to 200 g/L, characterized by high cellular PHA contents, above 90% g PHA/g CDW. Manipulation of genes related to the oxygen uptake, quorum sensing, and PHA biosynthetic mechanisms may enhance PHA production [65]. For example, oxygen limitation may occur, after obtaining high cell densities, in order to initiate/promote PHB production. In a relevant study, anaerobic metabolic pathways were designed in *E. coli* (over-expressing hydrogenase 3 and acetyl-CoA synthatase) to facilitate production of both hydrogen and PHB. In that way, the formation of toxic compounds such as formate and acetate was avoided by driving carbon fluxes towards the production of PHB. The engineered strain showed improved hydrogen and PHB production. In addition, PHB pathway optimization has been also investigated in *E. coli* by adjusting expression levels of the three genes *phbC*, *phbA*, and *phbB* [66]. *phbCAB* operon was cloned from the native PHA producing strain *Ralstonia eutropha*. Rational designed Ribosomal Binding Sites (RBS) libraries were constructed based on high or low copy number plasmids in a one-pot reaction by an Oligo-Linker Mediated Assembly method (OLMA). Bacterial strains accumulating cellular contents of 0 to 92% g PHB/g CDW were designed and a variety of molecular weights ranging between 2.7–6.8 × 10^6 was achieved. The certain study demonstrated that this semirational approach combining library design, construction, and proper screening is an efficient tool in order to optimize PHB production.

Another example where synthetic biology has been implemented are halophilic bacteria, which allow for PHA production under continuous mode and unsterile conditions. These features increase the competitiveness of industrial PHA production. In addition, halophilic bacteria have been proven easy for genetic manipulation, thus allowing for the construction of a hyper-producing strain [65,67]. For example, both recombinant and wild type *Halomonas campaniensis* LS21were able to grow on mixed substrates (kitchen wastes) in the presence of 26.7 g/L NaCl, at pH 10 and temperature of 37 °C continuously, for 65 days, without any contamination. Recombinant *H. campaniensis* produced almost 70% PHB compared to wild type strain that in which PHA accounted for 26% of its CDW [60].

Engineering the morphology of bacteria, in terms of cell size increase, has been recently investigated. Apart from PHA granules, several bacteria may accumulate polyphosphates, glycogen, sulfur, or proteins within their cells that limit cell space availability. In order to increase cell size, approaches such as deletion or weak expression of on actin-like protein gene mreB in recombinant

E. coli resulted in increasing PHB accumulation by 100% [68–70]. In addition, manipulating PHA granule-associated proteins leads to an increase in PHA granule size allowing for easier separation [71].

Intracellular accumulation of PHA necessitates several extraction and purification steps. Synthetic biology approaches have been developed to control and facilitate the release of PHA granules to the medium. For example, the programmed self-disruptive strain *P. putida* BXHL has been constructed in a recent study, deriving from the prototype *P. putida* KT2440 which is a well-known mcl-PHA producer [72]. This was based on a controlled autolysis system utilizing endolysin Ejl and holing Ejh isolated from EJ-1 phage and in order to improve the efficiency of the lytic system this was tested in *P. putida* tol-pal mutant strains with alterations in outer membrane integrity. According to results, it was shown that the engineered lytic system of *P. putida* BXHL provided a novel approach to inducing controlled cell lysis under PHA producing conditions, either produce PHA accumulating cells that were more susceptible to lytic treatments. The certain study demonstrated a new perspective on engineered cells facilitating PHA extraction in a more environmentally friendly and economic way.

PHA structures include PHA homopolymers, random and block copolymers, and also different monomer molar fractions in copolymers. Block copolymers have been reported regarding their resistance against polymer aging. This is of crucial importance since slower degradation of polymer occurs leading to better performance and consistent polymer properties [73]. It has been observed that downregulating of β-oxidation cycle in *P. putida* and *P. entomophila* may be used for controlling PHA structure when fatty acids are used as precursors for PHA production [73–76]. In the case of fatty acids, containing functional groups are consumed by bacteria, introduction of those functional groups into PHA polymer chains occurs [77]. In addition, recombinant strains of *E. coli* have been constructed for the synthesis of block polymers with superior properties [78–81]. PHA diversity is possible by engineering basic biosynthesis pathways (acetoacetyl-CoA pathway, in situ fatty acid synthesis, and/or β-oxidation cycles) as well as through the specificity of PHA synthase [82].

3. PHA Production by Mixed Microbial Consortia (MMC)

Currently, industrial PHA production is conducted using natural isolates or engineered strains and pure substrates [83,84]. An alternative scenario that would contribute to the reduction of the PHA production cost is to employ mixed culture biotechnology [85,86]. This approach uses open (under non-sterile conditions) mixed microbial consortia (MMC) and ecological selection principles, where microorganisms able to accumulate PHA are selected by the operational conditions imposed on the biological system. Thus, the principle is to engineer the ecosystem, rather than the strains, combining the methodology of environmental biotechnology with the goals of industrial biotechnology [87]. The cost reduction derives mainly from operations being performed under non-sterile conditions, and their consequent energy savings, and the higher adaptability of MMC to utilize waste streams as substrates.

Processes for PHA production in mixed cultures are usually performed in two steps (Figure 1). In the first step, SBR reactors (sequential batch reactors) are used to select and enrich a microbial population with high PHA production capacity by applying transient conditions. In the second step, the culture from the SBR is subjected to conditions maximizing the PHA accumulation, from where cells are harvested for PHA extraction and purification when they reach maximum PHA content [88,89].

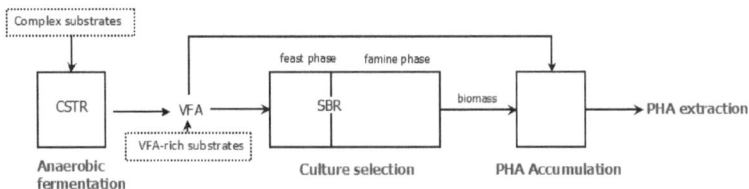

Figure 1. PHA production process by mixed microbial cultures. Modified from [88]. CSTR: continuous stirred tank reactor, SBR: Sequential Batch Reactor.

Unlike pure cultures, where glucose is mostly used as a substrate for PHA production, mixed culture biotechnology makes use of volatile fatty acids (VFA) as the precursors for PHA production [90]. The main reason is that carbohydrates in MMCs, as well as other substrates such as glycerol, tend to form glycogen besides PHA [91,92]. For those substrates, a previous step is generally included (Figure 1), during which they are fermented into volatile fatty acids (VFA) in continuous mode (CSTR). Moreover, this is also applied to complex substrates, such as olive mill wastewater [10,93,94], cheese whey [95], and other food wastes [96] in order to obtain a more homogeneous readily available feed for the PHA production. VFA conversion into PHA require few steps, and usually presents very high yields and rapid uptake rates [88,97]. This is especially the case for butyric acid, which has now been reported in many studies as the VFA presenting the highest yields (up to 0.94 Cmol PHA/Cmol S) and being the one preferably up-taken by MMCs [2,98–100]. As a matter of fact, butyric acid preference has been observed even in cultures that were not exposed to it during the enrichment [93].

It is worth mentioning that the distribution of VFA is known to affect the PHA monomer composition, where VFA with an even number of carbon atoms tend to produce PHB while VFA with odd carbon atoms tend to produce PHBV copolymers with different % HV molar fractions [101]. Based on this fact, many studies have suggested the possibility of regulating the PHA composition by manipulating the fermentation conditions in the preceding acidogenesis step [2,95,100].

This section provides an overview of the different types of existing enrichment techniques, performed within the last 10 years. Each one of the enrichment strategies presents different advantages; either related to the cost of the process or to increased cellular PHA content. Thus, they all present opportunities to further improve economic and sustainable PHA production. Such opportunities are commented for each of the enrichment techniques. This section is followed by a compilation of recent advances regarding the PHA accumulation stage, aiming at increasing the productivity. Finally, recent attempts to bring MMC to pilot scale are described, followed by a section highlighting the main challenges and bottlenecks of the MMC.

3.1. Types of Enrichments

Until the late 2000s, two types of enrichments dominated the research panorama related to PHA mixed culture biotechnological production, namely the anaerobic/aerobic selection and aerobic dynamic feeding. These types of enrichments have already been previously reviewed in other articles [88,102–104], thus apart from a general description of the mechanisms, only the recent trends are described in the respective Sections 3.1.1 and 3.1.2. The following sections are dedicated to recent (and still less widespread) types of enrichments developed within the last decade. The main characteristics of all types of enrichments described are summarized in Table 3.

Table 3. Summary on the main characteristics of the enrichment techniques applied for MMCs.

Anaerobic-Aerobic Enrichment (AN/AE) (Section 3.1.1)

	Feast phase	Famine phase
Aeration	No	Yes
e− acceptor	– (PHA)	Oxygen
Energy source	Glycogen/polyphosphate	Oxidation of PHA
Carbon source	External substrate	PHA
Driving force for PHA accumulation	• Lack of electron acceptor * • Transient presence of substrate ***	

Aerobic Dynamic Feeding (ADF) (Sections 3.1.2 and 3.1.3)

Classical Aerobic Dynamic Feeding (Section 3.1.2)

	Feast phase	Famine phase
Aeration	Yes	Yes
e− acceptor	Oxygen	Oxygen
Energy source	Oxidation of substrate	Oxidation of PHA
Carbon source	External substrate	PHA
Nitrogen availability	Yes **	Yes **
Driving force for PHA accumulation	• Transient presence of substrate ***	

Aerobic Dynamic Feeding (ADF) with Intermediate Settling Phase (Section 3.1.3)

	Feast phase	Famine phase
Aeration	Yes	Yes
e− acceptor	Oxygen	Oxygen
Energy source	Oxidation of substrate	Oxidation of PHA
Carbon source	External substrate	PHA
Nitrogen availability	Yes	Yes
Driving force for PHA accumulation	• Transient presence of substrate *** • Higher settling capacity of PHA rich cells • Elimination of residual COD after feast phase prevents growth of non-PHA accumulating bacteria	

Aerobic Dynamic Feeding (ADF) with Nitrogen Limitation in the Feast-Phase (Section 3.1.3)

	Feast phase	Famine phase
Aeration	Yes	Yes
e− acceptor	Oxygen	Oxygen
Energy source	Oxidation of substrate	Oxidation of PHA
Carbon source	External substrate	PHA
Nitrogen availability	No	Yes
Driving force for PHA accumulation	• Transient presence of substrate *** • Nitrogen limitation during the feast phase	

Table 3. *Cont.*

Photosynthetic Enrichment (Section 3.1.5)

Photosynthetic Enrichments—Illuminated SBR

	Feast phase	Famine phase
Aeration	No	No
e^- acceptor	– (PHA)	Oxygen produced by algae
Energy source	Light	Oxidation of PHA + Light
Carbon source	External substrate	PHA
Driving force for PHA accumulation	• Lack of external electron acceptor with presence of light • Transient presence of substrate ***	

Photosynthetic Enrichment—Dark Feast Phase

	Feast phase	Famine phase
Aeration	No	No
e^- acceptor	– (PHA)	Oxygen produced by algae
Energy source	Glycogen	Oxidation of PHA + Light
Carbon source	External substrate	PHA
Driving force for PHA accumulation	• Lack of external electron acceptor with presence of light *** • Transient presence of substrate ***	

Photosynthetic Enrichment—Permanent Feast Phase

	Feast phase	Famine phase
Aeration	No	No famine phase
e^- acceptor	– (PHA)	
Energy source	Light	
Carbon source	External substrate	
Driving force for PHA accumulation	• Lack of external electron acceptor with presence of light	

Aerobic-Anoxic Enrichment (Section 3.1.6)

	Feast phase	Famine phase
Aeration	Yes	No
e^- acceptor	Oxygen	NO_3 / NO_2
Energy source	Oxidation of substrate	Oxidation of PHA
Carbon source	External substrate	PHA
Driving force for PHA accumulation	• Transient presence of substrate ***	

Table 3. *Cont.*

	Anoxic-Aerobic Enrichment (Section 3.1.7)	
	Feast phase	**Famine phase**
Aeration	No	Yes
e⁻ acceptor	NO_3/NO_2	Oxygen
Energy source	Oxidation of substrate	Oxidation of PHA
Carbon source	External substrate	PHA
Driving force for PHA accumulation	• Transient presence of substrate ***	

	Microaerophilic Enrichment (Section 3.1.8)	
	Feast phase	**Famine phase**
Aeration	Yes	Yes
e⁻ acceptor	Oxygen	Oxygen
Energy source	Oxidation of substrate	Oxidation of PHA
Carbon source	External substrate	PHA
Driving force for PHA accumulation	• Transient presence of substrate ***	
	• Limitation of electron acceptor	

* Even though the lack of electron acceptor is the driving force of the enrichment, this limitation is not mandatory for these cultures to produce PHA, which can also be produced aerobically. ** various C/N ratios have been applied resulting in a limitation of nitrogen in the famine or late feast phase. Nevertheless, most wide-spread configuration provides nitrogen in both phases *** Transient presence of substrate leads to the following effects in all cases mentioned in the table: Growth during famine phase consuming the PHA accumulated; Limitation of internal growth factors; Higher responsiveness of PHA producers to substrate addition.

3.1.1. Anaerobic/Aerobic Enrichments (AN/AE)

PHA production from MMC was first observed in biological phosphate removal by activated sludge in wastewater treatment plants, where aerobic and anaerobic steps alternate [90]. Thus, the first attempts of enriching a PHA storing community were performed by replicating those conditions. In these cases, Polyphosphate Accumulating Organisms (PAOs) and Glycogen Accumulating Organisms (GAOs) were described to accumulate PHA during the anaerobic phase, where the electron acceptor becomes limiting. In the aerobic phase, these microbes would consume the internally stored PHA, using the available oxygen, obtaining thus a higher adenosine triphosphate (ATP) yield compared to the substrate being metabolized anaerobically [102,104]. Nevertheless, as substrate catabolism and PHA formation, during the anaerobic phase, requires ATP and reducing equivalents, these microbes would also depend on the accumulation of glycogen or polyphosphate from the stored PHA during respiration, that limit the PHA production capacity of the cultures. Maximum PHA contents of around 20% had been reported from these consortia enriched under AN/AE conditions. However, in the late 1990s the aerobic dynamic feeding (ADF) enrichment was introduced, obtaining higher PHA contents [102–104], and thus, research efforts in AN/AE remained limited. Nevertheless, recent developments in PHA production following AN/AE enrichment, demonstrated that PHA might be also accumulated aerobically, without depending on glycogen and phosphate reserves, and high PHA storage capacities (up to 60%) could be obtained [105]. Thus, further research could prove the feasibility of AN/AE that would demonstrate benefits of saving energy costs, in terms of aeration requirements, compared to ADF.

3.1.2. Aerobic Dynamic Feeding (ADF)

In enrichment under ADF conditions the limiting factor, promoting PHA accumulation, is carbon substrate availability rather than the electron acceptor [104]. This process relies on subsequent feast/famine cycles where the culture is subjected initially to an excess of carbon source, and then submitted to carbon deficiency under aerobic conditions. Bacteria that are able to convert carbon to PHA during the feast phase have a competitive advantage towards the rest of the microbial population, as they utilize PHA as a carbon and energy reserve during the famine phase, allowing them to grow over non-PHA storing microorganisms [104,105]. Moreover, a limitation of internal factors, such as RNA and enzymes required for growth, seems to be crucial [102]. In order for cells to grow, a considerable amount of RNA and enzymes are needed, which might not be available after long starvation periods. Nevertheless PHA synthase, the key enzyme for PHA polymerization, is active during PHA production and degradation, generating a futile cycle that wastes ATP but enables the PHA mechanism to be ready when a sudden addition of carbon occurs, providing them with higher responsiveness [106–108]. In this way, a new competitive advantage of PHA producers arises, given that they can use PHA to regulate substrate consumption and growth [106]. PHA contents up to 90% of the dry cell weight have been reported using this strategy [89,109], higher than the ones reached following AN/AE enrichment [102–104].

Most of the studies performed within the last 10 years have been based on ADF enrichment. Apart from investigating the feasibility of this strategy using a variety of substrates, recently reviewed by Valentino and colleagues [97], the main focus has been on evaluating the impact of different parameters during the enrichment process. An overview on how those parameters influence PHA production from MMCs has been recently reported on, including the effect of hydraulic retention time (HRT), solids retention time (SRT), pH, temperature, nitrogen concentration, dissolved oxygen concentration (DO), cycle length, influent concentration, feast/famine ratio, and food/microbe ratio [110]. Regarding process configuration, a continuous system has been proposed where instead of an SBR, the feast and the famine phases were operated in separate CSTR [111]. Although no significant improvements were observed with respect to the conventional SBR configuration, the study demonstrated that successful enrichment was also possible in continuous mode, which is considered advantageous in the case of putative coupling to the following PHA accumulation step under continuous mode.

3.1.3. Variations of the ADF Enrichments

Even though microbial communities with high PHA-storing capacity have been obtained with several parameter combinations, the presence of non-PHA accumulating microorganisms is still not completely avoided. This is partially due to the presence of organic content other than VFA present in waste streams, allowing the growth of non-PHA accumulating bacteria [112]. A possible way to overcome the presence of such bacteria has been recently proposed where the culture was settled and the supernatant was discharged just after VFA depletion. In this way, consumption of the remaining organic matter—measured as chemical oxygen demand (COD)—and growth of side-population was prevented while the fraction of PHA-producers was considerably increased as verified by molecular techniques [113]. An increase in the PHA content from 48% to 70% was observed by applying this strategy. The authors suggested that, apart from the role of the remaining COD, also the increased cell density of PHA packed cells enhanced the enrichment, as those cells would have a higher tendency to settle after the feast phase. This observation coincided with the results obtained using only acetate as a carbon source, where the settling after the feast phase also increased the PHA accumulation capacity of the culture from 41% to 64–74% [114]. As no residual COD was present in those experiments, the effect could be entirely attributed to differences in the cell density that led to an additional physical selection.

Similarly, the growth of such non-PHA accumulating bacteria was observed to be restricted by applying nitrogen limitation [112]. Following this reasoning, a variation of the ADF was recently proposed, where also a nitrogen deficiency was imposed during the feast phase, while providing nitrogen during the famine phase to enable growth from the accumulated PHA. Thus, an uncoupling of the carbon and nitrogen took place in the SBR [2,86,115]. This strategy resulted in higher PHA contents at the end of the feast phase using synthetic VFA [2,115], cheese whey [86], and 1,3-propanediol from fermented crude glycerol [116] as a substrates. When a separate PHA accumulation was performed with the cultures of such enrichments, higher PHA yields and productivities were obtained with this strategy compared to carbon-nitrogen coupled ADF [86]. This strategy also allowed for a more stable system for long term operation. Nevertheless, given that the enrichment is already performed at conditions maximizing PHA accumulation, it was also suggested that a separate PHA accumulation might no longer be required if part of the biomass is already harvested after the feast phase, something that would significantly contribute towards the reduction of the costs of the process [115]. Moreover, given that there is already a selective pressure in the feast phase, the duration of the famine phase would be less important, so as to achieve an effective selection, and this could enable a reduction of its duration leading to enhanced process productivity [86].

3.1.4. ADF Enrichments in Halophilic Conditions

As previously discussed, the use of halophilic bacteria comes with various advantages. PHA production using halophilic bacterial populations can be performed using seawater instead of fresh or distilled water or using high salinity wastewater produced by several industries, namely food processing industries. In addition, halophilic bacteria can be lysed in distilled water thus reducing downstream processing costs due to lower quantities of solvents required. Enrichment of a halophilic PHA accumulating consortium under ADF conditions has been recently investigated using different carbon sources as substrate, which resulted in cellular PHA contents reaching up to 65% and 61% PHA using acetate and glucose respectively, demonstrating the potential of this strategy in MMC [117]. Recently, a previously enriched MMC fed with a mixture of VFAs containing 0.8 g/L Na^+ was examined regarding its PHA accumulation capacity under transient concentrations of 7, 13, and 20 g/L NaCl [118]. Since the particular MMC was not adapted to saline conditions, PHA accumulation capacities and rates decreased with higher NaCl concentrations while biopolymer composition was affected in terms of HB:HV ratio.

3.1.5. Mixed Photosynthetic Consortia

A new approach relying on the photosynthetic activity of mixed consortia has been explored recently [119–122]. Based on the previous observation of PHA production in photosynthetic strains, an illuminated SBR operating without aeration was proposed, eliminating the costly need for aeration during ADF enrichments. In such a system, photosynthetic bacteria uptake an external carbon source, in the form of acetate, during the feast phase using light as an energy source. PHB was produced at the same time as a sink of NADH, given that no electron acceptor was present. During the famine phase PHB was consumed using oxygen as an electron acceptor, which was not provided though aeration but produced by algae also present in the SBR. Under these conditions, up to 20% PHB was attained during the PHA accumulation step [119], which was also possible by utilizing other VFAs such as propionate and butyrate [121]. However, it was observed that a dark feast phase could also be envisioned, given that similarly to AN/AE enrichments, glycogen accumulation occurred during the famine phase, which was subsequently used as a complementary energy source to uptake acetate. With this SBR configuration, which would considerably reduce the need of illumination, a 30% PHA was accomplished [122].

The best PHA productivity though, was obtained in a system operating in a permanent illuminated feast phase (instead of successive feast-famine cycles) without oxygen supply [120]. This was based on the fact that photosynthetic accumulating bacteria out-compete other bacteria and algae without the need of transient presence of carbon source. Therefore, productivity was significantly increased due to the elimination of the famine phase and since there was no need for a separate PHA accumulation reactor. However, considerable input of light was required in order to obtain cultures with high PHA content (up to 60% of the dry weight), so the economic advantages of such systems should be further explored. Nevertheless, this process will allow significant savings in energy, since no sterilization and aeration are required, which will have an impact on the final price of the polymer.

3.1.6. Aerobic–Anoxic Enrichment Coupled with Nitrification/Denitrification

Basset and colleagues [123] developed a novel scheme for the treatment of municipal wastewater integrating nitrification/denitrification with the selection of PHA storing biomass, under an aerobic/anoxic and feast/famine regime. The process took place in a SBR (where NH_4^+ is converted into NO_3^- with a simultaneous selection of PHA storing biomass—and with COD being converted to PHA) and the subsequent PHA accumulation in a batch reactor (where PHA is consumed to allow denitrification, under famine–anoxic conditions, without the need of external addition of organic matter). The carbon source added during the selection and accumulation steps consisted of fermentation liquid from the Organic Fraction of Municipal Solids Waste (OFMSW) and primary sludge fermentation liquor.

The advantage of this approach is that the potential for recovering biopolymers from wastewater presents particular interest, when the latter is integrated within the normal operation of the plant. An important benefit of this strategy is that anoxic denitrification usually requires a carbon source, which at that point is usually low and already consumed in a wastewater treatment plant (WWTP). In this case however, it can occur without external addition of carbon source by using those stored internally in the form of PHA.

Results showed that during SBR operation ammonium oxidation to nitrite reached on average $93.4 \pm 5.25\%$. The overall nitrogen removal was 98% (resulting in an effluent with only 0.8 mg NH_4^- N L^{-1}). Similar results were obtained by Morgan-Sagastume et al. [124]. Denitrification efficiency and rate did not seem to be affected by the carbon source. When sufficient PHA amount was available, denitrification of all available nitrate was observed. COD removal reached up to 70% when DO level was higher (2–3 mg L^{-1}). PHA content decreased during nitrification due to the lack of external COD. However, after complete nitrification, there was enough PHA to carry out the denitrification process. Even though biomass was rich in PHA storing bacteria, PHA accumulation reached only 6.2% during the feast phase (first 10 min) and it was progressively consumed before the initiation of the anoxic phase to 2.3%, which was enough to complete the subsequent denitrification. Nevertheless, the selection of PHA storing biomass under feast (aerobic)–famine (anoxic) conditions required less DO compared to the typical feast–famine regime carried out under continuous aerobic conditions, leading to a reduction of 40% of the energy demand.

The PHA accumulation capacity of the biomass, previously selected in the SBR, was further evaluated in accumulation batch reactors with the use of OFMSW, and primary sludge fermentation liquid. After 8 h of accumulation with OFMSW, the stored PHA was 10.6% (wt.). In the case where fermented sludge liquid and OFMSW was used as carbon source, the contribution to growth was higher, due to the elevated nutrient content, and the lower VFA/COD ratio but only 8.6% PHA was accumulated after 8 h. The carbon source was proven to play an important role in the PHA accumulation step as the presence of non-VFA COD contributed to the growth of non-PHA-storing biomass [124,125]. PHA storage yields could be potentially improved with a more efficient solid–liquid separation after the fermentation process.

3.1.7. Anoxic–Aerobic Strategy Coupled with Nitrification/Denitrification

Anoxic–aerobic enrichments coupled with denitrification, where nitrate is used as electron acceptor during PHA accumulation, has been already explored using synthetic VFA since early 2000s [103]. A recent study applying this strategy has been reported which investigated the use of the condensate and wash water from a sugar factory [126]. Furthermore, they considered that, in combination with harvesting enriched biomass from the process water treatment, side-streams could be exploited as a substrate for PHA accumulation. This approach (together with the aerobic–anoxic strategy shown in Section 3.1.6) would have the big advantage of significantly reducing the PHA production costs, through the integration of already existing full-scale WWTP and reduction of aeration needs.

In that study [126], they used parallel SBRs fed alternatively with condensate and wash water, developing a microbial consortium that removes inorganic nitrogen by aerobic and anoxic bioprocess steps of nitrification and denitrification. Alternating bioprocess conditions of anoxic feast (supporting denitrification) and aerobic famine (supporting nitrification) in mixed open cultures was expected to furnish a biomass with stable PHA accumulating potential characterized by its ability to remove carbon, nitrogen, and phosphorus in biological processes. In more detail, one laboratory SBR was operated with suspended activated sludge (AS) and long SRT, similar to the full-scale (SRT > 6 days), while the other SBR was a hybrid suspended activated sludge and moving bed biofilm reactor (MBBR) with short SRT of 4–6 days. MBBR technology employs thousands of polyethylene biofilm carriers operating in mixed motion within an aerated wastewater treatment basin. Therefore, the MBBR-SBR was used as a means for maintaining nitrifying activity while enabling enrichment of biomass at relatively low SRT.

The results showed a COD removal performance of 94% and 96% for AS- and MBBR-SBRs, respectively. Full nitrification was achieved in both systems, with exception of periods showing phosphorous or mineral trace element limitation. Soluble nitrogen removal reached 80 ± 21% and 83 ± 11% for AS- and MBBR-SBRs, respectively. MBBR-SBR showed more stable performance under lab-scale operation. The process achieved a PHA content of 60% g PHA/g VSS in both cases. A significant advantage was the possibility of lowering the SRT while maintaining a robust nitrification activity and improving the removal of soluble phosphorus from the process water.

3.1.8. Microaerophilic Conditions

In 1998, Satoh and colleagues [127] investigated the feasibility of activated sludge (from laboratory scale anaerobic–aerobic reactors) for the production of PHA, by optimizing the DO concentration provided to the system. They were able to obtain a PHA content of around 20% in anaerobic conditions and 33% under aerobic conditions. When applying a microaerophilic–aerobic process, by supplying a limited amount of oxygen into the anaerobic zone, they were able to increase the PHA accumulation to 62% of sludge dry weight.

PHA production using palm oil mill effluent (POME) was investigated by Din and colleagues [128], using a laboratory SBR system under aerobic feeding conditions. The microorganisms were grown in serial configuration under non-limiting conditions for biomass growth, whereas in the parallel configuration the nutrient presence was controlled so as to minimize biomass growth in favor of intracellular PHA production. PHA production under aerobic, anoxic, and microaerophilic conditions was investigated and it was shown that PHA concentration and content increased rapidly at the early stages of oxygen limitation while the production rate was reduced at a later stage implying that oxygen limitation would be more advantageous in the PHA accumulation step.

Another interesting study was published by Pratt and colleagues [129], where the effect of microaerophilic conditions was evaluated during the accumulation phase, using an enriched PHA culture, harvested from a SBR fed with fermented dairy waste. Batch experiments were conducted to examine the effect of DO on PHA storage and biomass growth. The results showed that in microaerophilic conditions a higher fraction of substrate was accumulated as PHA, compared to high DO conditions. Also, the intracellular PHA content was 50% higher during early accumulation phase. Interestingly, the accumulation capacity was not affected by the DO, despite its influence on biomass growth. The PHA content in both low and high DO concentrations reached approximately 35%. However, the time needed to achieve maximum PHA content at low DO level was three times longer than in the case of high DO concentration. The reason why PHA accumulation was proven to be less sensitive to DO, compared to its effect on biomass growth, was explained by the fact that low DO levels limit the availability of ATP, while high DO supply provides surplus ATP and high growth rates (and consequently reduced PHA yield). In addition, when MMCs were fed with multiple VFAs (acetate, propionate, butyrate, and valerate) it was also shown that, during PHA accumulation, high DO concentration is required to reach maximum PHA accumulation rates due to low specific VFA uptake rates under low DO levels [130].

The effect of dual nitrogen and DO limitation has been also investigated in MMCs fed with a VFA mixture of acetate, propionate, and butyrate and acidified OMW [10,99]. As discussed above, it was shown that during the PHA accumulation step, under batch mode, lower substrate uptake and PHA production rates were obtained compared to assays performed under nitrogen limitation. Moreover, PHA accumulation percentages and the yield coefficient $Y_{PHA/S}$ was lower in the case of dual limitation, while the accumulation of non-PHA polymers within the cells was indicated.

Those reports demonstrate that manipulating oxygen concentration could influence growth and PHA storage. Manipulating DO instead of limiting nitrogen or phosphate availability could represent a significant opportunity for PHA production processes that utilize nutrient rich feedstocks. A major advantage of operating at low DO is the reduced aeration requirements leading to reduction of operating costs. However, this advantage can be countered by the fact that PHA accumulation in low DO environments can be significantly slower.

3.1.9. PHA Accumulation without Previous Enrichment

The three-step process described in Figure 1, consisting in an enrichment step followed by an accumulation step, has been proven efficient to obtain cultures with high PHA contents. Nevertheless, several authors have put in doubt the alternative of having a separate enrichment step. The main reason is that during the enrichment, PHA is produced, but it is also allowed to be consumed to drive the selection. Thus, this step consumes substrate without leading to any net production of PHA, lowering the overall PHA/Substrate yields of the system [97,131]. Thus, skipping this step could imply considerable improvements on those parameters.

Already in 2002, PHA accumulation without a previous enrichment step using activated sludge was reported to obtain PHA contents up to 30% [132]. Later on, fed-batch cultivation under nitrogen limiting conditions was reported to obtain up to 57% PHA [133]. Cavaillé and colleagues performed fed-batch PHA accumulation experiment as well using activated sludge without previous enrichment, but applying phosphorous limitation and achieved up to 70% PHA [134]. Substrate to PHA yield reached up to 0.2 Cmol PHA/Cmol S using acetic acid. This yield was considerably lower than that obtained in PHA accumulation steps submitted to a previous enrichment step (up to 0.9 Cmol PHA/Cmol S), but comparable to the overall yields of enrichment and accumulation strategies summed up [134]. Moreover, they further developed the system into a continuous process, and attained a stable operation where the cells in the effluent contained 74% PHA [135]. Their findings also evidenced that the continuous system was not stable at severely phosphorous limited experiments, because the growth rate could not be maintained and the cells were washed out of the reactor. The key was that differently from when a separate enrichment is performed, the continuous PHA accumulation without previous enrichment relied on the occurrence of both growth and storage responses. Moreover, the authors suggested that phosphorous limitation might offer more flexibility than nitrogen limitation when both PHA formation and growth are a goal, given that phosphorous is less needed than nitrogen for growth-related metabolism [135].

3.2. PHA Accumulation

As in the enrichment step, several operational parameters such the temperature [136] and the pH have an impact on PHA accumulation [2,137–139]. Nonetheless, the most critical aspect during PHA accumulation experiments is the cultivation strategy employed. High substrate concentrations supplied under batch mode should be avoided since they can cause inhibition and thus limit PHA productivity [105,140]. In order to circumvent that, several fed-batch strategies have been suggested. Pulsed fed-batch cultivation has been suggested when synthetic VFA mixtures are used [140,141]. However, due to an increase occurring in the working volume after the addition of substrate the feed should be very concentrated. Nevertheless, this is rarely the case with fermentation effluents, since they usually do not exceed 20 g COD/L [97]. Discharge of the exhaust supernatant has been suggested as an alternative [137,142,143] yet, this approach requires a settling step between batches, which severely limits the productivity [137].

Continuous feeding processes have shown the best results until now, given that they can attain a sustained productivity [89,137,143,144]. Pulsed fed-batch production may result in high PHA productivity but as the substrate is being consumed, PHA productivity eventually decreases. This phenomenon might be avoided by supplying substrate in a continuous manner. Continuous substrate addition has been successfully performed using the pH as an indicator, given the pH increases with VFA consumption [89,137,143,144]. On the other hand, less successful results have been obtained when the substrate was supplied, taking into account previously observed substrate uptake rates resulting in either accumulation or limitation of substrate in the reactor [140,141]. Alternatively, an on-demand continuous addition of substrate, based on change of DO, has been proven efficient to maintain optimal amounts of carbon substrate in the reactor [145].

High productivities up to 1.2 g PHA/L h combined with high PHA yields, 0.8 Cmol PHA/Cmol S, have been achieved with continuous-feeding systems [137]. However, similarly to the pulsed fed-batch, such values have been reported only when synthetic substrates of high concentration were present in the feed. Much lower values are reported in real substrates due to the diluted nature of these substrates and the consequent increment in reactor volumes [97]. A way to overcome this could be the development of a continuous feeding scheme for PHA accumulation under low biomass loading rates (3.5–5.5 Cmol VFA/Cmol X/d). So far, this venture has only been investigated once using MMC [139]. The authors suggested a PHA accumulation reactor operating under continuous mode, were the effluent was allowed to settle and the resulting sludge recycled back to the reaction vessel. The system worked with a rather diluted effluent in the feed (around 100 Cmol VFA) and was shown to obtain higher specific productivities than the pulse-fed-batch. Nevertheless, overall productivity of the system was not reported in that study. It is worth noting that the system was not coupled to an SBR operating in continuous mode, so the operation of the reactor was also for a limited period of time.

Regarding the nutrient availability, nitrogen limitation or deficiency is usually reported to improve the PHA yield and content [2,105,146,147]. On the other hand, several studies have concluded that the role of nitrogen during the accumulation step is secondary since PHA storage was preferred over growth regardless of the nitrogen concentration [144,148]. Moreover, in another study it was observed that nitrogen limitation did not enhance the PHA accumulation [149]. As a matter of fact, the main reason for limiting the amount nitrogen is to prevent bacterial growth of non PHA-accumulating bacteria [112]. Hence, different observations from different cultures do not imply contradictions but highlight the fact that the requirement for nitrogen limitation, in order to obtain high PHA contents, is highly dependent on the composition of the enriched culture and the type of substrate fed. In terms of productivity though, nutrient limitation rather than deficiency was reported to show higher productivities [145]. According to a certain study, the absence of an essential nutrient for growth leads to cellular PHA saturation, while nutrient limitation allows cells to duplicate prolonging PHA accumulation without enabling excessive growth response. In addition, it was shown that the best productivities were obtained from dual limitation of nitrogen and phosphorous.

3.3. Pilot Scale Experiences

Several industrial/agro-industrial effluents and residues have been investigated so far as potential feedstocks for PHA production. Effluents, rich in sugars, glycerol, and/or fatty acids were either directly used for the selection of PHA accumulating MMCs or were previously fermented for the conversion of carbohydrates to VFAs, the preferable precursors for PHA production using MMCs. Numerous studies on PHA production from industrial effluents have been performed, that have been recently reviewed by Valentino et al. [97]. On the other hand, studies on PHA production in pilot-scale by MMCs are rather scarce and relevant efforts have recently started, in 2010s, while no full-scale production of PHA by MMC exists yet.

A common feature in all pilot-scale studies is that effluents/feedstocks were always fermented prior to PHA production. Also, most efforts on PHA production in pilot-scale focus on integrating and combining PHA production with existing processes in wastewater treatment plants so as to reduce the production cost by exploiting the available infrastructure as much as possible. In this context, anoxic/aerobic MMC selection regimes can be coupled to nitrification and denitrification activities despite the fact that the highest PHA yields and cell content in PHA are usually reported for aerobic dynamic feeding selection regimes. Another tendency in pilot-scale studies is the use of a different effluent for the MMC enrichment than the one fed during the PHA accumulation step. The first pilot-scale study, concerning MMC PHA production, was performed from pre-fermented milk and ice-cream processing wastewater, as reported by Chakravarty et al. [150], with a PHA content of 43% and a PHA yield of 0.25 kg PHA/kg COD being obtained. In 2014, Jia et al. [125] studied the production of PHA in pilot-scale with pre-hydrolyzed and fermented raw excess sludge. In both studies, activated sludge was used as the raw material for its enrichment to PHA accumulating microorganisms. A series of pilot-scale studies was conducted and published with the participation of Anoxkaldnes and Veolia Water Technologies [126,151,152]. In the study of Morgan-Sagastume et al. [151], the potential of waste sludge, generated in wastewater treatment plants as a feedstock for PHA production was evaluated. This was done in the context of integrating PHA production in existent WWTP valorizing at the same time the excess sludge that in general represents a burden for further treatment and disposal. A very interesting point in the study of Bengtsson et al. [152] is that denitrifying microbial biomass was also selected towards high PHA producing potential by applying an anoxic-feast and aerobic-famine selection pattern and therefore the process comprised nitrification and denitrification steps followed by accumulation of PHA. Tamis et al. [153] investigated PHA production from fermented wastewater deriving from a candy bar factory. Activated sludge was enriched in PHA accumulating microorganisms under aerobic feast and famine regime and the obtained PHA content was the highest reported so far in pilot-scale at 70–76%. Table 4 summarizes the main features of the pilot-scale studies published so far. Overall, as it regards pilot-scale studies, their performance cannot really be compared to respective lab-scale studies but they provide valuable information on PHA formation under variable feedstock characteristics and allow production of significant amounts of polymer that can be processed for a full characterization. The variation of feedstock composition combined with the oxygen mass transfer limitations occurring at a larger scale, could be the main reason why PHA yields and cell content in pilot-scale studies are in general lower than the ones reported in lab-scale experiments.

Table 4. Main characteristics of PHA production in pilot-scale.

Pilot Plant, Location	Feedstock	Origin of MMC and Enrichment Strategy	$Y_{P/S}$ (g/g) *	PHA % (%mol HB: %mol HV)	mg PHA/g X/h	Ref.
Nagpur, India	Pre-fermented milk and ice cream processing wastewater	Activated sludge	0.425 *	39–43		[150]
Lucun WWTP in Wuxi, China	Hydrolyzed and acidified raw excess sludge	Activated sludge/synthetic mixture of VFA, ADF feast famine with carbon limitation and inhibitor of nitrification	0.044–0.29 *		2.06–39.31	[125]
Eslöv, Sweden	Beet process water, 38% in VFA	PHA producing MMC from pre-fermented effluent of Procordia Foods		60 (85:15 HB:HV)		[126]
Brussels North WWTP (Aquiris, Belgium)	Pre-hydrolyzed and fermented WWTP sludge	Sludge fed with municipal WW under aerobic feast famine	0.25–0.38	27–38 (66–74:26–34 HB:HV)	100–140	[151]
Leeuwarden WWTP, Friesland, Netherlands	Fermented residuals from green-house tomato production	Sludge fed with municipal WW under anoxic feast/aerobic famine	0.30–0.39	34–42 (51–58:42:49 HB:HV)	28–35	[152]
Mars company, Veghel, Netherlands	Fermented wastewater from a candy bar factory	Activated sludge from a WWTP fed with the fermented wastewater under aerobic feast/famine with inhibitor of nitrification	0.30	70–76 (84:16 HB:HV)		[153]

* Yield calculated on a COD basis by using the coefficients: for HB: 1.67 g COD PHA/g PHA and for HV: 1.92 g COD PHA/g PHA.

3.4. Challenges and Perspectives Regarding PHA Production by Mixed Microbial Consortia

PHA process brought to an industrial scale using pure substrates and strains has broadly surpassed cell densities of 150 g/L with PHA contents up to 90% and productivities in the range from 1 to 3 g PHA/L h [4]. Comparable productivities, of up to 1.2 g PHA/L h, combined with high PHA yields, up to 0.8 Cmol PHA/Cmol S, have already been attained from MMCs using synthetic substrates [137]. Likewise, PHA contents up to 90% of the CDW have been reported [109]. Thus, MMC have proven to be able to achieve comparable results to pure cultures in synthetic media. Nevertheless, PHA productivities could be further increased if higher cell densities were obtained, a parameter that is usually below 10 g/L in MMC. In pure cultures, the PHA accumulation phase is preceded by a biomass growth phase in order to achieve high biomass densities [154]. In the accumulation phase, the feeding strategy is modified accordingly (usually limiting the nitrogen) to obtain a high PHA content. Nevertheless, in processes for PHA production from MMC, the biomass generation step is also the enrichment step. Thus, two objectives, which might not have the same optimal conditions, are combined in the same process unit. A future direction could be to test if adopting a microbial biomass generation step—leading to higher cell densities before the PHA accumulation—could maintain the high PHA content of the cells in MMC.

In order to achieve a sustainable production of PHA, both in economic and environmental terms, high productivities should also be obtained in waste substrates. In the current state of art,

Bioengineering **2017**, *4*, 55

neither pure strains nor MMC have achieved high productivities when waste streams are used as substrates [83,97,155]. Thus, the challenge is common.

One of the main issues that compromises productivity when using waste streams is their diluted nature [155,156]. This applies for the case of MMC where an anaerobic fermentation step is usually performed in order to convert sugars to fatty acids. However, it is also the case with other industrial wastes used both in pure and mixed cultures, such as whey or lignocellulosic biomass, that requires pre-processing to release its sugars [83,155]. When these effluents are provided as feed in fed-batch PHA accumulations, they provoke substantial increases in the reactor operating volume, thus reducing the productivity.

A promising way to obtain high cell densities and productivities would be the use of cell recycling systems coupled to fed-batch processes [156]. This strategy has been scarcely applied to PHA production until now, with only one report using an MMC [139] and one study in pure strains [157]. The latter obtained cell densities up to 200 g/L with a productivity of 4.6 g PHA/L/h by using external cross-flow membranes to recirculate cells into the fed-batch reactor using *C. necator* [157]. Thus, the strategy seems to offer good opportunities to increase the cell density and PHA concentration. Likewise, reactor designs preventing the cells from escaping the system, while allowing supernatant removal, could result in high cell densities and reduced reactor volumes.

With the same scope, other research groups have proposed influent concentration. Although evaporation has been suggested [157], other less energy intensive methods such as forward osmosis membranes could more likely be applied [158,159]. A very interesting approach was recently published where forward osmosis Aquaporin® membranes, mimicking biological protein channels, were suggested to concentrate fermentation effluents from glycerol and wheat straw [160]. The novelty lied in the fact that the concentrated feedstock could be used as the water draw solution with the diluted fermentation effluent being the water feed solution. This enabled the recirculation of water from the effluent to the influent in an energy efficient way since the process was based solely on the use of forward osmosis membranes without the need of the costly regeneration of the draw solution (usually performed by applying the energy intensive reverse osmosis). Integration of such systems in the PHA production process could also enhance its productivity.

4. PHA Recovery

The development of new strategies and methodologies for PHA recovery is one of the main factors associated with the feasibility of a PHA production bio-refinery using microbial mixed fermentation. As demonstrated in Figure 2, PHA recovery uses a variation of different techniques. However, PHA purification generally requires five steps: biomass-harvesting, pre-treatment, PHA recovery, PHA accumulation, polishing, and drying. Biomass harvesting is the concentration of biomass using techniques such as filtration or centrifugation. As PHAs are intracellular polymers, it is necessary to concentrate the biomass prior PHA recovery. Nevertheless, some researchers have evaluated PHA recovery without biomass harvesting to facilitate process scale-up and to reduce costs. Pre-treatment's main objective is to facilitate PHA retrieval from the microbial biomass; these techniques include drying techniques (lyophilization and thermal drying), grinding, chemical and biochemical pre-treatments, etc. The pre-treatment step can combine two or more methods. PHA retrieval phase utilizes two principal methods: PHA solubilization and the disruption of non-PHA cell mass (NPCM). In some cases, NPCM disruption precedes a PHA solubilization step. The PHA accumulation step is dependent on the retrieval technique utilized. In PHA solubilization, the PHA is concentrated by using alcohols precipitation. On the other hand, in NPCM disruption, recovery is performed by collecting the PHA granules. As final steps, recovered PHAs can be polished by removing residues, from the previous steps, or can be dried; depending on the separation steps utilized. As of 2013, two reviews specialized in PHA recovery were published [161,162]. This section primarily focuses in the most recent developments in pre-treatments and PHA retrieval steps; additionally, this section describes a more industrial opinion on the PHA recovery methods.

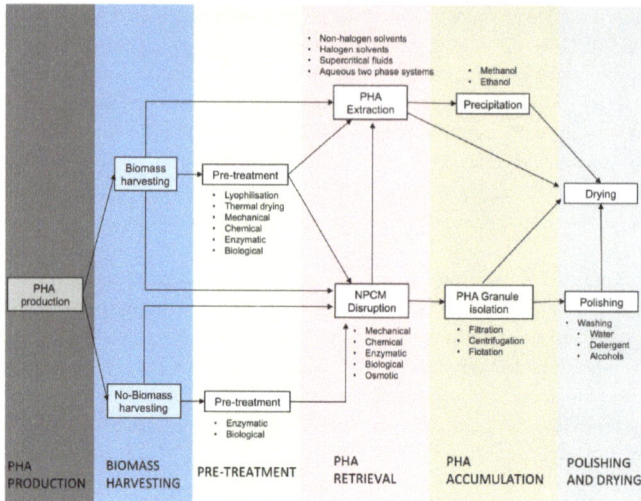

Figure 2. PHA recovery strategies.

4.1. Pre-Treatments

Pre-treatments are chemical, physical, or biological methods employed to facilitate the retrieval of PHA. These methods focus on weakening the cell structure that protects and surrounds the PHA granules. After biomass harvesting, drying is the traditional pre-treatment used in PHA recovery, which includes heat drying and lyophilization; the latter is the most employed pre-treatment in PHA recovery processes. This technique removes the majority of water molecules in the biomass facilitating the posterior PHA extraction. Although lyophilization has interesting features, it has economic and technical difficulties that reduce its future application in an industrial PHA recovery process. In recent years, these industrial difficulties increased the interest in PHA recovery from wet biomass instead of dry biomass [163–165]. Lyophilization can be a unique pre-treatment or it can precede a further pre-treatment. Lyophilization has preceded thermal, mechanical, and chemical pre-treatments (see Table 5). All these treatments included a further retrieval technique associated with solvent extraction [166,167]. Samori et al. [166], described chemical pre-treatment with NaClO as the best pre-treatment for PHA recovery compared with thermal and mechanical pre-treatments; however, this pre-treatment generated an important reduction in the molecular weight of the polymer.

Pre-treatments without lyophilization consisted of chemical and physical methods, with chemical methods including sodium chloride (NaCl) and sodium hypochlorite (NaClO). Anis et al. [163], employed NaCl as a pre-treatment for a NaClO digestion. The additional pre-treatment step generated an increment in purity and yield. NaCl pre-treatment modifies the osmotic conditions in the medium leading the cells to dehydrate and shrink, this destabilizes the cell membranes facilitating the PHA granules' liberation. Physical methods include high temperature and ultrasonication methods both methods anteced NPCM digestion methods. Neves and Muller [168] evaluated three temperatures (121, 100, and 80 °C) during 15 min as pre-treatment conditions. The 121 °C treatment achieved the best results; whereas, 100 and 80 °C treatments recovered significantly lower amounts of PHA. The heat treatment improves the PHA removal by denaturizing proteins, DNA and RNA, destabilizing the microbial cell wall and inactivating the PHB depolymerase [168]. Ultrasonication employs sound waves to create disruption in the cell wall and open the cytosolic material to the aqueous medium. Leong et al. [164] utilized ultrasonication prior an aqueous two-phase extraction. The advantages of this method are the lack of any previous cell harvesting method and the fast pace in which is performed.

Table 5. Pre-treatment techniques applied for PHA recovery.

Pre-Treatment	Further PHA Retrieval Treatment	Pre-Treatment Conditions	Purity (%)	Recovery (%)	Ref.
Sodium chloride (NaCl)	NaOH digestion	NaCl (8 g/L), 30 °C, 3 h	97.7	97.5	[163]
Ultrasonication	Aqueous two-phase extraction	Ultrasonication at 30 kHz per cycle 15 min		-	[164]
Sodium hypochlorite (NaClO)	Non-halogenated solvent extraction	NaClO (10%), 37 °C, 1 h		-	[165]
Thermal pre-treatment	Enzymatic digestion and chloroform extraction	Autoclave, 15 min 121 °C		94.1	[168]
Thermal pre-treatment [1]	Non-halogenated solvent extraction	150 °C, 24 h		50	[166]
Ultrasonication and glass beds [1]	Non-halogenated solvent extraction	Glass beads (0.5 mm) and Ultrasonication (10 pulses of 2 min)		50	[166]
Sodium hypochlorite [1]	Non-halogenated solvent extraction	NaClO (5%), 100 °C, 15 min	93	82	[166]
Hot acetone [1]	Non-halogenated solvent extraction	Acetone, 100 °C, 30 min		-	[167]

[1] The pre-treatment included a previous lyophilization step.

Pre-treatments can increase the recovery and purity of the PHA extracted from a fermentation broth. However, their implementation in a PHA industrial recovery process needs to be evaluated using economic and technical analysis. The yield and purity increment should counterbalance with the additional cost associated with the introduction of this step in the purification line. Additionally, it is important to remark that even though several pre-treatments can be used, not all of them are suitable for industrial applications. Industrially suitable pre-treatments require the use of wet biomass or unharvested biomass in order to reduce the number of purification steps and the costs associated with the purification process. Lyophilization use should concentrate on PHA chemical analyses or be replaced with more suitable drying techniques for process scale up. The selection of a pre-treatment is dependent of the bacterial strain, fermentation broth characteristics, and further PHA application; therefore, each PHA process needs to be analyzed individually from an economic, environmental, and technical point of view.

4.2. Retrieval Techniques

4.2.1. Non-PHA Cell Mass (NPCM) Disruption

In recent years, the use of NPCM disruption as a tool for retrieving PHA from bacterial biomass has increased. This increment is associated with the necessity of environmental and safe options to replace the use of halogenated solvents used in the traditional PHA extraction methods. Additionally, some of the NPCM techniques for the PHA extraction have used wet biomass or unharvested biomass, which is an advance through the reduction of purification steps and costs. This review grouped the novel NPCM disruption techniques into chemical, enzymatic, and biological disruption.

Chemical Disruption

NPCM chemical disruption includes methods that utilize chemical compounds to disrupt bacterial cell wall and the denaturalization or degradation of cytosolic material. The three principal methods for the chemical disruption are sodium hydroxide (NaOH), sodium hypochlorite (NaClO), and sodium dodecyl sulphate (SDS). Additional chemical treatments include water and acid treatments. NaOH treatment destabilizes the cell wall by reacting with the lipid layer in a saponification reaction and increases the cell membrane permeability [163]. NaOH treatment has obtained high recovery and purification percentage using pre-treated and unpretreated biomass (see Table 6). However, pre-treated biomass aided the NaOH treatments to achieve improvements in purity and recovery [163,169]. The combination of NaCl pre-treatment and NaOH digestion increased the purity and recovery of NaOH treatment by approximately 10% [163]. Similarly, lyophilization and freezing helped to increase the PHA purity using NaOH; however, these pre-treatments did not increase the recovery percentage [169]. López-Abelairas et al. [170] described a recovery reduction in treatments with high biomass and NaOH concentrations. Biomass concentration above 2.5% created a constant reduction in recovery and purity; however, biomass concentrations of 7.5% and 10% yielded the greatest recovery reductions. NaOH concentrations over 0.5 N affected recovery, although, the purity percentage at high percentage was constant. Likewise, Villano et al. [171] achieved recovery between 80–87% utilizing high NaOH (1 M), high biomass concentration (6:1 biomass:chemical solution) and extended extraction times (6–24 h). However, the purity achieved by this treatment was below 60%.

Sodium hypochlorite (NaClO) has confirmed its positive aspects as an NCMP disruption treatment [161,162]. Therefore, in recent years, the assessments of this treatment advanced to larger scales or continuous processes [171,172]. A continuous sequential process for PHA production and recovery obtained high polymer recovery (100%) and purity (98%). This continuous process contains three steps: a production step using microbial mixed culture, a PHA accumulation step, and a PHA extraction step using NaClO (5%) disruption for 24 h. This approach produced 1.43 g PHA/L·d and was stable for four months [171].

Another chemical disruption treatment is sodium dodecyl sulphate (SDS). This surfactant is a well-known detergent used in the recovery of genetic material. SDS treatment obtained recovery and purity values comparable with other chemical disruption techniques; moreover, this disruption technique obtained similar retrieval with and without biomass pre-treatment [169,173]. The amount of SDS varied between 0.025% and 0.2%, higher concentrations of SDS generated higher purity as result of SDS micelles formation. High micelles production is also associated with the solubilization of PHA granules generating a reduction in the recovery yield. SDS has complemented NaOH disruption, this combination exhibited superior levels of purity, especially in the removal of hydrophobic impurities [169].

Besides the previous treatments, other authors have described water and acid disruption methods as effective treatments for NPCM disruption. Distilled water achieved high purity (94%) and recovery (98%) percentages; however, the process needed 18 h and lyophilized biomass to reach these high percentages. The process duration improved by adding SDS (0.1%) into the mixture [173]. In contrast, distilled water disruption treatment with wet biomass obtained recovery (80%) and purity (58%) percentages lower than lyophilized biomass. Mohammadi et al. [174], described a higher purity and recovery yield using distilled water disruption with recombinant bacteria instead of wild bacteria. Recombinant bacterial cell wall is thinner than in wild type bacteria, which facilitates the cell wall breaking by osmotic pressure. Acid treatments have demonstrated their capability to disrupt NPCM; López-Abelairas et al. [170] described a recovery and purity percentage using a sulphuric acid solution (0.64 M) similar to alkaline treatments (NaOH, NaClO). They selected acid disruption as the best recovery method focused in operational and environmental factors. The authors chose acid disruption because this process had lower cost, environmental impact (greenhouse gas emissions), and polymer degradation than alkaline treatments [170].

Chemical disruption treatments are a significant option in PHA recovery since they present environmental and economic advantages over the traditional PHA extraction using halogenated solvents. Environmentally, chemical disruption avoids the use of toxic solvents such as chloroform. Economic advantages include liquid current recycling, the use of wet biomass, and the reagent cost. The principal drawback for chemical disruption has been polymer degradation; however, the use of mixtures of chemicals and process optimization has reduced polymer degradation. The selection of a chemical disruption method for PHA recovery from mixed microbial cultures needs an all-around evaluation of technical, economic, and environmental factors that consider the positive and negative effects and how they can affect the feasibility of a PHA bio-refinery plant.

Enzymatic NPCM Disruption

Enzymatic NPCM disruption utilizes purified enzymes or crude extracts to disrupt the bacterial cell wall. Proteases are the principal enzymatic activities employed in enzymatic disruption; however, other types of enzymes or enzymatic cocktail have effectively degraded NCMP. Enzymatic disruption advantages include their low energy requirements, aqueous recovery, and low capital investment; in contrast, the enzymes production cost is the principal disadvantage for industrial implementation [175]. Gutt et al. [176], evaluated the recovery of P3HBHV from *Cupriavidus necator* by several methods including enzymatic disruption. Simple enzymatic treatment (lysozyme) obtained low recovery and purity (Table 6). The authors attributed these low percentages to the absence of additional chemical or mechanical treatments, which have proved necessary for achieving high recovery and purity. Martino et al. [177] evaluated enzymatic disruption using simultaneously enzymatic (Alcalase) and chemical treatments (SDS and EDTA). SDS and EDTA contributed to cell wall and membrane lysis whereas Alcalase solubilized the cytosolic material. This enzymatic/chemical digestion treatment eliminated the requirement of heat pre-treatment used in previous enzymatic disruption researchers [161,162]. Kachrimanidou et al. [175], developed a novel enzymatic disruption method by using crude enzymes from solid-state fermentation of *Aspergillus awamori*. This method achieved good recovery (98%) and purity (97%) without using additional chemicals; however, it required heat and lyophilization as pre-treatments. Approaches focused in the reduction of enzyme costs are necessary to facilitate the industrial application of enzymatic disruption including the use of immobilized enzymes, integration of enzymes production as part of a PHA biorefinery, and genetically engineering enzymes.

Biological NPCM Disruption

Biological disruption utilizes biological agents (virus) or organisms to liberate PHA from bacterial cells. The first biological disruption technique used viral particles to break bacterial cells. Bacteriophages were included in bacterial lines to utilize the viruses' lytic cycle to liberate PHA granules. When the lytic cycle is completed, the virus escapes from the host cell by breaking down the cell wall; this breaking down also liberates the PHA particles allowing their recovery [162]. In recent years, biological disruption methods included bacteria predators, rats, and mealworms [178–180].

Martinez et al. [181], proposed the use of obligate predatory bacteria *Bdellovibrio bacteriovorus* as an innovative cell lytic agent suitable to recover intracellular bioproducts such as PHA. *B. bacteriovorus* achieved a PHA recovery of 60%, the recovery percentage obtained was attributed to PhaZ depolymerase activity which hydrolyses PHA and expresses during all the stages of *B. bacteriovorus*'s life cycle [181]. To improve the use of *B. bacteriovorus* in PHA recovery, Martinez et al. [180], developed *B. bacteriovorus* mutant strains, one with an inactive medium-chain-length PHA depolymerase (*B. bacteriovorus* Bd3709) and another with inactive short-chain-length-PHA depolymerase (*B. bacteriovorus* Bd2637). *B. bacteriovorus* Bd3709 increased PHA recovery from 60% to 80% when predating *Pseudomonas putida*, whereas, *B. bacteriovorus* Bd2637 increased PHB recovery from 48% to 63% when predating PHB-accumulating *E. coli* ML35 [180]. Biological recovery using *B. bacteriovorus* has advantages such as avoiding the use of cell harvesting and pre-treatments. However, it has disadvantages such as the use of organic solvent steps, processing time, and low recovery. In the future, this biological method can be complemented with other recovery treatments to avoid solvents' usage and to increase PHA recovery.

In recent years, authors have used complex organisms' digestive system as a NPCM disruption technique. In these treatments, different organisms were fed with PHA-rich bacteria; afterwards, PHA granules were recovered from these organism's feces. This process selectively digested the NPCM without reducing the PHA molecular weight. Kunasundari et al. [182], fed Sprague Dawley rats with lyophilized cells of *Cupriavidus necator* in a single cell protein diet during several days. The fecal pellets were whitish and rich in PHA (82–97%). The authors also demonstrated the safety and tolerability of the Sprague Dawley rats to a *Cupriavidus necator* diet. Kunasundari et al. [179], studied the purification of the Sprague Dawley PHA-rich fecal pellets using water and surfactants. The use of SDS 2% as a further purification step increased the purity of the PHA biological recovered at levels similar to solvent extraction. Similar to Sprague Dawley rats, Murugan et al. [180] fed lyophilized bacteria to mealworms and recovered PHA granules from their fecal pellets. The PHA granules had an 89% purity when washed with water and reached almost 100% purity when treated with SDS. The authors reported higher protein content in mealworms fed with *C. necator* cells than mealworms fed with oats. Mealworms fed with *C. necator* can be an alternative protein source in aquaculture and poultry diets. Biological NPCM disruption is an alternative to other disruption methods; it does not require expensive instrumentation, solvents, or strong chemicals, and the organisms doing the PHA recovery can be a marketable product too. However, the biological recovery process takes longer than any other recovery process and needs biomass pre-treatment. Depending on the final use, PHA biological recovery can be an integrated process for renewable production of feed, food, and materials.

Table 6. NPCM chemical disruption treatments.

NPCM Digestion Type	NPCM Disruption Method	Pre-Treatment	PHA Accumulation Method	Disruption Conditions	Microbial Strain	PHA Content in Biomass (%)	Purity (%)	Recovery (%)	Ref.
Chemical	NaOH	Chemical Pre-treatment	Centrifugation	NaOH (0.1 M), 30 °C, 1 h, 350 rpm	*C. necator*	68	90.8	95.3	[163]
Chemical	NaOH		Centrifugation	NaOH (0.1 M), 30 °C, 1 h, 350 rpm	*C. necator*	68	82.7	94.4	[163]
Chemical	NaOH	Lyophilization	Centrifugation	NaOH (0.1M), 30 °C, 1 h,	*C. necator*	68	80–90	80–90	[183]
Chemical	NaOH	Lyophilization	Centrifugation	NaOH 0.05 M, 3 h, 0 rpm, 4 °C	*C. necator*	30	98.6	96.9	[174]
Chemical	NaOH	Lyophilization and milling	Centrifugation	NaOH (0.5 N), 4 h, 37 °C, 500 rpm	*C. necator*	65	93	80	[170]
Chemical	NaOH		Centrifugation	NaOH (0.2 M), 200 rpm, 30 °C, 1 h	*Mixed Culture*	62–72	87	97	[169]
Chemical	NaClO		Centrifugation	NaClO (5%) 24 h	*Mixed Culture*	46	90	~100	[171]
Chemical	NaClO	Mechanical pre-treatment	Precipitation	NaClO 13% (v/v), room temperature, 1 h.	*Ralstonia eutropha*	65.2	95.6	91.3	[172]
Chemical	NaClO	Lyophilization and Milling	Centrifugation	NaClO (13%), 37 °C, 500 rpm,4 h.	*C. necator*	65	97	82	[170]
Chemical	NaOH and SDS		Centrifugation	NaOH (0.2 M) and SDS (0.2 %), 200 rpm, 30°C, 1 h	*Mixed Culture*	62–72	99	91	[169]
Chemical	SDS		NaClO and Centrifugation	SDS (0.1%), 24 h	*H. mediterranei*	70	~100	97	[55]
Chemical	SDS		Centrifugation	SDS (0.1%), 24 h	*H. mediterranei*	71.2	~100	97	[56]
Chemical	SDS		Centrifugation	SDS (0.1%), 60 °C, 2 h	*Halomonas sp.* SK5	48	94	98	[173]
Chemical	SDS		Centrifugation	SDS (0.2 %), 200 rpm, 30 °C, 1 h	*Mixed Culture*	62–72	79	63.5	[169]
Chemical	H_2SO_4	Lyophilization and Milling	Chemical treatment and Centrifugation	H_2SO_4 (0.64 M), 6 h, 80 °C	*C. necator*	65	98	79	[170]
Chemical	Water	Lyophilization	Centrifugation	dH_2O, 30 °C, 1 h,	*Comamonassp.*	30	80.6	96	[174]
Chemical	Water	Lyophilization	Centrifugation	dH_2O, 30 °C, 18 h	*Halomonas sp.*	48	94	98	[173]
Enzymatic	Alcalase, SDS and EDTA		Centrifugation	Alcalase (0.3 U g−1), SDS (0.3 g g−1), EDTA (0.01 g g−1), Na2HPO4 buffer, 150 rpm, 55 °C, 1 h	*C. necator*	37	94		[177]
Enzymatic	Crude extract	Heat treatment and lyophilization	Centrifugation	*Aspergillus oryzae* crude extract, Na2HPO4–citric acid buffer and 47 °C	*C. necator*	78.9	98	97	[175]
Enzymatic	Lysozyme		Centrifugation	Lysozyme solution (2 mg/mL), 1 h, 3 °C	*C. necator*	41	41	75	[176]
Biological	Mealworm (*Tenebrio molitor*)	Lyophilization	Chemical treatment, centrifugation	50 g of mealworms fed 5% of their body weight per day for 16 days.	*C. necator*	37	89%		[178]
Biological	Sprague Dawley rats	Lyophilization and grinding	Chemical treatment, centrifugation	150–200 g rats were feed 15 g/day/animal, 28 days 25 °C	*C. necator*	37	89.3	100	[179]
Biological	Sprague Dawley rats	Lyophilization	Water	150–200 g rats were feed 15 g/day/animal, 28 days 25 °C	*C. necator*	54	82–97	40–47	[182]
Biological	*Bdellovibrio bacteriovorus* HD100		Centrifugation	*P. putida* was inoculated with *B. bacteriovorus* strains 48 h. 30 °C	*P. putida*	55		60	[181]
Biological	*Bdellovibrio bacteriovorus* HD100 and Bd3709		Centrifugation	*P. putida* and *E. coli* cultures were inoculated with *B. bacteriovorus* strains 48 h. 30 °C	*P. putida*	55		80	[180]

4.2.2. PHA Extraction

As PHA are produced inside the cellular biomass, in order to be retrieved they have to be separated from the non-PHA cell mass (NPCM). The simplest, least destructive to biopolymer and most direct way for PHA is to be extracted from the biomass; significant quantities of hazardous solvents and energy input are required for this, creating a potential counterbalance to sustainability and economics towards commercialization [184].

Several studies have proposed various solvent extraction methods for PHA recovery that improve parameters such as yield, purity, and cost of extraction, while at the same time maintaining the physicochemical properties of the biopolymer [185].

Non-Halogenated Solvents

Although several types of extraction systems exist for the production of biopolymers, the majority of extraction methods for PHA still involve the use of organic solvents in which the polymer is soluble. Fei et al. [186], aimed to develop an effective and environmentally-friendly solvent system so as to extract PHB from bacterial biomass. In order to accomplish that, they used a solvent mixture of acetone/ethanol/propylene carbonate (A/E/P, 1:1:1 v/v/v) for extracting PHB from *Cupriavidus necator*. When the A/E/P mixture was used at high temperature, it could recover 92% pure PHB with 85% yield from dry biomass, and 90% purity with 83% PHB yield from wet biomass. Additionally, if hexane was added, it could further enhance the purity and recovery quantities of PHB.

Bacterial PHA could be used for medical applications due to its biocompatibility. However, using inappropriate solvents or techniques during extraction of PHA from bacterial biomass could result in contamination by pyrogenic compounds (e.g., lipopolysaccharides), which eventually leads to rejection of the material for medical use. This problem could be overcome by using a temperature-controlled method for the recovery of poly(3-hydroxyoctanoate-co-3-hydroxyhexanoate) from *Pseudomonas putida* GPo1. Non-chlorinated solvents were found to be the optimal solvents for such tests and, specifically, n-hexane and 2-propanol. The purity reached more than 97% (w/w) and the endotoxicity between 10–15 EU/g PHA. Further re-dissolution in 2-propanol at 45 °C and precipitation at 10 °C resulted in a purity of nearly 100% (w/w) and endotoxicity equal to 2 EU/g PHA [187].

Another approach, however, is that the use of aqueous solvents could benefit the integration into a biorefinery scheme. A study aimed at connecting the exploitation of a raw biowaste, such as used cooking oil (UCO), and producing a desired final product (i.e., amorphous granules of PHA) based on aqueous solvents in a way that will prove the effective reliability of an overall biotechnological approach. Used cooking oil was utilized as the only carbon source for the production of PHB by cultivating *Cupriavidus necator* DSM 428 in batch reactors. The PHB granules were extracted from the biomass using sodium dodecyl sulfate (SDS), ethylenediaminetetraacetic acid (EDTA), and the enzyme Alcalase in an aqueous medium. The PHB granule recovery reached more than 90% and highly pure amorphous polymer was finally obtained [177].

A different alternative to the use of halogen-free and environmental-friendly methods implemented the use of water and ethanol for the recovery of PHA from recombinant *Cupriavidus necator*, in comparison to the well-established chloroform extraction method. Comparing the results obtained from experiments under different incubation times (1, 3, and 5 h) and temperatures (4 and 30 °C) showed that the optimized halogen-free method produced a PHA with 81% purity and 96% recovery yield, whereas the chloroform extraction system resulted in a highly pure PHA with 95% recovery yield. This method could potentially be developed as an alternative and more environmentally-friendly method for industrial application [188].

Aqueous Two-Phase Extraction Systems (ATPS)

In addition to the conventional isolation and purification methods, such as solvent extraction, aqueous two-phase systems (ATPS) have many advantages and important characteristics that attract the attention of researchers and industries. The main advantages are that because ATPS comprise of high water content (70–90% w/w), thus they provide a beneficial environment for separation of sensitive biomaterials. Also, the materials that form the different phases/layers of ATPS are, in principle, safe and environmentally-friendly compared to conventional solvent extraction methods; additionally, an intricate ability for high capacity processing which leads to reduced purification steps; finally, large-scale purification using ATPS can be easily and reliably predicted from laboratory experimental data. ATPS is a feasible solution for industrial demand of cost-effective and highly efficient large-scale bioseparation technologies with short processing times [189,190].

Regarding PHA retrieval methods, one interesting approach is the use of aqueous two-phase systems with the aim of enhancing the accumulation of PHA in one phase using environmentally-friendly layer-forming constituents. This method offers advantages of supporting a beneficial environment for bioseparation, capability of handling high operating capacity, and reducing downstream processing volume, thus proving extractive bioconversion via ATPS can be an optimum solution. Leong et al. [189], examined the effect of pH and salts' addition in *Ralstonia eutropha* H16 cultures, using an ATPS system as a mechanism for PHA extraction. The optimum result obtained in this study was a PHA concentration of 0.139 g/L (purification factor: 1.2–1.63) and recovery yield 55–65% using ATPS of polyethylene glycol (PEG) 8000/sodium sulphate adjusted to pH 6 and the addition of 0.5 M NaCl.

In another study using ATPS, the thermal separation of the phases and how it affected the PHA extraction was studied. The most important ATPS parameters (type and concentration of thermoseparating polymer, salt addition, feedstock load, and thermoseparating temperature) were optimized in order to achieve high PHA retrieval from the bacterial lysate. By taking advantage of the properties of the thermo-responsive polymer (whose solubility decreases in its aqueous solution as the temperature increases), cloud point extraction (CPE) is an ATPS technique that offers the capability to its phase-forming component to be reused. Extraction of PHA from *Cupriavidus necator* H16 via CPE was investigated. The best conditions for PHA extraction (recovery yield of 94.8% and purification factor 1.42) were reached under the following conditions: 20% w/w ethylene oxide-propylene oxide (EOPO), 10 mM NaCl addition, and a thermoseparating temperature of 60 °C with crude feedstock limit of 37.5% w/w. Another benefit of this process is the ability to recycle and reuse EOPO 3900 at least twice, achieving a satisfying yield and purification factor [164].

5. Conclusions

Wide production, commercialization, and thus application of PHAs as a biodegradable alternative to conventional plastics is still limited due to high production cost. Bioprocess technologies are still being developed while bacterial resources are still being explored.

Pure cultures are constantly investigated for their potential to valorize waste byproducts as a low-cost feedstock. The ability of certain bacteria to directly utilize lignocellulosic biomass as the carbon source for PHA production is a huge bioprocess advantage. In this case, the need for chemicals, energy, and labor is minimized. PHA production can be also used as a tool for the bioremediation of oil contaminated sites, as bacterial strains can degrade environmental pollutants, minimizing their toxicity and environmental impact. Moreover, halophilic PHA producers combine a series of advantages, such as growth on seawater and possibility of continuous processes under non-sterile conditions. Those features will significantly contribute to the PHA cost reduction and minimize the requirements for fresh water. Synthetic biology tools are expected to aid in the enhancement of PHA production efficiency, simplify downstream process, and regulate PHA composition providing customized materials for specific applications. However, robust strains are yet to be developed.

Efforts have been also made towards the PHA production using mixed microbial cultures (MMC). MMC-based processes, apart from offering a reduction in the operational costs and the possibility to adapt to a wider range of waste substrates, they could be integrated in current wastewater treatment plants. Recent developments regarding different enrichment strategies and the PHA production

Bioengineering **2017**, *4*, 55

step offer new opportunities to make the PHA production more feasible. Cell densities and derived productivities attained with MMC are the main current bottleneck.

The development of economic and simple downstream processes is crucial for the recovery of PHAs. Methods based on the utilization of environmentally friendlier techniques are constantly being investigated, with enzymatic methods advancing the bio-based profile of the process. Reduction of large amounts of chemicals, used per cell dry mass, is going to benefit the economics of the process as well as society and the environment.

Acknowledgments: Constantina Kourmentza would like to thank the European Commission for providing financial support through the 'SimPHAsRLs' project (Grant Agreement no. 625774) funded by the scheme FP7-PEOPLE-2013-IEF-Marie Curie Action: Intra-European Fellowships for Career Development. Jersson Plácido would like to thank the financial support provided by the European Regional Development Fund/Welsh Government funded BEACON research program (Swansea University). Anna Burniol-Figols, Christiano Varrone, and Hari Gavala thank the European Commission for the financial support of this work under FP7 Grant Agreement no. 613667 (acronym: GRAIL). Maria Reis would like to acknowledge and thank all partners and researchers working on 'INCOVER' (Grant Agreement no. 689242) and 'NoAW' (Grant Agreement no. 688338) projects and the European Union for providing funding under H2020 research and innovation program.

Author Contributions: Constantina Kourmentza conceived the concept of the study and reviewed the literature regarding PHA production using pure culture biotechnology. Anna Burniol-Figols, Christiano Varrone, and Hari Gavala prepared the section referring to PHA production by mixed microbial consortia. Nikolaos Venetsaneas and Jersson Plácido reviewed the part of PHA recovery techniques. Maria Reis revised and edited the manuscript.

Conflicts of Interest: The authors declare no conflict of interest.

References

1. Plastics Europe. *Plastics Europe Plastics—The Facts 2016*; Plastics Europe: Brussels, Belgium, 2016; pp. 1–38.
2. Kourmentza, C.; Kornaros, M. Biotransformation of volatile fatty acids to polyhydroxyalkanoates by employing mixed microbial consortia: The effect of pH and carbon source. *Bioresour. Technol.* **2016**, *222*, 388–398. [CrossRef] [PubMed]
3. Singh Saharan, B.; Grewal, A.; Kumar, P. Biotechnological Production of Polyhydroxyalkanoates: A Review on Trends and Latest Developments. *Chin. J. Biol.* **2014**, *2014*, 1–18. [CrossRef]
4. Chen, G.-Q. A microbial polyhydroxyalkanoates (PHA) based bio- and materials industry. *Chem. Soc. Rev.* **2009**, *38*, 2434–2446. [CrossRef] [PubMed]
5. López, N.I.; Pettinari, M.J.; Nikel, P.I.; Méndez, B.S. Polyhydroxyalkanoates: Much More than Biodegradable Plastics. *Adv. Appl. Microbiol.* **2015**, *93*, 73–106. [PubMed]
6. Kourmentza, C.; Koutra, E.; Venetsaneas, N.; Kornaros, M. Integrated Biorefinery Approach for the Valorization of Olive Mill Waste Streams Towards Sustainable Biofuels and Bio-Based Products. In *Microbial Appications Vol. 1—Bioremediation and Bioenergy*; Kalia, V.C., Kumar, P., Eds.; Springer: Berlin, Germany, 2017; Volume 1, pp. 211–238.
7. Plastics Technology. Available online: http://www.ptonline.com/articles/prices-bottom-out-for-polyolefins-pet-ps-pvc-move-up (accessed on 26 May 2017).
8. Eno, R.; Hill, J. Metabolix Bio-industrial Evolution. In Proceedings of the Jefferies 11th Global Clean Technology Conference, New York, NY, USA, 23–24 February 2011.
9. Markets and Markets Polyhydroxyalkanoate (PHA) Market by Type (Monomers, Co-Polymers, Terpolymers), Manufacturing Technology (Bacterial Fermentation, Biosynthesis, Enzymatic Catalysis), Application (Packaging, Bio Medical, Food Services, Agriculture)-Global Forecast to 202. Available online: http://www.marketsandmarkets.com/Market-Reports/pha-market-395.html (accessed on 26 May 2017).
10. Kourmentza, C.; Ntaikou, I.; Lyberatos, G.; Kornaros, M. Polyhydroxyalkanoates from *Pseudomonas* sp. using synthetic and olive mill wastewater under limiting conditions. *Int. J. Biol. Macromol.* **2015**, *74*, 202–210. [CrossRef] [PubMed]
11. Fraiberg, M.; Borovok, I.; Weiner, R.M.; Lamed, R. Discovery and characterization of cadherin domains in *Saccharophagus degradans* 2-40. *J. Bacteriol.* **2010**, *192*, 1066–1074. [CrossRef] [PubMed]
12. Weiner, R.M.; Taylor, L.E.; Henrissat, B.; Hauser, L.; Land, M.; Coutinho, P.M.; Rancurel, C.; Saunders, E.H.; Longmire, A.G.; Zhang, H.; et al. Complete genome sequence of the complex carbohydrate-degrading marine bacterium, *Saccharophagus degradans* strain 2-40T. *PLoS Genet.* **2008**, *4*, e1000087. [CrossRef] [PubMed]

13. Ekborg, N.A.; Taylor, L.E.; Longmire, A.G.; Henrissat, B.; Weiner, R.M.; Steven, W.; Hutcheson, S.W. Genomic and Proteomic Analyses of the Agarolytic System Expressed by *Saccharophagus degradans* 2-40. *Appl. Environ. Microbiol.* **2006**, *72*, 3396–3405. [CrossRef] [PubMed]

14. Andrykovitch, G.; Marx, I. Isolation of a new polysaccharide-digesting bacterium from a salt marsh. *Appl. Environ. Microbiol.* **1988**, *54*, 1061–1062. [PubMed]

15. Taylor, L.E.; Henrissat, B.; Coutinho, P.M.; Ekborg, N.A.; Hutcheson, S.W.; Weiner, R.M. Complete cellulase system in the marine bacterium *Saccharophagus degradans* strain 2-40T. *J. Bacteriol.* **2006**, *188*, 3849–3861. [CrossRef] [PubMed]

16. Howard, M.B.; Ekborg, N.A.; Taylor, L.E.; Weiner, R.M.; Hutcheson, S.W. Genomic Analysis and Initial Characterization of the Chitinolytic System of *Microbulbifer degradans* Strain 2-40. *J. Bacteriol* **2003**, *185*, 3352–3360. [CrossRef] [PubMed]

17. Howard, M.B.; Ekborg, N.A.; Taylor, L.E.; Weiner, R.M.; Hutcheson, S.W. Chitinase B of "*Microbulbifer degradans*" 2-40 Contains Two Catalytic Domains with Different Chitinolytic Activities. *J. Bacteriol.* **2004**, *186*, 1297–1303. [CrossRef] [PubMed]

18. Suvorov, M.; Kumar, R.; Zhang, H.; Hutcheson, S. Novelties of the cellulolytic system of a marine bacterium applicable to cellulosic sugar production. *Biofuels* **2011**, *2*, 59–70. [CrossRef]

19. Munoz, L.E.A.; Riley, M.R. Utilization of cellulosic waste from tequila bagasse and production of polyhydroxyalkanoate (pha) bioplastics by *Saccharophagus degradans*. *Biotechnol. Bioeng.* **2008**, *100*, 882–888. [CrossRef] [PubMed]

20. González-García, Y.; Nungaray, J.; Córdova, J.; González-Reynoso, O.; Koller, M.; Atlic, A.; Braunegg, G. Biosynthesis and characterization of polyhydroxyalkanoates in the polysaccharide-degrading marine bacterium *Saccharophagus degradans* ATCC 43961. *J. Ind. Microbiol. Biotechnol.* **2008**, *35*, 629–633. [CrossRef] [PubMed]

21. González-García, Y.; Rosales, M.A.; González-Reynoso, O.; Sanjuán-Dueñas, R.; Córdova, J. Polyhydroxybutyrate production by *Saccharophagus degradans* using raw starch as carbon source. *Eng. Life Sci.* **2011**, *11*, 59–64. [CrossRef]

22. Sheu, D.S.; Chen, W.M.; Yang, J.Y.; Chang, R.C. Thermophilic bacterium *Caldimonas taiwanensis* produces poly(3-hydroxybutyrate-co-3-hydroxyvalerate) from starch and valerate as carbon sources. *Enzyme Microb. Technol.* **2009**, *44*, 289–294. [CrossRef]

23. Sawant, S.S.; Tran, T.K.; Salunke, B.K.; Kim, B.S. Potential of *Saccharophagus degradans* for production of polyhydroxyalkanoates using cellulose. *Process Biochem.* **2017**. [CrossRef]

24. Sawant, S.S.; Salunke, B.K.; Taylor, L.E.; Kim, B.S. Enhanced agarose and xylan degradation for production of polyhydroxyalkanoates by co-culture of marine bacterium, *Saccharophagus degradans* and its contaminant, *Bacillus cereus*. *Appl. Sci.* **2017**, *7*, 225. [CrossRef]

25. Sawant, S.S.; Salunke, B.K.; Kim, B.S. A Laboratory Case Study of Efficient Polyhydoxalkonates Production by *Bacillus cereus*, a Contaminant in *Saccharophagus degradans* ATCC 43961 in Minimal Sea Salt Media. *Curr. Microbiol.* **2014**, *69*, 832–838. [CrossRef] [PubMed]

26. Salamanca-Cardona, L.; Ashe, C.S.; Stipanovic, A.J.; Nomura, C.T. Enhanced production of polyhydroxyalkanoates (PHAs) from beechwood xylan by recombinant *Escherichia coli*. *Appl. Microbiol. Biotechnol.* **2014**, *98*, 831–842. [CrossRef] [PubMed]

27. Chen, W.M.; Chang, J.S.; Chiu, C.H.; Chang, S.C.; Chen, W.C.; Jiang, C.M. *Caldimonas taiwanensis* sp. nov., a amylase producing bacterium isolated from a hot spring. *Syst. Appl. Microbiol.* **2005**, *28*, 415–420. [CrossRef] [PubMed]

28. Elleuche, S.; Antranikian, G. Starch-Hydrolyzing Enzymes from Thermophiles. In *Thermophilic Microbes in Environmental and Industrial Biotechnology: Biotechnology of Thermophiles*; Satyanarayana, T., Littlechild, J., Kawarabayasi, Y., Eds.; Springer: Dordrecht, The Netherlands, 2013; pp. 509–533.

29. Ward, P.G.; De Roo, G.; O'Connor, K.E. Accumulation of polyhydroxyalkanoate from styrene and phenylacetic acid by *Pseudomonas putida* CA-3. *Appl. Environ. Microbiol.* **2005**, *71*, 2046–2052. [CrossRef] [PubMed]

30. Thompson, R.C.; Moore, C.J.; Vom Saal, F.S.; Swan, S.H. Plastics, the environment and human health: Current consensus and future trends. *Philos. Trans. R. Soc. B* **2009**, *364*, 2153–2166. [CrossRef] [PubMed]

31. Khandare, R.V.; Govindwar, S.P. Phytoremediation of textile dyes and effluents: Current scenario and future prospects. *Biotechnol. Adv.* **2015**, *33*, 1697–1714. [CrossRef] [PubMed]

32. Goudarztalejerdi, A.; Tabatabaei, M.; Eskandari, M.H.; Mowla, D.; Iraji, A. Evaluation of bioremediation potential and biopolymer production of pseudomonads isolated from petroleum hydrocarbon-contaminated areas. *Int. J. Environ. Sci. Technol.* **2015**, *12*, 2801–2808. [CrossRef]
33. Hori, K.; Abe, M.; Unno, H. Production of triacylglycerol and poly(3-hydroxybutyrate-co-3-hydroxyvalerate) by the toluene-degrading bacterium *Rhodococcus aetherivorans* IAR1. *J. Biosci. Bioeng.* **2009**, *108*, 319–324. [CrossRef] [PubMed]
34. Hori, K.; Kobayashi, A.; Ikeda, H.; Unno, H. *Rhodococcus aetherivorans* IAR1, a new bacterial strain synthesizing poly(3-hydroxybutyrate-co-3-hydroxyvalerate) from toluene. *J. Biosci. Bioeng.* **2009**, *107*, 145–150. [CrossRef] [PubMed]
35. Nikodinovic-Runic, J.; Casey, E.; Duane, G.F.; Mitic, D.; Hume, A.R.; Kenny, S.T.; O'Connor, K.E. Process analysis of the conversion of styrene to biomass and medium chain length polyhydroxyalkanoate in a two-phase bioreactor. *Biotechnol. Bioeng.* **2011**, *108*, 2447–2455. [CrossRef] [PubMed]
36. Tan, G.-Y.A.; Chen, C.-L.; Ge, L.; Li, L.; Tan, S.N.; Wang, J.-Y. Bioconversion of styrene to poly(hydroxyalkanoate) (PHA) by the new bacterial strain *Pseudomonas putida* NBUS12. *Microbes Environ.* **2015**, *30*, 76–85. [CrossRef] [PubMed]
37. Goff, M.; Ward, P.G.; O'Connor, K.E. Improvement of the conversion of polystyrene to polyhydroxyalkanoate through the manipulation of the microbial aspect of the process: A nitrogen feeding strategy for bacterial cells in a stirred tank reactor. *J. Biotechnol.* **2007**, *132*, 283–286. [CrossRef] [PubMed]
38. Tamboli, D.P.; Kurade, M.B.; Waghmode, T.R.; Joshi, S.M.; Govindwar, S.P. Exploring the ability of *Sphingobacterium* sp. ATM to degrade textile dye Direct Blue GLL, mixture of dyes and textile effluent and production of polyhydroxyhexadecanoic acid using waste biomass generated after dye degradation. *J. Hazard. Mater.* **2010**, *182*, 169–176. [CrossRef] [PubMed]
39. Tamboli, D.P.; Gomare, S.S.; Kalme, S.S.; Jadhav, U.U.; Govindwar, S.P. Degradation of Orange 3R, mixture of dyes and textile effluent and production of polyhydroxyalkanoates from biomass obtained after degradation. *Int. Biodeterior. Biodegrad.* **2010**, *64*, 755–763. [CrossRef]
40. Kahlon, R.S. *Pseudomonas: Molecular and Applied Biology*; Springer: Berlin, Germany, 2016.
41. Nikodinovic, J.; Kenny, S.T.; Babu, R.P.; Woods, T.; Blau, W.J.; O'Connor, K.E. The conversion of BTEX compounds by single and defined mixed cultures to medium-chain-length polyhydroxyalkanoate. *Appl. Microbiol. Biotechnol.* **2008**, *80*, 665–673. [CrossRef] [PubMed]
42. Ni, Y.Y.; Kim, D.Y.; Chung, M.G.; Lee, S.H.; Park, H.Y.; Rhee, Y.H. Biosynthesis of medium-chain-length poly(3-hydroxyalkanoates) by volatile aromatic hydrocarbons-degrading *Pseudomonas fulva* TY16. *Bioresour. Technol.* **2010**, *101*, 8485–8488. [CrossRef] [PubMed]
43. Narancic, T.; Kenny, S.T.; Djokic, L.; Vasiljevic, B.; O'Connor, K.E.; Nikodinovic-Runic, J. Medium-chain-length polyhydroxyalkanoate production by newly isolated *Pseudomonas* sp. TN301 from a wide range of polyaromatic and monoaromatic hydrocarbons. *J. Appl. Microbiol.* **2012**, *113*, 508–520. [CrossRef] [PubMed]
44. Ward, P.G.; Goff, M.; Donner, M. A Two Step Chemo—Biotechnological Conversion of Polystyrene to a Biodegradable Thermoplastic. *Environ. Sci. Technol.* **2006**, *40*, 2433–2437.
45. Kenny, S.T.; Runic, J.N.; Kaminsky, W.; Woods, T.; Babu, R.P.; Keely, C.M.; Blau, W.; O'Connor, K.E. Up-cycling of PET (Polyethylene Terephthalate) to the biodegradable plastic PHA (Polyhydroxyalkanoate). *Environ. Sci. Technol.* **2008**, *42*, 7696–7701. [CrossRef] [PubMed]
46. Guzik, M.W.; Kenny, S.T.; Duane, G.F.; Casey, E.; Woods, T.; Babu, R.P.; Nikodinovic-Runic, J.; Murray, M.; O'Connor, K.E. Conversion of post consumer polyethylene to the biodegradable polymer polyhydroxyalkanoate. *Appl. Microbiol. Biotechnol.* **2014**, *98*, 4223–4232. [CrossRef] [PubMed]
47. Tamboli, D.P.; Kagalkar, A.N.; Jadhav, M.U.; Jadhav, J.P.; Govindwar, S.P. Production of polyhydroxyhexadecanoic acid by using waste biomass of *Sphingobacterium* sp. ATM generated after degradation of textile dye Direct Red 5B. *Bioresour. Technol.* **2010**, *101*, 2421–2427. [CrossRef] [PubMed]
48. Maheshwari, D.K.; Saraf, M. *Halophiles: Biodiversity and Sustainable Exploitation*; Springer: Berlin, Germany, 2015; Volume 6.
49. Konstantinidis, G. *Elsevier'S Dictionary of Medicine and Biology Greek German Italian Latin*; Elsevier: Amsterdam, The Netherlands, 2005.
50. Setati, M. Diversity and industrial potential of hydrolaseproducing halophilic/halotolerant eubacteria. *Afr. J. Biotechnol.* **2010**, *9*, 1555–1560.

51. Yin, J.; Chen, J.C.; Wu, Q.; Chen, G.Q. Halophiles, coming stars for industrial biotechnology. *Biotechnol. Adv.* **2015**, *33*, 1433–1442. [CrossRef] [PubMed]

52. Quillaguamán, J.; Guzmán, H.; Van-Thuoc, D.; Hatti-Kaul, R. Synthesis and production of polyhydroxyalkanoates by halophiles: Current potential and future prospects. *Appl. Microbiol. Biotechnol.* **2010**, *85*, 1687–1696. [CrossRef] [PubMed]

53. Kirk, R.G.; Ginzburg, M. Ultrastructure of two species of halobacterium. *J. Ultrastruct. Res.* **1972**, *41*, 80–94. [CrossRef]

54. Rodriguez-Valera, F.; Ruiz-Berraquero, F.; Ramos-Cormenzana, A. Isolation of Extremely Halophilic Bacteria Able to Grow in Defined Inorganic Media with Single Carbon Sources. *Microbiology* **1980**, *119*, 535–538. [CrossRef]

55. Bhattacharyya, A.; Pramanik, A.; Maji, S.K.; Haldar, S.; Mukhopadhyay, U.K.; Mukherjee, J. Utilization of vinasse for production of poly-3-(hydroxybutyrate-co-hydroxyvalerate) by *Haloferax mediterranei*. *AMB Express* **2012**, *2*, 34. [CrossRef] [PubMed]

56. Bhattacharyya, A.; Saha, J.; Haldar, S.; Bhowmic, A.; Mukhopadhyay, U.K.; Mukherjee, J. Production of poly-3-(hydroxybutyrate-co-hydroxyvalerate) by *Haloferax mediterranei* using rice-based ethanol stillage with simultaneous recovery and re-use of medium salts. *Extremophiles* **2014**, *18*, 463–470. [CrossRef] [PubMed]

57. Pais, J.; Serafim, L.S.; Freitas, F.; Reis, M.A.M. Conversion of cheese whey into poly(3-hydroxybutyrate-co-3-hydroxyvalerate) by *Haloferax mediterranei*. *New Biotechnol.* **2016**, *33*, 224–230. [CrossRef] [PubMed]

58. Alsafadi, D.; Al-Mashaqbeh, O. A one-stage cultivation process for the production of poly-3-(hydroxybutyrate-co-hydroxyvalerate) from olive mill wastewater by *Haloferax mediterranei*. *New Biotechnol.* **2017**, *34*, 47–53. [CrossRef] [PubMed]

59. Tan, D.; Xue, Y.-S.; Aibaidula, G.; Chen, G.-Q. Unsterile and continuous production of polyhydroxybutyrate by *Halomonas* TD01. *Bioresour. Technol.* **2011**, *102*, 8130–8136. [CrossRef] [PubMed]

60. Yue, H.; Ling, C.; Yang, T.; Chen, X.; Chen, Y.; Deng, H.; Wu, Q.; Chen, J.; Chen, G.-Q. A seawater-based open and continuous process for polyhydroxyalkanoates production by recombinant *Halomonas campaniensis* LS21 grown in mixed substrates. *Biotechnol. Biofuels* **2014**, *7*, 108. [CrossRef]

61. Salgaonkar, B.B.; Mani, K.; Braganca, J.M. Characterization of polyhydroxyalkanoates accumulated by a moderately halophilic salt pan isolate *Bacillus megaterium* strain H16. *J. Appl. Microbiol.* **2013**, *114*, 1347–1356. [CrossRef] [PubMed]

62. Moorkoth, D.; Nampoothiri, K.M. Production and characterization of poly(3-hydroxy butyrate-co-3 hydroxyvalerate) (PHBV) by a novel halotolerant mangrove isolate. *Bioresour. Technol.* **2016**, *201*, 253–260. [CrossRef] [PubMed]

63. Rodríguez-Contreras, A.; Koller, M.; Braunegg, G.; Marqués-Calvo, M.S. Poly[(R)-3-hydroxybutyrate] production under different salinity conditions by a novel *Bacillus megaterium* strain. *New Biotechnol.* **2016**, *33*, 73–77. [CrossRef] [PubMed]

64. Chen, G.-Q. New challenges and opportunities for industrial biotechnology. *Microb. Cell Fact.* **2012**, *11*, 111. [CrossRef] [PubMed]

65. Wang, Y.; Yin, J.; Chen, G.Q. Polyhydroxyalkanoates, challenges and opportunities. *Curr. Opin. Biotechnol.* **2014**, *30*, 59–65. [CrossRef] [PubMed]

66. Li, T.; Ye, J.; Shen, R.; Zong, Y.; Zhao, X.; Lou, C.; Chen, G.-Q. Semirational Approach for Ultrahigh Poly(3-hydroxybutyrate) Accumulation in *Escherichia coli* by Combining One-Step Library Construction and High-Throughput Screening. *ACS Synth. Biol.* **2016**, *5*, 1308–1317. [PubMed]

67. Fu, X.-Z.; Tan, D.; Aibaidula, G.; Wu, Q.; Chen, J.-C.; Chen, G.-Q. Development of *Halomonas* TD01 as a host for open production of chemicals. *Metab. Eng.* **2014**, *23*, 78–91. [CrossRef] [PubMed]

68. Wang, Y.; Wu, H.; Jiang, X.; Chen, G.-Q. Engineering *Escherichia coli* for enhanced production of poly(3-hydroxybutyrate-co-4-hydroxybutyrate) in larger cellular space. *Metab. Eng.* **2014**, *25*, 183–193. [CrossRef] [PubMed]

69. Jiang, X.-R.; Wang, H.; Shen, R.; Chen, G.-Q. Engineering the bacterial shapes for enhanced inclusion bodies accumulation. *Metab. Eng.* **2015**, *29*, 227–237. [CrossRef] [PubMed]

70. Jiang, X.-R.; Chen, G.-Q. Morphology engineering of bacteria for bio-production. *Biotechnol. Adv.* **2016**, *34*, 435–440. [CrossRef] [PubMed]

71. Pfeiffer, D.; Jendrossek, D. Localization of poly(3-Hydroxybutyrate) (PHB) granule-associated proteins during PHB granule formation and identification of two new phasins, phap6 and phap7, in *Ralstonia eutropha* H16. *J. Bacteriol.* **2012**, *194*, 5909–5921. [CrossRef] [PubMed]

72. Martínez, V.; García, P.; García, J.L.; Prieto, M.A. Controlled autolysis facilitates the polyhydroxyalkanoate recovery in *Pseudomonas putida* KT2440. *Microb. Biotechnol.* **2011**, *4*, 533–547. [CrossRef] [PubMed]

73. Tripathi, L.; Wu, L.-P.; Chen, J.; Chen, G.-Q. Synthesis of Diblock copolymer poly-3-hydroxybutyrate -block-poly-3-hydroxyhexanoate [PHB-b-PHHx] by a β-oxidation weakened *Pseudomonas putida* KT2442. *Microb. Cell Fact.* **2012**, *11*, 44. [CrossRef] [PubMed]

74. Tripathi, L.; Wu, L.-P.; Dechuan, M.; Chen, J.; Wu, Q.; Chen, G.-Q. *Pseudomonas putida* KT2442 as a platform for the biosynthesis of polyhydroxyalkanoates with adjustable monomer contents and compositions. *Bioresour. Technol.* **2013**, *142*, 225–231. [CrossRef] [PubMed]

75. Li, S.; Cai, L.; Wu, L.; Zeng, G.; Chen, J.; Wu, Q.; Chen, G.-Q. Microbial Synthesis of Functional Homo-, Random, and Block Polyhydroxyalkanoates by β-Oxidation Deleted *Pseudomonas entomophila*. *Biomacromolecules* **2014**, *15*, 2310–2319. [CrossRef] [PubMed]

76. Shen, R.; Cai, L.W.; Meng, D.C.; Wu, L.P.; Guo, K.; Dong, G.X.; Liu, L.; Chen, J.C.; Wu, Q.; Chen, G.Q. Benzene containing polyhydroxyalkanoates homo- and copolymers synthesized by genome edited *Pseudomonas entomophila*. *Sci. China Life Sci.* **2014**, *57*, 4–10. [CrossRef] [PubMed]

77. Meng, D.C.; Shen, R.; Yao, H.; Chen, J.C.; Wu, Q.; Chen, G.Q. Engineering the diversity of polyesters. *Curr. Opin. Biotechnol.* **2014**, *29*, 24–33. [CrossRef] [PubMed]

78. Tripathi, L.; Wu, L.-P.; Meng, D.; Chen, J.; Chen, G.-Q. Biosynthesis and Characterization of Diblock Copolymer of P(3-Hydroxypropionate)-block-P(4-hydroxybutyrate) from Recombinant *Escherichia coli*. *Biomacromolecules* **2013**, *14*, 862–870. [CrossRef] [PubMed]

79. Zhuang, Q.; Wang, Q.; Liang, Q.; Qi, Q. Synthesis of polyhydroxyalkanoates from glucose that contain medium-chain-length monomers via the reversed fatty acid β-oxidation cycle in *Escherichia coli*. *Metab. Eng.* **2014**, *24*, 78–86. [CrossRef] [PubMed]

80. Wang, Q.; Luan, Y.; Cheng, X.; Zhuang, Q.; Qi, Q. Engineering of *Escherichia coli* for the biosynthesis of poly(3-hydroxybutyrate-co-3-hydroxyhexanoate) from glucose. *Appl. Microbiol. Biotechnol.* **2015**, *99*, 2593–2602. [CrossRef] [PubMed]

81. Meng, D.-C.; Wang, Y.; Wu, L.-P.; Shen, R.; Chen, J.-C.; Wu, Q.; Chen, G.-Q. Production of poly(3-hydroxypropionate) and poly(3-hydroxybutyrate-co-3-hydroxypropionate) from glucose by engineering *Escherichia coli*. *Metab. Eng.* **2015**, *29*, 189–195. [CrossRef] [PubMed]

82. Chen, G.Q.; Hajnal, I.; Wu, H.; Lv, L.; Ye, J. Engineering Biosynthesis Mechanisms for Diversifying Polyhydroxyalkanoates. *Trends Biotechnol.* **2015**, *33*, 565–574. [CrossRef] [PubMed]

83. Koller, M.; Maršálek, L.; de Sousa Dias, M.M.; Braunegg, G. Producing microbial polyhydroxyalkanoate (PHA) biopolyesters in a sustainable manner. *New Biotechnol.* **2017**, *37*, 24–38. [CrossRef] [PubMed]

84. Możejko-Ciesielska, J.; Kiewisz, R. Bacterial polyhydroxyalkanoates: Still fabulous? *Microbiol. Res.* **2016**, *192*, 271–282. [CrossRef] [PubMed]

85. Bugnicourt, E.; Cinelli, P.; Lazzeri, A.; Alvarez, V. Polyhydroxyalkanoate (PHA): Review of synthesis, characteristics, processing and potential applications in packaging. *Express Polym. Lett.* **2014**, *8*, 791–808. [CrossRef]

86. Oliveira, C.S.S.; Silva, C.E.; Carvalho, G.; Reis, M.A. Strategies for efficiently selecting PHA producing mixed microbial cultures using complex feedstocks: Feast and famine regime and uncoupled carbon and nitrogen availabilities. *New Biotechnol.* **2016**, *37*, 69–79. [CrossRef] [PubMed]

87. Kleerebezem, R.; van Loosdrecht, M.C. Mixed culture biotechnology for bioenergy production. *Curr. Opin. Biotechnol.* **2007**, *18*, 207–212. [CrossRef] [PubMed]

88. Serafim, L.S.; Lemos, P.C.; Albuquerque, M.G.E.; Reis, M.A.M. Strategies for PHA production by mixed cultures and renewable waste materials. *Appl. Microbiol. Biotechnol.* **2008**, *81*, 615–628. [CrossRef] [PubMed]

89. Johnson, K.; Jiang, Y.; Kleerebezem, R.; Muyzer, G.; Van Loosdrecht, M.C.M. Enrichment of a mixed bacterial culture with a high polyhydroxyalkanoate storage capacity. *Biomacromolecules* **2009**, *10*, 670–676. [CrossRef] [PubMed]

90. Van Loosdrecht, M.C.M.; Pot, M.A.; Heijnen, J.J. Importance of bacterial storage polymers in bioprocesses. *Water Sci. Technol.* **1997**, *35*, 41–47. [CrossRef]

91. Dircks, K.; Beun, J.J.; Van Loosdrecht, M.; Heijnen, J.J.; Henze, M. Glycogen metabolism in aerobic mixed cultures. *Biotechnol. Bioeng.* **2001**, *73*, 85–94. [CrossRef] [PubMed]

92. Moralejo-Gárate, H.; Mar'Atusalihat, E.; Kleerebezem, R.; Van Loosdrecht, M.C.M. Microbial community engineering for biopolymer production from glycerol. *Appl. Microbiol. Biotechnol.* **2011**, *92*, 631–639. [CrossRef] [PubMed]

93. Dionisi, D.; Carucci, G.; Petrangeli Papini, M.; Riccardi, C.; Majone, M.; Carrasco, F. Olive oil mill effluents as a feedstock for production of biodegradable polymers. *Water Res.* **2005**, *39*, 2076–2084. [CrossRef] [PubMed]

94. Ntaikou, I.; Valencia Peroni, C.; Kourmentza, C.; Ilieva, V.I.; Morelli, A.; Chiellini, E.; Lyberatos, G. Microbial bio-based plastics from olive-mill wastewater: Generation and properties of polyhydroxyalkanoates from mixed cultures in a two-stage pilot scale system. *J. Biotechnol.* **2014**, *188C*, 138–147. [CrossRef] [PubMed]

95. Duque, A.F.; Oliveira, C.S.S.; Carmo, I.T.D.; Gouveia, A.R.; Pardelha, F.; Ramos, A.M.; Reis, M.A.M. Response of a three-stage process for PHA production by mixed microbial cultures to feedstock shift: Impact on polymer composition. *New Biotechnol.* **2014**, *31*, 276–288. [CrossRef] [PubMed]

96. Amulya, K.; Jukuri, S.; Venkata Mohan, S. Sustainable multistage process for enhanced productivity of bioplastics from waste remediation through aerobic dynamic feeding strategy: Process integration for up-scaling. *Bioresour. Technol.* **2015**, *188*, 231–239. [CrossRef] [PubMed]

97. Valentino, F.; Morgan-Sagastume, F.; Campanari, S.; Villano, M.; Werker, A.; Majone, M. Carbon recovery from wastewater through bioconversion into biodegradable polymers. *New Biotechnol.* **2016**, *37*, 9–23. [CrossRef] [PubMed]

98. Kourmentza, C.; Mitova, E.; Stoyanova, N.; Ntaikou, I.; Kornaros, M. Investigation of PHAs production from acidified olive oil mill wastewater (OOMW) by pure cultures of *Pseudomonas* spp. strains. *New Biotechnol.* **2009**, *25*, S269. [CrossRef]

99. Kourmentza, C.; Ntaikou, I.; Kornaros, M.; Lyberatos, G. Production of PHAs from mixed and pure cultures of *Pseudomonas* sp. using short-chain fatty acids as carbon source under nitrogen limitation. *Desalination* **2009**, *248*, 723–732. [CrossRef]

100. Marang, L.; Jiang, Y.; van Loosdrecht, M.C.M.; Kleerebezem, R. Butyrate as preferred substrate for polyhydroxybutyrate production. *Bioresour. Technol.* **2013**, *142*, 232–239. [CrossRef] [PubMed]

101. Shen, L.; Hu, H.; Ji, H.; Cai, J.; He, N.; Li, Q.; Wang, Y. Production of poly(hydroxybutyrate-hydroxyvalerate) from waste organics by the two-stage process: Focus on the intermediate volatile fatty acids. *Bioresour. Technol.* **2014**, *166*, 194–200. [CrossRef] [PubMed]

102. Dias, J.M.L.; Lemos, P.C.; Serafim, L.S.; Oliveira, C.; Eiroa, M.; Albuquerque, M.G.E.; Ramos, A.M.; Oliveira, R.; Reis, M.A.M. Recent advances in polyhydroxyalkanoate production by mixed aerobic cultures: From the substrate to the final product. *Macromol. Biosci.* **2006**, *6*, 885–906. [CrossRef] [PubMed]

103. Salehizadeh, H.; Van Loosdrecht, M.C.M. Production of polyhydroxyalkanoates by mixed culture: Recent trends and biotechnological importance. *Biotechnol. Adv.* **2004**, *22*, 261–279. [CrossRef] [PubMed]

104. Reis, M.A.M.; Serafim, L.S.; Lemos, P.C.; Ramos, A.M.; Aguiar, F.R.; Van Loosdrecht, M.C.M. Production of polyhydroxyalkanoates by mixed microbial cultures. *Bioprocess Biosyst. Eng.* **2003**, *25*, 377–385. [CrossRef] [PubMed]

105. Albuquerque, M.G.E.; Eiroa, M.; Torres, C.; Nunes, B.R.; Reis, M.A.M. Strategies for the development of a side stream process for polyhydroxyalkanoate (PHA) production from sugar cane molasses. *J. Biotechnol.* **2007**, *130*, 411–421. [CrossRef] [PubMed]

106. Ren, Q.; De Roo, G.; Ruth, K.; Witholt, B.; Zinn, M. Simultaneous Accumulation and Degradation of Polyhydroxyalkanoates: Futile Cycle or Clever Regulation? *Biomacromolecules* **2009**, *10*, 916–922. [CrossRef] [PubMed]

107. Frigon, D.; Muyzer, G.; Van Loosdrecht, M.; Raskin, L. rRNA and poly-β-hydroxybutyrate dynamics in bioreactors subjected to feast and famine cycles. *Appl. Environ. Microbiol.* **2006**, *72*, 2322–2330. [CrossRef] [PubMed]

108. Prieto, A.; Escapa, I.F.; Martínez, V.; Dinjaski, N.; Herencias, C.; de la Peña, F.; Tarazona, N.; Revelles, O. A holistic view of polyhydroxyalkanoate metabolism in *Pseudomonas putida*. *Environ. Microbiol.* **2016**, *18*, 341–357. [CrossRef] [PubMed]

109. Jiang, Y.; Marang, L.; Kleerebezem, R.; Muyzer, G.; van Loosdrecht, M.C.M. Polyhydroxybutyrate production from lactate using a mixed microbial culture. *Biotechnol. Bioeng.* **2011**, *108*, 2022–2035. [CrossRef] [PubMed]

110. Pardelha, F.A. *Constraint-Based Modelling of Mixed Microbial Populations: Application to Polyhydroxyalkanoates Production*; Faculdade de Ciências e Tecnologia: Caparica, Portugal, 2013.

111. Albuquerque, M.G.E.; Concas, S.; Bengtsson, S.; Reis, M.A.M. Mixed culture polyhydroxyalkanoates production from sugar molasses: The use of a 2-stage CSTR system for culture selection. *Bioresour. Technol.* **2010**, *101*, 7112–7122. [CrossRef] [PubMed]

112. Marang, L.; Jiang, Y.; Van Loosdrecht, M.C.M.; Kleerebezem, R. Impact of non-storing biomass on PHA production: An enrichment culture on acetate and methanol. *Int. J. Biol. Macromol.* **2014**, *71*, 74–80. [CrossRef] [PubMed]

113. Korkakaki, E.; van Loosdrecht, M.C.M.; Kleerebezem, R. Survival of the fastest: Selective removal of the side population for enhanced PHA production in a mixed substrate enrichment. *Bioresour. Technol.* **2016**, *216*, 1022–1029. [CrossRef] [PubMed]

114. Chen, Z.; Guo, Z.; Wen, Q.; Huang, L.; Bakke, R.; Du, M. A new method for polyhydroxyalkanoate (PHA) accumulating bacteria selection under physical selective pressure. *Int. J. Biol. Macromol.* **2015**, *72*, 1329–1334. [CrossRef] [PubMed]

115. Silva, F.; Campanari, S.; Matteo, S.; Valentino, F.; Majone, M.; Villano, M. Impact of nitrogen feeding regulation on polyhydroxyalkanoates production by mixed microbial cultures. *New Biotechnol.* **2017**, *37*, 90–98. [CrossRef] [PubMed]

116. Burniol-Figols, A.; Varrone, C.; Daugaard, A.E.; Skiadas, I.V.; Gavala, H.N. Polyhydroxyalkanoates (PHA) production from fermented crude glycerol by mixed microbial cultures. In Proceedings of the Sustain ATV Conference, Book of Abstracts, Kobenhavn, Denmark, 30 November 2016.

117. Cui, Y.-W.; Zhang, H.-Y.; Lu, P.-F.; Peng, Y.-Z. Effects of carbon sources on the enrichment of halophilic polyhydroxyalkanoate-storing mixed culture in an aerobic dynamic feeding process. *Sci. Rep.* **2016**, *6*, 30766. [CrossRef] [PubMed]

118. Palmeiro-Sánchez, T.; Fra-Vázquez, A.; Rey-Martínez, N.; Campos, J.L.; Mosquera-Corral, A. Transient concentrations of NaCl affect the PHA accumulation in mixed microbial culture. *J. Hazard. Mater.* **2016**, *306*, 332–339. [CrossRef] [PubMed]

119. Fradinho, J.C.; Domingos, J.M.B.; Carvalho, G.; Oehmen, A.; Reis, M.A.M. Polyhydroxyalkanoates production by a mixed photosynthetic consortium of bacteria and algae. *Bioresour. Technol.* **2013**, *132*, 146–153. [CrossRef] [PubMed]

120. Fradinho, J.C.; Reis, M.A.M.; Oehmen, A. Beyond feast and famine: Selecting a PHA accumulating photosynthetic mixed culture in a permanent feast regime. *Water Res.* **2016**, *105*, 421–428. [CrossRef] [PubMed]

121. Fradinho, J.C.; Oehmen, A.; Reis, M.A.M. Photosynthetic mixed culture polyhydroxyalkanoate (PHA) production from individual and mixed volatile fatty acids (VFAs): Substrate preferences and co-substrate uptake. *J. Biotechnol.* **2014**, *185*, 19–27. [CrossRef] [PubMed]

122. Fradinho, J.C.; Oehmen, A.; Reis, M.A.M. Effect of dark/light periods on the polyhydroxyalkanoate production of a photosynthetic mixed culture. *Bioresour. Technol.* **2013**, *148*, 474–479. [CrossRef] [PubMed]

123. Basset, N.; Katsou, E.; Frison, N.; Malamis, S.; Dosta, J.; Fatone, F. Integrating the selection of PHA storing biomass and nitrogen removal via nitrite in the main wastewater treatment line. *Bioresour. Technol.* **2016**, *200*, 820–829. [CrossRef] [PubMed]

124. Morgan-Sagastume, F.; Karlsson, A.; Johansson, P.; Pratt, S.; Boon, N.; Lant, P.; Werker, A. Production of polyhydroxyalkanoates in open, mixed cultures from a waste sludge stream containing high levels of soluble organics, nitrogen and phosphorus. *Water Res.* **2010**, *44*, 5196–5211. [CrossRef] [PubMed]

125. Jia, Q.; Xiong, H.; Wang, H.; Shi, H.; Sheng, X.; Sun, R.; Chen, G. Production of polyhydroxyalkanoates (PHA) by bacterial consortium from excess sludge fermentation liquid at laboratory and pilot scales. *Bioresour. Technol.* **2014**, *171*, 159–167. [CrossRef] [PubMed]

126. Anterrieu, S.; Quadri, L.; Geurkink, B.; Dinkla, I.; Bengtsson, S.; Arcos-Hernandez, M.; Alexandersson, T.; Morgan-Sagastume, F.; Karlsson, A.; Hjort, M.; et al. Integration of biopolymer production with process water treatment at a sugar factory. *New Biotechnol.* **2014**, *31*, 308–323. [CrossRef] [PubMed]

127. Satoh, H.; Iwamoto, Y.; Mino, T.; Matsuo, T. Activated sludge as a possible source of biodegradable plastic. *Water Sci. Technol.* **1998**, *38*, 103–109. [CrossRef]

128. Din, M.F.; Mohanadoss, P.; Ujang, Z.; van Loosdrecht, M.; Yunus, S.M.; Chelliapan, S.; Zambare, V.; Olsson, G. Development of Bio-PORec®system for polyhydroxyalkanoates (PHA) production and its storage in mixed cultures of palm oil mill effluent (POME). *Bioresour. Technol.* **2012**, *124*, 208–216. [CrossRef] [PubMed]

129. Pratt, S.; Werker, A.; Morgan-Sagastume, F.; Lant, P. Microaerophilic conditions support elevated mixed culture polyhydroxyalkanoate (PHA) yields, but result in decreased PHA production rates. *Water Sci. Technol.* **2012**, *65*, 243–246. [CrossRef]

130. Wang, X.; Oehmen, A.; Freitas, E.B.; Carvalho, G.; Reis, M.A.M. The link of feast-phase dissolved oxygen (DO) with substrate competition and microbial selection in PHA production. *Water Res.* **2017**, *112*, 269–278. [CrossRef] [PubMed]

131. Gurieff, N.; Lant, P. Comparative life cycle assessment and financial analysis of mixed culture polyhydroxyalkanoate production. *Bioresour. Technol.* **2007**, *98*, 3393–3403. [CrossRef] [PubMed]

132. Takabatake, H.; Satoh, H.; Mino, T.; Matsuo, T. PHA (polyhydroxyalkanoate) production potential of activated sludge treating wastewater. *Water Sci. Technol.* **2002**, *45*, 119–126. [PubMed]

133. Mengmeng, C.; Hong, C.; Qingliang, Z.; Shirley, S.N.; Jie, R. Optimal production of polyhydroxyalkanoates (PHA) in activated sludge fed by volatile fatty acids (VFAs) generated from alkaline excess sludge fermentation. *Bioresour. Technol.* **2009**, *100*, 1399–1405. [CrossRef] [PubMed]

134. Cavaillé, L.; Grousseau, E.; Pocquet, M.; Lepeuple, A.S.; Uribelarrea, J.L.; Hernandez-Raquet, G.; Paul, E. Polyhydroxybutyrate production by direct use of waste activated sludge in phosphorus-limited fed-batch culture. *Bioresour. Technol.* **2013**, *149*, 301–309. [CrossRef] [PubMed]

135. Cavaillé, L.; Albuquerque, M.; Grousseau, E.; Lepeuple, A.S.; Uribelarrea, J.L.; Hernandez-Raquet, G.; Paul, E. Understanding of polyhydroxybutyrate production under carbon and phosphorus-limited growth conditions in non-axenic continuous culture. *Bioresour. Technol.* **2016**, *201*, 65–73. [CrossRef] [PubMed]

136. Johnson, K.; van Geest, J.; Kleerebezem, R.; van Loosdrecht, M.C.M. Short- and long-term temperature effects on aerobic polyhydroxybutyrate producing mixed cultures. *Water Res.* **2010**, *44*, 1689–1700. [CrossRef] [PubMed]

137. Albuquerque, M.G.E.; Martino, V.; Pollet, E.; Avérous, L.; Reis, M.A.M. Mixed culture polyhydroxyalkanoate (PHA) production from volatile fatty acid (VFA)-rich streams: Effect of substrate composition and feeding regime on PHA productivity, composition and properties. *J. Biotechnol.* **2011**, *151*, 66–76. [CrossRef] [PubMed]

138. Villano, M.; Beccari, M.; Dionisi, D.; Lampis, S.; Miccheli, A.; Vallini, G.; Majone, M. Effect of pH on the production of bacterial polyhydroxyalkanoates by mixed cultures enriched under periodic feeding. *Process Biochem.* **2010**, *45*, 714–723. [CrossRef]

139. Chen, Z.; Huang, L.; Wen, Q.; Guo, Z. Efficient polyhydroxyalkanoate (PHA) accumulation by a new continuous feeding mode in three-stage mixed microbial culture (MMC) PHA production process. *J. Biotechnol.* **2015**, *209*, 68–75. [CrossRef]

140. Serafim, L.S.; Lemos, P.C.; Oliveira, R.; Reis, M.A.M. Optimization of polyhydroxybutyrate production by mixed cultures submitted to aerobic dynamic feeding conditions. *Biotechnol. Bioeng.* **2004**, *87*, 145–160. [CrossRef] [PubMed]

141. Moita, R.; Freches, A.; Lemos, P.C. Crude glycerol as feedstock for polyhydroxyalkanoates production by mixed microbial cultures. *Water Res.* **2014**, *58*, 9–20. [CrossRef] [PubMed]

142. Pardelha, F.; Albuquerque, M.G.E.; Reis, M.A.M.; Dias, J.M.L.; Oliveira, R. Flux balance analysis of mixed microbial cultures: Application to the production of polyhydroxyalkanoates from complex mixtures of volatile fatty acids. *J. Biotechnol.* **2012**, *162*, 336–345. [CrossRef] [PubMed]

143. Chen, H.; Meng, H.; Nie, Z.; Zhang, M. Polyhydroxyalkanoate production from fermented volatile fatty acids: Effect of pH and feeding regimes. *Bioresour. Technol.* **2013**, *128*, 533–538. [CrossRef]

144. Jiang, Y.; Marang, L.; Tamis, J.; van Loosdrecht, M.C.M.; Dijkman, H.; Kleerebezem, R. Waste to resource: Converting paper mill wastewater to bioplastic. *Water Res.* **2012**, *46*, 5517–5530. [CrossRef] [PubMed]

145. Valentino, F.; Karabegovic, L.; Majone, M.; Morgan-Sagastume, F.; Werker, A. Polyhydroxyalkanoate (PHA) storage within a mixed-culture biomass with simultaneous growth as a function of accumulation substrate nitrogen and phosphorus levels. *Water Res.* **2015**, *77*, 49–63. [CrossRef]

146. Johnson, K.; Kleerebezem, R.; van Loosdrecht, M.C.M. Influence of ammonium on the accumulation of polyhydroxybutyrate (PHB) in aerobic open mixed cultures. *J. Biotechnol.* **2010**, *147*, 73–79. [CrossRef] [PubMed]

147. Venkateswar Reddy, M.; Venkata Mohan, S. Effect of substrate load and nutrients concentration on the polyhydroxyalkanoates (PHA) production using mixed consortia through wastewater treatment. *Bioresour. Technol.* **2012**, *114*, 573–582. [CrossRef] [PubMed]

148. Moralejo-Garate, H.; Palmeiro-Sanchez, T.; Kleerebezem, R.; Mosquera-Corral, A.; Campos, J.L.; van Loosdrecht, M.C.M. Influence of the cycle lenght on the production of PHA from Glycerol by Bacterial Enrichments in Sequencing Batch Reactors. *Biotechnol. Bioeng.* **2013**, *110*, 3148–3155. [CrossRef] [PubMed]

149. Dionisi, D.; Majone, M.; Vallini, G.; Di Gregorio, S.; Beccari, M. Effect of the applied organic load rate on biodegradable polymer production by mixed microbial cultures in a sequencing batch reactor. *Biotechnol. Bioeng.* **2006**, *93*, 76–88. [CrossRef]

150. Chakravarty, P.; Mhaisalkar, V.; Chakrabarti, T. Study on poly-hydroxyalkanoate (PHA) production in pilot scale continuous mode wastewater treatment system. *Bioresour. Technol.* **2010**, *101*, 2896–2899. [CrossRef] [PubMed]

151. Morgan-Sagastume, F.; Hjort, M.; Cirne, D.; Gérardin, F.; Lacroix, S.; Gaval, G.; Karabegovic, L.; Alexandersson, T.; Johansson, P.; Karlsson, A.; et al. Integrated production of polyhydroxyalkanoates (PHAs) with municipal wastewater and sludge treatment at pilot scale. *Bioresour. Technol.* **2015**, *181*, 78–89. [CrossRef] [PubMed]

152. Bengtsson, S.; Karlsson, A.; Alexandersson, T.; Quadri, L.; Hjort, M.; Johansson, P.; Morgan-Sagastume, F.; Anterrieu, S.; Arcos-Hernandez, M.; Karabegovic, L.; et al. A process for polyhydroxyalkanoate (PHA) production from municipal wastewater treatment with biological carbon and nitrogen removal demonstrated at pilot-scale. *New Biotechnol.* **2017**, *35*, 42–53. [CrossRef] [PubMed]

153. Tamis, J.; Lužkov, K.; Jiang, Y.; van Loosdrecht, M.C.M.; Kleerebezem, R. Enrichment of *Plasticicumulans acidivorans* at pilot-scale for PHA production on industrial wastewater. *J. Biotechnol.* **2014**, *192*, 161–169. [CrossRef] [PubMed]

154. Lee, S.Y. Bacterial polyhydroxyalkanoates. *Biotechnol. Bioeng.* **1996**, *49*, 1–14. [CrossRef]

155. Obruca, S.; Benesova, P.; Marsalek, L.; Marova, I. Use of Lignocellulosic Materials for PHA Production. *Chem. Biochem. Eng. Q.* **2015**, *29*, 135–144. [CrossRef]

156. Ienczak, J.L.; Schmidell, W.; De Aragão, G.M.F. High-cell-density culture strategies for polyhydroxyalkanoate production: A review. *J. Ind. Microbiol. Biotechnol.* **2013**, *40*, 275–286. [CrossRef] [PubMed]

157. Ahn, W.S.; Park, S.J.; Lee, S.Y. Production of poly (3-hydroxybutyrate) from whey by cell recycle fed-batch culture of recombinant *Escherichia coli*. *Biotechnol. Lett.* **2001**, *23*, 235–240. [CrossRef]

158. Shaffer, D.L.; Werber, J.R.; Jaramillo, H.; Lin, S.; Elimelech, M. Forward osmosis: Where are we now? *Desalination* **2015**, *356*, 271–284. [CrossRef]

159. Jung, K.; Choi, J.D.R.; Lee, D.; Seo, C.; Lee, J.; Lee, S.Y.; Chang, H.N.; Kim, Y.C. Permeation characteristics of volatile fatty acids solution by forward osmosis. *Process Biochem.* **2015**, *50*, 669–677. [CrossRef]

160. Kalafatakis, S.; Braekevelt, S.; Vilhelmsen Carlsen, N.S.; Lange, L.; Skiadas, I.V.; Gavala, H.N. On a novel strategy for water recovery and recirculation in biorefineries through application of forward osmosis membranes. *Chem. Eng. J.* **2017**, *311*, 209–216. [CrossRef]

161. Koller, M.; Niebelschütz, H.; Braunegg, G. Strategies for recovery and purification of poly[(*R*)-3-hydroxyalkanoates] (PHA) biopolyesters from surrounding biomass. *Eng. Life Sci.* **2013**, *13*, 549–562. [CrossRef]

162. Madkour, M.H.; Heinrich, D.; Alghamdi, M.A.; Shabbaj, I.I.; Steinbüchel, A. PHA recovery from biomass. *Biomacromolecules* **2013**, *14*, 2963–2972. [CrossRef] [PubMed]

163. Anis, S.N.S.; Md Iqbal, N.; Kumar, S.; Amirul, A.A. Effect of different recovery strategies of P(3HB-co-3HHx) copolymer from *Cupriavidus necator* recombinant harboring the PHA synthase of Chromobacterium sp. USM2. *Sep. Purif. Technol.* **2013**, *102*, 111–117. [CrossRef]

164. Leong, Y.K.; Lan, J.C.-W.; Loh, H.-S.; Ling, T.C.; Ooi, C.W.; Show, P.L. Cloud-point extraction of green-polymers from *Cupriavidus necator* lysate using thermoseparating-based aqueous two-phase extraction. *J. Biosci. Bioeng.* **2017**, *123*, 370–375. [CrossRef] [PubMed]

165. Aramvash, A.; Gholami-Banadkuki, N.; Seyedkarimi, M.-S. An efficient method for the application of PHA-poor solvents to extract polyhydroxybutyrate from *Cupriavidus necator*. *Biotechnol. Prog.* **2016**, *32*, 1480–1486. [CrossRef] [PubMed]

166. Samorì, C.; Abbondanzi, F.; Galletti, P.; Giorgini, L.; Mazzocchetti, L.; Torri, C.; Tagliavini, E. Extraction of polyhydroxyalkanoates from mixed microbial cultures: Impact on polymer quality and recovery. *Bioresour. Technol.* **2015**, *189*, 195–202. [CrossRef] [PubMed]
167. Yang, Y.H.; Jeon, J.M.; Yi, D.H.; Kim, J.H.; Seo, H.M.; Rha, C.K.; Sinskey, A.J.; Brigham, C.J. Application of a non-halogenated solvent, methyl ethyl ketone (MEK) for recovery of poly(3-hydroxybutyrate-co-3-hydroxyvalerate) [P(HB-co-HV)] from bacterial cells. *Biotechnol. Bioprocess Eng.* **2015**, *20*, 291–297. [CrossRef]
168. Neves, A.; Müller, J. Use of enzymes in extraction of polyhydroxyalkanoates produced by *Cupriavidus necator*. *Biotechnol. Prog.* **2012**, *28*, 1575–1580. [CrossRef] [PubMed]
169. Jiang, Y.; Mikova, G.; Kleerebezem, R.; van der Wielen, L.A.; Cuellar, M.C. Feasibility study of an alkaline-based chemical treatment for the purification of polyhydroxybutyrate produced by a mixed enriched culture. *AMB Express* **2015**, *5*, 5. [CrossRef] [PubMed]
170. López-Abelairas, M.; García-Torreiro, M.; Lú-Chau, T.; Lema, J.M.; Steinbüchel, A. Comparison of several methods for the separation of poly(3-hydroxybutyrate) from *Cupriavidus necator* H16 cultures. *Biochem. Eng. J.* **2015**, *93*, 250–259. [CrossRef]
171. Villano, M.; Valentino, F.; Barbetta, A.; Martino, L.; Scandola, M.; Majone, M. Polyhydroxyalkanoates production with mixed microbial cultures: From culture selection to polymer recovery in a high-rate continuous process. *New Biotechnol.* **2014**, *31*, 289–296. [CrossRef] [PubMed]
172. Heinrich, D.; Madkour, M.H.; Al-Ghamdi, M.A.; Shabbaj, I.I.; Steinbüchel, A. Large scale extraction of poly(3-hydroxybutyrate) from *Ralstonia eutropha* H16 using sodium hypochlorite. *AMB Express* **2012**, *2*, 59. [CrossRef] [PubMed]
173. Rathi, D.-N.; Amir, H.G.; Abed, R.M.M.; Kosugi, A.; Arai, T.; Sulaiman, O.; Hashim, R.; Sudesh, K. Polyhydroxyalkanoate biosynthesis and simplified polymer recovery by a novel moderately halophilic bacterium isolated from hypersaline microbial mats. *J. Appl. Microbiol.* **2013**, *114*, 384–395. [CrossRef] [PubMed]
174. Mohammadi, M.; Hassan, M.A.; Phang, L.Y.; Shirai, Y.; Man, H.C.; Ariffin, H. Intracellular polyhydroxyalkanoates recovery by cleaner halogen-free methods towards zero emission in the palm oil mill. *J. Clean. Prod.* **2012**, *37*, 353–360. [CrossRef]
175. Kachrimanidou, V.; Kopsahelis, N.; Vlysidis, A.; Papanikolaou, S.; Kookos, I.K.; Monje Martínez, B.; Escrig Rondán, M.C.; Koutinas, A.A. Downstream separation of poly(hydroxyalkanoates) using crude enzyme consortia produced via solid state fermentation integrated in a biorefinery concept. *Food Bioprod. Process.* **2016**, *100*, 323–334. [CrossRef]
176. Gutt, B.; Kehl, K.; Ren, Q.; Boesel, L.F. Using ANOVA Models to Compare and Optimize Extraction Protocols of P3HBHV from *Cupriavidus necator*. *Ind. Eng. Chem. Res.* **2016**, *55*, 10355–10365. [CrossRef]
177. Martino, L.; Cruz, M.V.; Scoma, A.; Freitas, F.; Bertin, L.; Scandola, M.; Reis, M.A.M. Recovery of amorphous polyhydroxybutyrate granules from *Cupriavidus necator* cells grown on used cooking oil. *Int. J. Biol. Macromol.* **2014**, *71*, 117–123. [CrossRef] [PubMed]
178. Murugan, P.; Han, L.; Gan, C.Y.; Maurer, F.H.J.; Sudesh, K. A new biological recovery approach for PHA using mealworm, *Tenebrio molitor*. *J. Biotechnol.* **2016**, *239*, 98–105. [CrossRef] [PubMed]
179. Kunasundari, B.; Arza, C.R.; Maurer, F.H.J.; Murugaiyah, V.; Kaur, G.; Sudesh, K. Biological recovery and properties of poly(3-hydroxybutyrate) from *Cupriavidus necator* H16. *Sep. Purif. Technol.* **2017**, *172*, 1–6. [CrossRef]
180. Martínez, V.; Herencias, C.; Jurkevitch, E.; Prieto, M.A. Engineering a predatory bacterium as a proficient killer agent for intracellular bio-products recovery: The case of the polyhydroxyalkanoates. *Sci. Rep.* **2016**, *6*, 24381. [CrossRef] [PubMed]
181. Martínez, V.; Jurkevitch, E.; García, J.L.; Prieto, M.A. Reward for *Bdellovibrio bacteriovorus* for preying on a polyhydroxyalkanoate producer. *Environ. Microbiol.* **2013**, *15*, 1204–1215. [CrossRef] [PubMed]
182. Kunasundari, B.; Murugaiyah, V.; Kaur, G.; Maurer, F.H.J.; Sudesh, K. Revisiting the Single Cell Protein Application of *Cupriavidus necator* H16 and Recovering Bioplastic Granules Simultaneously. *PLoS ONE* **2013**, *8*, 1–15. [CrossRef] [PubMed]
183. Anis, S.N.S.; Iqbal, N.M.; Kumar, S.; Al-Ashraf, A. Increased recovery and improved purity of PHA from recombinant *Cupriavidus necator*. *Bioengineered* **2013**, *4*, 115–118. [CrossRef] [PubMed]

184. Koller, M.; Bona, R.; Chiellini, E.; Braunegg, G. Extraction of short-chain-length poly-acetone under elevated temperature and pressure. *Biotechnol. Lett.* **2013**, *35*, 1023–1028. [CrossRef] [PubMed]
185. Aramvash, A.; Gholami-Banadkuki, N.; Moazzeni-Zavareh, F.; Hajizadeh-Turchi, S. An Environmentally Friendly and Efficient Method for Extraction of. *J. Microbiol. Biotechnol.* **2015**, *25*, 1936–1943. [CrossRef] [PubMed]
186. Fei, T.; Cazeneuve, S.; Wen, Z.; Wu, L.; Wang, T. Effective Recovery of Poly-β-Hydroxybutyrate (PHB) Biopolymer from *Cupriavidus necator* Using a Novel and Environmentally Friendly Solvent System. *Biotechnol. Prog.* **2016**, *32*, 678–685. [CrossRef] [PubMed]
187. Furrer, P.; Panke, S.; Zinn, M. Efficient recovery of low endotoxin medium-chain-length poly ([*R*]-3-hydroxyalkanoate) from bacterial biomass. *J. Microbiol. Methods* **2007**, *69*, 206–213. [CrossRef] [PubMed]
188. Mohammadi, M.; Ali Hassan, M.; Phang, L.-Y.; Ariffin, H.; Shirai, Y.; Ando, Y. Recovery and purification of intracellular polyhydroxyalkanoates from recombinant *Cupriavidus necator* using water and ethanol. *Biotechnol. Lett.* **2012**, *34*, 253–259. [CrossRef] [PubMed]
189. Leong, Y.K.; Koroh, F.E.; Show, P.L.; Lan, C.-W.J.; Loh, H.-S. Optimisation of Extractive Bioconversion for Green Polymer via Aqueous Two-Phase System Optimisation of Extractive Bioconversion for Green Polymer. *Chem. Eng. Trans.* **2015**, *45*, 1495–1500.
190. Iqbal, M.; Tao, Y.; Xie, S.; Zhu, Y.; Chen, D.; Wang, X.; Huang, L.; Peng, D.; Sattar, A.; Shabbir, M.A.B.; et al. Aqueous two-phase system (ATPS): An overview and advances in its applications. *Biol. Proced. Online* **2016**, *18*, 18. [CrossRef] [PubMed]

bioengineering

MDPI

Article

Prospecting for Marine Bacteria for Polyhydroxyalkanoate Production on Low-Cost Substrates

Rodrigo Yoji Uwamori Takahashi, Nathalia Aparecida Santos Castilho, Marcus Adonai Castro da Silva, Maria Cecilia Miotto and André Oliveira de Souza Lima *

Centro de Ciências Tecnológicas da Terra e do Mar, Universidade do Vale do Itajaí, R. Uruguai 458, Itajaí-SC 88302-202, Brazil; rodrigo.aquicultura@gmail.com (R.Y.U.T.); nathi_zuca@hotmail.com (N.A.S.C.); marcus.silva@univali.br (M.A.C.S.); cecilia.miotto@gmail.com (M.C.M.)
* Correspondence: andreolima@gmail.com; Tel.: +55-(47)-3341-7716

Academic Editor: Martin Koller
Received: 5 May 2017; Accepted: 17 June 2017; Published: 23 June 2017

Abstract: Polyhydroxyalkanoates (PHAs) are a class of biopolymers with numerous applications, but the high cost of production has prevented their use. To reduce this cost, there is a prospect for strains with a high PHA production and the ability to grow in low-cost by-products. In this context, the objective of this work was to evaluate marine bacteria capable of producing PHA. Using Nile red, 30 organisms among 155 were identified as PHA producers in the medium containing starch, and 27, 33, 22 and 10 strains were found to be positive in media supplemented with carboxymethyl cellulose, glycerol, glucose and Tween 80, respectively. Among the organisms studied, two isolates, LAMA 677 and LAMA 685, showed strong potential to produce PHA with the use of glycerol as the carbon source, and were selected for further studies. In the experiment used to characterize the growth kinetics, LAMA 677 presented a higher maximum specific growth rate ($\mu max = 0.087$ h^{-1}) than LAMA 685 ($\mu max = 0.049$ h^{-1}). LAMA 677 also reached a D-3-hydroxybutyrate (P(3HB)) content of 78.63% (dry biomass), which was 3.5 times higher than that of LAMA 685. In the assay of the production of P(3HB) from low-cost substrates (seawater and biodiesel waste glycerol), LAMA 677 reached a polymer content of 31.7%, while LAMA 685 reached 53.6%. Therefore, it is possible to conclude that the selected marine strains have the potential to produce PHA, and seawater and waste glycerol may be alternative substrates for the production of this polymer.

Keywords: biopolymer; seawater; waste glycerol; deep sea

1. Introduction

PHAs (polyhydroxyalkanoates) are a class of polyesters produced by prokaryotic microorganisms that are accumulated inside cells as carbon and energy reserves [1,2]. These biopolymers have drawn great interest due to their biodegradability, biocompatibility, the possibility of biosynthesis from renewable resources, and similar physical and chemical characteristics to the main petrochemical polymers [3,4]. Despite the environmental benefits and the potential of using this raw material in several areas, the production costs are relatively high compared to those of conventional polymers [5].

There are bacteria known to be PHA producers, such as *Cupriavidus necator*, *Azohydromonas lata* and *Azotobacter vinelandii*. However, the other organisms capable of using low-cost substrates, accumulating a high PHA content, presenting high productivity, producing copolymers from single carbon sources, and producing other PHA monomer compositions have high economic importance [3,4,6,7]. According to Quillaguamán et al. [8], the marine environment has been poorly explored in terms of prospecting for PHA-producing organisms. However, recent research on halophiles indicates a strong potential for biotechnological production of PHAs, based on a study of the bacterium *Halomonas hydrothermalis*, which was able to accumulate a polyhydroxybutyrate (P(3HB)) content of 75.8% when cultivated in residual glycerol as the only source of carbon [9]. Two other *Halomonas* spp. have been isolated and showed great potential for low-cost PHA production. *Halomonas* sp. TD01 grew rapidly to over 80 g/L cell dry weight (CDW) in a lab fermentor and accumulated a P(3HB) content of over 80% on a glucose–salt medium [10]. *H. campaniensis* LS21 was able to grow in artificial seawater and kitchen-waste-like mixed substrates consisting of cellulose, proteins, fats, fatty acids and starch [11].

There are low-cost ways to synthesize PHAs with the use of halophilic bacteria [12] whose salinity requirements may inhibit the growth of non-halophilic microorganisms, allowing for growth in non-sterile conditions [8,12]. This result was noted in a study by Tan et al. [10], who found evidence of the production of P(3HB) using a non-sterile fermentation process in cultures of *Halomonas* TD01. Moreover, a study by Kawata and Aiba [13] reported the growth of *Halomonas* sp. KM-1 in unsterilized medium cultures. Additionally, seawater can be used as a source of minerals and low-cost nutrients for cultivation, according to a study conducted by Pandian et al. [14] which illustrated the production of PHAs in cultures of *Bacillus megaterium* SRKP-3 (an organism isolated from the marine environment) from seawater, milk residues and rice bran. According to Yin et al. [15], due to their characteristics, using halophiles to produce PHAs can reduce the costs of the fermentation and recovery processes, making them a promising alternative for PHA production.

Therefore, this study describes a search for marine bacteria for PHA production using low-cost substrates, as well as the growth, productivity, and PHA characteristics of the selected producers.

2. Materials and Methods

2.1. Marine Bacteria

In the present study, marine bacteria maintained in the culture collection of the Laboratory of Applied Microbiology (LAMA) of the University of Vale do Itajaí (UNIVALI) were used. The bacteria were obtained through the South Atlantic MAR-ECO Patterns and Processes of the Ecosystems of The Southern Mid-Atlantic projects. The organisms were isolated from sediment and water samples collected between the surface and a depth of 5500 m from the Mid-South Atlantic ridge, the Rio Grande Rise and the Walvis Ridge.

2.2. Qualitative Screening of PHA Producers

All isolates were evaluated based on their ability to synthesize PHAs according to the method described by Spiekermann et al. [16]. Thus, the organisms were cultivated in: (a) Zobell Marine Agar 2216 (AM); (b) AM with added glucose (0.5% w/v); and (c) a mineral medium (MM) with added starch (0.5%), carboxymethyl cellulose (CMC), glycerol, glucose, or Tween 80. The MM composition described by Baumann et al. [17] for 1 L of the medium with a pH of 7.5 was 11.7 g of NaCl, 12.32 g of $MgSO_4 \cdot 7H_2O$, 0.745 g of KCl, 1.47 g of $CaCl_2 \cdot 2H_2O$, 6.05 g of Tris (hydroxymethyl) aminomethane ($C_4H_{11}NO_3$), 6.65 g of NH_4Cl, 0.062 g of $K_2HPO_4 \cdot 3H_2O$, and 0.026 g of $FeSO_4 \cdot 7H_2O$. In order to evaluate the potential of the bacteria to produce PHA, all media (except the control) were supplemented with Nile red (final concentration: 0.5 µg/mL), using a stock solution (0.025% m/v in dimethyl sulfoxide). Methods using Nile red are fast and can detect PHA inside intact cells [18,19]; however, the fluorophore can also stain other lipophilic compounds [20]. Organisms were cultivated in an incubator at 28 °C until evident growth occurred (1–9 days). After growth, the colonies were inspected by direct exposure to ultraviolet light (λ = 312 nm; transilluminator UV-TRANS, model UVT-312) and photographed with a digital camera (Canon, EOS Rebel model, Canon Inc., Tokyo, Japan). The fluorescence intensity of isolates is proportional to the PHA content in the cells, as reported by Degelau et al. [21] and Reddy et al. [22]. In this context, the fluorescence intensity was used (Image-Pro Plus, version 6.0, Media Cybernetics Inc., Rockville, MD, USA) as an indicator of polymer content, and the colony area was associated with the growth capacity. The fluorescence intensity and colony area data were statistically analyzed, and the organisms in each test were compared using a Kruskal-Wallis test. In cases of significance ($p < 0.05$), Dunn's test was applied. Statistical analyses were performed using BioEstat (version 5.3, Instituto Mamirauá; Tefé, Brasil) and Statistica (version 8.0, StatSoft Inc., Oklahoma, OK, USA) software.

2.3. Production of PHA in Semi-Solid and Liquid Mediums

The preculture was prepared by inoculating isolated bacterial colonies in Zobell Marine Broth 2216 (CM) or the MM supplemented with glycerol (5%), and cultivating (28 °C; 150 rpm) for 24–48 h. After growth, the preinoculum (10% v/v) was precipitated via centrifugation, and resuspended to assay the PHA production media. Then, the broth was divided into three aliquots (each 100 mL), and cultivated (68 h; 28 °C; 150 rpm; in 250 mL Erlenmeyer flasks) to assay the PHA production. In the first evaluation round, the bacteria were cultured in a MM supplemented with one of the following carbon sources: (a) starch (2% w/v), (b) CMC (2% w/v), (c) glycerol (5% v/v), (d) glucose (5% w/v), or (e) Tween 80 (5% v/v). To evaluate the growth kinetics, cells were cultivated (28 °C; 150 rpm; in triplicate) for 143 h in a MM supplemented with glycerol (5% v/v). Periodically, samples were collected to determine the biomass using gravimetry.

When evaluating the PHA synthesis from the low-cost carbon sources, different media formulations were used (Table 1). For those experiments, residual glycerol from biodiesel production was used, which was provided by a Brazilian company from Rio Grande do Sul (not identified). Selected bacteria were cultivated (28 °C; 150 rpm; 69 h), and the produced biomass was recovered by centrifugation, washed (with distilled water), lyophilized, weighed, and frozen for further analysis by gas chromatography (GC). The PHA quantification analysis was performed using a complex sample that constituted of equal amounts of each replicate. Thus, the PHA content results represent the average of the three cultures.

Assays were also conducted in semi-solid media. Organisms were inoculated (streak) and incubated (28 °C; 6 days) on the MM with 1.8 g· L^{-1} of agar, and separately supplemented with the following carbon sources: starch (2% w/v), CMC (2% w/v), glycerol (5% v/v), glucose (5% v/v) and Tween 80 (5% v/v). After culturing, cells were resuspended in a saline solution, recovered, washed (with distilled water), lyophilized, weighed, and frozen for further analysis by GC (Shimadzu, modelo 17A, Kyoto, Japan).

Table 1. Culture media formulations used in PHA synthesis assays from different sources of low-cost carbon, minerals and nutrients (residual glycerol and seawater).

Culture Media	Composition
1	90% MM medium * + 5% (v/v) glycerol + 5% distilled water
2	90% MM medium + 10% (v/v) glycerol
3	90% seawater + 5% (v/v) glycerol + 5% distilled water
4	90% seawater + 5% (v/v) residual glycerol + 5% distilled water
5	90% seawater + 10% (v/v) residual glycerol
6	90% MM medium + 5% (v/v) residual glycerol + 5% distilled water

* Marine mineral medium.

2.4. Quantification of PHA by Gas Chromatography

For PHA quantification, approximately 50 mg of freeze-dried cells were weighed and transferred to a screw-cap tube. Then, to extract the polymer and submit it to the methanolysis reaction to form monomers of methyl esters, 2 mL of H_2SO_4/methanol (5:95) was added, along with 2 mL of chloroform and 250 µL of internal standard solution, which was composed of 20 mg of benzoic acid in 1 mL of methanol. Due to the reduced biomass in some assays, 1 mL of chloroform was added during methanolysis. The tubes were then heated (100 °C) in a dry block for 3 h with occasional stirring. After the reaction, the tubes were cooled to room temperature, 1 mL of distilled water was added, and the tubes were vortexed for 30 s and left for phase separation. The chloroform phase was transferred to chromatography vials for analysis. A Shimadzu gas chromatograph (model 17A) equipped with a flame ionization detector (FID) adjusted to 280 °C was used. The temperature program was set to an initial temperature of 50 °C for 2 min, then increased from 50 °C to 110 °C at a rate of 20 °C/min, and finally increased to 250 °C at a rate of 20 °C/min. The injector was maintained at 250 °C, and the oven was maintained at 120 °C. The column used was a VB-WAX (VICI) column, 30 m long, 0.25 mm in diameter, and with 0.25 mm film thickness. The volume injected was 1 μL, with the helium flow set at 1 mL/min, with a total run time of 12 min. The standard curve was made using P(3HB) (Sigma-Aldrich, St. Louis, MO, USA) as an external standard, with a mass ranging from 0.005 to 0.020 g. The standards were submitted to the methanolysis process, as described previously for the bacterial samples.

3. Results and Discussion

3.1. Screening for PHA-Producing Marine Bacteria

Among the 155 isolates evaluated, 40.6% (63 isolates) presented fluorescence indicative of PHA production in at least one of the growth conditions tested. The results were distributed as follows: 19.4% (30 isolates) were active in one culture condition, 5.8% (9 isolates) were active in two mediums, 5.2% (8 isolates) were active in three mediums, 7.1% (11 isolates) were active in four mediums, 0.6% (1 isolate) was active in five mediums, 1.9% (3 isolates) were active in six mediums, and 0.6% (1 isolate) presented fluorescence in every medium tested (Table 2). Regarding the substrates added in the MM, the number of isolates capable of producing PHA was higher when starch was added, with 30 representatives, followed by 27, 23, 22 and 10 organisms classified as producers in assays supplied with CMC, glycerol, glucose and Tween 80, respectively. In the MM supplied with glycerol, 23 positive isolates (14.7%) were identified as polymer producers, a percentage similar to that observed by Shrivastav et al. [9], who selected PHA producers from the marine environment using residual glycerol as the carbon source, and identified 14% positives using the Nile red method.

Table 2. Results of the qualitative analysis of the growth and PHA production of marine bacteria isolates when exposed to two culture media (marine agar and mineral marine) with different formulations.

Isolate LAMA [3]	MA [1]		MM [2]					Isolate LAMA	MA		MM				
	NS	GU	ST	CM	GL	GU	TW		NS	GU	ST	CM	GL	GU	TW
570	−	−	wg	wg	wg	wg	wg	671	−	−	−	+	−	−	−
571	−	−	wg	wg	wg	−	wg	672	−	−	−	−	−	−	−
572	−	−	+	wg	−	−	−	673	−	−	+	+	+	+	−
573	−	−	wg	wg	wg	−	wg	674	+	−	+	+	+	+	+
574	−	−	wg	wg	wg	wg	−	675	−	−	wg	wg	wg	wg	−
575	−	−	wg	wg	wg	wg	wg	677	−	−	+	+	+	−	+
576	−	−	wg	wg	wg	wg	wg	679	+	+	+	+	+	+	+
577	−	−	wg	wg	−	−	wg	680	−	−	−	−	−	−	−
580	−	−	wg	−	−	−	wg	681	−	−	wg	−	−	−	wg
582	−	−	wg	wg	−	−	wg	683	−	−	wg	wg	wg	wg	+
583	−	−	−	wg	−	−	−	684	−	−	wg	wg	wg	wg	−
584	+	−	wg	wg	wg	wg	wg	685	−	−	−	+	+	+	−
585	−	−	wg	wg	−	−	wg	687	−	−	wg	wg	wg	wg	wg
587	−	−	−	−	−	−	−	688	−	−	wg	wg	wg	wg	wg
592	+	−	−	−	−	−	−	689	−	−	−	−	wg	−	−
593	−	−	wg	wg	wg	wg	wg	690	−	−	wg	−	wg	−	−
594	−	−	+	+	+	−	−	691	−	−	wg	wg	wg	wg	−
595	−	−	wg	−	wg	wg	wg	692	−	−	wg	wg	wg	wg	wg
597	−	−	wg	wg	−	wg	wg	693	−	−	−	−	−	−	−
598	−	−	wg	wg	wg	−	−	694	−	−	wg	wg	wg	wg	wg
599	−	−	+	+	−	−	−	695	−	−	−	−	−	−	−
600	−	−	−	−	−	−	−	696	−	−	wg	wg	wg	wg	wg
601	−	−	+	+	−	−	−	697	+	+	−	−	−	+	−
604	−	−	+	−	wg	−	−	698	−	−	wg	wg	wg	wg	wg
606	−	−	wg	wg	wg	wg	wg	699	−	−	wg	wg	wg	−	−

Table 2. Cont.

Isolate LAMA [3]	MA [1]		MM [2]					Isolate LAMA	MA		MM				
	NS	GU	ST	CM	GL	GU	TW		NS	GU	ST	CM	GL	GU	TW
607	−	−	wg	wg	wg	−	−	700	−	−	−	wg	−	−	−
608	−	−	wg	wg	wg	wg	wg	701	−	−	wg	wg	−	wg	wg
610	−	−	wg	−	−	−	wg	702	−	−	+	+	+	+	−
611	−	−	wg	wg	wg	wg	−	703	−	−	+	−	−	−	−
612	−	−	+	+	+	+	−	704	−	−	+	+	−	−	−
613	−	−	wg	wg	wg	wg	wg	705	−	−	−	wg	−	−	−
614	−	−	wg	wg	+	−	wg	706	−	−	wg	wg	wg	wg	+
615	−	−	−	−	−	−	−	707	+	+	−	−	−	−	wg
616	−	−	wg	wg	wg	−	−	708	−	−	wg	−	−	−	wg
617	−	−	wg	wg	wg	wg	−	709	−	−	wg	wg	−	wg	wg
618	−	+	wg	wg	wg	wg	wg	710	−	−	+	wg	−	−	−
619	−	−	wg	wg	wg	−	−	711	+	+	+	+	+	+	−
644	−	+	−	−	−	−	−	712	−	−	wg	wg	wg	wg	−
647	−	−	−	−	−	−	−	713	−	−	wg	wg	wg	wg	wg
650	−	−	wg	−	−	+	−	715	−	−	wg	wg	wg	−	wg
653	+	−	−	−	−	−	+	716	−	−	wg	wg	wg	wg	wg
659	−	−	+	+	+	+	−	717	−	+	−	wg	wg	wg	wg
667	−	−	−	wg	−	−	−	718	−	−	wg	wg	−	−	wg
669	−	−	−	−	−	+	−	719	−	−	+	−	−	−	−
720	−	−	wg	wg	wg	wg	wg	759	−	−	wg	wg	wg	wg	wg
722	−	−	wg	+	−	+	−	760	−	−	+	+	+	−	−
723	−	−	−	−	−	−	−	761	+	+	−	+	+	+	+
725	−	−	−	wg	wg	−	wg	762	−	−	wg	wg	−	−	−
726	+	−	−	wg	−	+	+	763	+	+	wg	wg	wg	−	−
727	−	−	−	wg	wg	−	wg	764	−	−	wg	wg	wg	wg	wg
728	−	−	wg	wg	−	−	wg	765	−	−	+	+	+	+	−
729	−	+	+	+	+	+	−	766	−	−	+	wg	−	−	−
730	−	−	wg	wg	−	−	−	767	−	−	−	−	−	−	−
731	−	−	−	−	−	−	wg	768	−	−	wg	wg	wg	wg	wg
732	+	−	−	wg	wg	−	−	769	−	−	wg	wg	wg	wg	wg
733	−	−	−	−	−	−	−	773	−	−	+	+	+	+	−
734	−	+	−	wg	wg	−	−	775	−	−	wg	wg	−	−	wg
735	−	−	wg	wg	−	+	−	778	−	−	+	−	−	−	−
736	+	−	−	wg	+	+	−	779	−	−	−	−	wg	−	−
737	−	−	+	+	+	+	−	781	−	−	wg	−	−	−	−
738	−	−	wg	wg	−	−	−	782	−	−	−	wg	wg	wg	wg
739	−	−	wg	wg	wg	wg	wg	786	−	−	−	−	−	−	−
741	−	−	wg	wg	wg	−	−	790	−	−	+	−	−	−	−
742	−	−	wg	−	wg	−	−	791	−	−	+	−	−	−	−
743	−	−	wg	wg	−	−	wg	M 112	−	−	wg	wg	wg	wg	wg
744	−	+	wg	wg	wg	−	−	M 135	+	+	wg	wg	wg	wg	wg
746	−	−	wg	wg	−	−	−	M 151	+	−	wg	wg	−	−	−
747	−	−	−	−	−	−	+	M 169	−	−	wg	wg	wg	−	wg
748	−	−	+	+	+	−	−	M 171	+	−	wg	wg	wg	−	−
749	−	−	wg	wg	wg	−	wg	M 173	−	+	wg	wg	−	wg	wg
750	−	−	wg	−	+	+	−	M 180	+	+	+	+	−	−	−
751	−	−	−	−	wg	−	wg	M 189	−	−	−	−	−	−	−
753	−	−	wg	wg	wg	−	wg	M 198	−	−	wg	wg	−	−	wg
754	−	−	+	+	+	−	−	M 199	−	−	+	+	+	−	+
755	−	−	wg	wg	wg	wg	−	M 211	−	−	−	+	−	−	−
756	−	−	wg	wg	wg	−	−	M 84	+	−	wg	wg	wg	wg	wg
757	−	−	wg	−	−	−	wg	M 97	−	−	wg	−	−	wg	wg
758	+	−	−	+	+	+	−								

(+) PHA producer; (−) non-PHA producer; (wg) without growth; [1] MA—commercial marine agar medium (NS—not supplemented, and GU—supplemented with glucose); [2] MM—mineral medium (ST—supplemented with starch, CM—supplemented with carboxymethylcellulose, GL—supplemented with glycerol, GU—supplemented with glucose, and TW—supplemented with Tween 80). [3] Marine bacteria of the Laboratory of Applied Microbiology of the University of Vale do Itajaí.

When considering the use of agro-industrial by-products as substrates to produce PHA, the great potential of some bacteria can be recognized, particularly the isolates LAMA 674, LAMA 677,

LAMA 679 and M 199. The statistical analyses of the colony area and fluorescence on semi-solid media were not significantly different among organisms (data not shown). However, it was possible to identify the organisms with higher fluorescence and growth (area) in each medium, and these were selected for further evaluation. LAMA 679 and LAMA 732 growing in the AM medium were among the highest producers. On the other hand, LAMA 711 and LAMA 644 performed better in the AM medium supplemented with glucose, and LAMA 748 and LAMA 737 performed well in the MM with added starch. In the MM with CMC, LAMA 748 and LAMA 674 performed the best; in the MM with added glycerol, LAMA 685 and LAMA 677 were also sufficient. Finally, when cultivated in the MM with added Tween 80, LAMA 726 and M 199 presented the highest indices.

According to the data provided by the MAR-ECO project, 49.7% of the isolates were obtained from water samples, and 50.3% were obtained from sediments. When considering the sea zonation defined by Hedgpeth [23], 76.6% of the bacteria were collected in the epipelagic zone, 16.9% in the mesopelagic zone, and 6.5% in the bathypelagic zone. In the benthic domain, 29.5% of bacteria were collected from the bathyal zone, and 70.5% were obtained from the abyssal zone. Regarding the sampling locations, 49.7% were from the Rio Grande Rise, 38.7% were from the Walvis Ridge, and 11.6% were from the Mid-South Atlantic ridge. Given this information, the most promising organisms for PHA production were taken from the epipelagic zone, with six organisms, while one isolate was taken from the mesopelagic zone. It is also important to emphasize that 93.5% of the studied bacteria were from the epi- and meso-pelagic zones. To produce PHA, microorganisms need substrates with excess carbon sources, as can be observed in the epi- and meso-pelagic zones [24]. This fact may explain why most of the organisms capable of accumulating PHA were from these zones.

3.2. PHA Production in Different Substrates

Among the bacteria screened in the MM assay, those with higher intensities and colony areas were chosen to be evaluated for PHA production in liquid media. As shown in Table 3, production of PHA was not detected in the assays where the media were supplemented with starch, CMC or Tween. However, it is believed that most of these samples were composed of the carbon source itself, so the GC did not detect PHA production. On the other hand, when evaluated in the medium supplemented with glycerol, LAMA 677 had a productivity of 0.0058 g· L^{-1}·h^{-1}, reaching 1.41 ± 0.18 g· L^{-1} of biomass with 28.28% of a polymer identified by GC as P(3HB). Additionally, LAMA 685 presented better indices with a biomass of 2.03 ± 0.27 g· L^{-1} (32.79% P(3HB)) with 0.0098 g· L^{-1}·h^{-1} productivity. A higher P(3HB) production was observed when the media was supplemented with glycerol as a carbon source. This condition was even better than when the media were supplemented with glucose, which is a readily assimilated source. For example, LAMA 685 accumulated 17% more biomass (2.03 g· L^{-1}) and 3.4 times more P(3HB) (32.79%) when growing in glycerol compared to glucose as the carbon source. Similar results have been reported previously. Chien et al. [25] evaluated the use of bacteria isolated from mangrove sediments to produce PHA and reported the best PHA content with the use of glycerol in comparison to glucose. According to Chien et al. [25], the organism named M11 produced 30.2% PHA in cells in glycerol culture and 8.1% in the supplied glucose culture. Mahishi et al. [26] used recombinant *Escherichia coli* cultures and reported PHA contents of up to 60% in relation to dry-weight PHA with the glycerol supply, and 38% when supplied with glucose as the carbon source. When evaluating different carbon sources, Mohandas et al. [27] reported that glycerol supported the maximum biomass yield and P(3HB) production, followed by glucose and fructose. The biomass yield and P(3HB) production in the presence of glycerol were 3 g· L^{-1} and 68% (w/w), respectively.

Due to the presence of insoluble substrates in some of the treatments tested, it was difficult to detect PHA production, as the biomass analyzed by GC consisted of both the microbial biomass and the residual substrate. Taking this into consideration, a new set of assays were carried out in a semi-solid medium, as an alternative to obtain cells free from the substrates. For instance, when using media supplemented with CMC, LAMA 674 and LAMA 748 produced PHA as 22.26% and 27.64% of biomass, respectively. Although the substrate used in this study was a source of carbon that is more easily assimilated by microorganisms compared to sources used in other studies, these results are comparable to a study by Van-Thuoc et al. [28]. Van-Thuoc et al. [28] used xylose as the carbon source for PHA production by *Halomonas boliviensis* LC1, and obtained polymer yields ranging from 23.1% to 33.8% of biomass. Bertrand et al. [29] reported a 19.6% accumulation of PHA in cultures of *Pseudomonas pseudoflava* supplied with xylose. Silva et al. [30] reported a PHA content of between 35% and 58.2% when using xylose, and contents of 15.39% to 23.22% when using hydrolyzed sugarcane. Additionally, the PHA accumulation results from this study where Tween 80 was added to the medium were interesting, as isolated LAMA 726 reached a biomass content of 55.24% PHA. In a study by Fernández et al. [31], who used cultures of *Pseudomonas aeruginosa* NCIB 40045, a PHA content of 29.4% was obtained when residual frying oil was added to the culture. On the other hand, He et al. [32] achieved a yield of 63% polymers by cultivating *Pseudomonas stutzeri* 1317 in soybean oil.

Table 3. Results of the parameters used to analyze the PHA production of the isolates grown in the mineral medium (MM) supplemented with starch, carboxymethylcellulose, glycerol, glucose and Tween 80.

Isolate	Carbon Source	Total Biomass $(g \cdot L^{-1})$	P(3HB) Concentration $(g \cdot L^{-1})$	P(3HB) Content in Total Biomass (%)	P(3HB) Productivity $(g \cdot L^{-1} \cdot h^{-1})$
LAMA 748	Starch	16.80 ± 0.97	0	0	0
LAMA 737	Starch	25.31 ± 1.09	0	0	0
LAMA 748	Carboxymethylcellulose	1.19 ± 0.19	0	0	0
LAMA 674	Carboxymethylcellulose	1.24 ± 0.20	0	0	0
LAMA 677	Glycerol	1.41 ± 0.18	0.4	28.28	0.0058
LAMA 685	Glycerol	2.03 ± 0.27	0.67	32.79	0.0098
LAMA 685	Glucose	1.73 ± 0.22	0.17	9.62	0.0025
LAMA 737	Glucose	0.68 ± 0.10	0.05	7.85	0.0008
LAMA 726	Tween 80	0.91 ± 0.08	0	0	0
M 199 A	Tween 80	0.56 ± 0.05	0	0	0

3.3. Growth Kinetics and P(3HB) Production

As the best productivity was obtained when using glycerol as the carbon source, this condition was further analyzed during bacterial growth. However, the biomass production was much lower than in the previous experiments. For example, LAMA 677 produced 0.16 g· L^{-1} of biomass, of which 22.74% was polymer, compared to the 28.28% of polymer produced in the previous liquid test. This fact can be explained by the longer cultivation time at this stage, allowing for the consumption of the polymer. Moreover, LAMA 685 in the same condition had a 78.63% yield of P(3HB) in the biomass. This value is higher than reported by Cavalheiro et al. [33], who reported a PHA content of 62% when evaluating the production of *Cupriavidus necator* DSM 545 (the main organism used in the industrial production of PHA) under controlled conditions and high cellular concentrations, with a supply of pure glycerol. Mothes et al. [34] cultured the bacteria *Cupriavidus necator* JMP 134 at low cell concentrations using pure glycerol and obtained a 70% accumulation. When compared to other studies, the PHA content of LAMA 685 was similar. Zhu et al. [35] reported a content of 81.9% in cultures of *Burkholderia cepacia* ATCC 17759 grown in a shaker using residual glycerol as the carbonic substrate. Kangsadan et al. [36] cultivated *Cupriavidus necator* ATCC 17699 with pure and residual glycerol and reached contents of 83.23% and 78.26%, respectively. Considering these results, the polymer content obtained in this experiment with LAMA 685 is relevant because it is comparable to that of the organisms used in the production of the polymer on a commercial scale.

Compared to the literature, the maximum specific growth rates obtained in this study were lower, possibly because the medium culture used was poorer in nutrients than those employed in other studies. The values of maximum specific growth rates (μmax) obtained were μmax = 0.087 h^{-1} for LAMA 677 and μmax = 0.049 h^{-1} for LAMA 685. Piccoli et al. [37] studied PHA-producing lines using pure glycerol and found specific speeds ranging from 0.155 to 0.222 h^{-1}. Rodrigues [38] reported values of 0.27, 0.24 and 0.23 h^{-1} in glucose, apple cake and starch residues, respectively, in cultures of *Cupriavidus necator* DSM 545. Nascimento et al. [39] cultured *Burkholderia sacchari* LFM 101 on glucose, sucrose and glycerol, but did not observe differences in the maximum specific growth rates obtained with glucose and sucrose (an average of 0.539 h^{-1}). However, the highest PHA productivity (0.054 g·L^{-1}·h^{-1}) was seen with glucose at 35 °C.

3.4. Production of P(3HB) in Seawater and Residual Biodiesel Glycerol

An increase in the concentration of pure glycerol in the MM led to a reduction in the dry biomass as well as the PHA content in the cells, as seen in Table 4. When 5% pure glycerol was used in the culture, LAMA 677 produced 1.11 ± 0.04 g· L^{-1} biomass, with 64.28% P(3HB) content, and LAMA 685 produced 0.95 ± 0.06 g· L^{-1} biomass including 43.64% of the biopolymer. When the carbon source increased to 10%, the biomass of LAMA 677 decreased to 0.85 ± 0.05 g· L^{-1}, corresponding to a 23.4% reduction in total biomass. In terms of the P(3HB) content, the numbers showed a small decrease when comparing the two concentrations of the carbon source (64.28% at 5% glycerol and 64.04% at 10% glycerol), corresponding to a reduction of 0.24% in the total biomass. The biomass produced by LAMA 685 in the 10% glycerol assay resulted in 0.36 ± 0.07 g· L^{-1}, presenting a 62% reduction, and 28.08% P(3HB), presenting a reduction of 15.56%. The negative effect of increasing the concentration of pure glycerol on the growth of organisms has already been reported in other studies. Zhu et al. [35] cultured *Burkholderia cepacia* ATCC 17759 and obtained lower biomass values in the assay of 9% pure glycerol (1.3 g· L^{-1}) compared to the assay of 3% of the same carbon source, which yielded 2.8 g· L^{-1}.

Table 4. Results of the parameters used to analyze the P(3HB) production of the isolates LAMA 677 and LAMA 685, in seawater and residual biodiesel glycerol, with different formulations.

Isolate	Culture Medium Composition	Total Biomass (g·L^{-1})	P(3HB) Concentration (g·L^{-1})	P(3HB) Content in Total Biomass (%)	P(3HB) Productivity (g·L^{-1}·h^{-1})
LAMA 677	90% mineral medium + 5% glycerol + 5% distilled water	1.11 ± 0.04	0.71	64.28	0.0103
	90% mineral medium + 10% glycerol	0.85 ± 0.05	0.55	64.04	0.0079
	90% seawater + 5% glycerol + 5% distilled water	0.10 ± 0.03	0.03	35.04	0.0005
	90% seawater + 5% residual glycerol + 5% distilled water	0.06 ± 0.01	0.02	31.70	0.0003
	90% seawater + 10% residual glycerol	*	*	*	*
	90% mineral medium + 5% residual glycerol + 5% distilled water	1.39 ± 0.05	0.74	52.94	0.0107
LAMA 685	90% mineral medium + 5% glycerol + 5% distilled water	0.95 ± 0.06	0.42	43.64	0.0060
	90% mineral medium + 10% glycerol	0.36 ± 0.07	0.10	28.08	0.0015
	90% seawater + 5% glycerol + 5% distilled water	0.27 ± 0.03	0.13	48.26	0.0019
	90% seawater + 5% residual glycerol + 5% distilled water	0.32 ± 0.01	0.17	53.60	0.0024
	90% seawater + 10% residual glycerol	0.10 ± 0.02	0.01	10.97	0.0002
	90% mineral medium + 5% residual glycerol + 5% distilled water	2.71 ± 0.96	1.22	44.95	0.0177

* Unobserved growth.

To evaluate the production of P(3HB) from low-cost minerals and nutrients, a medium composed of seawater and pure glycerol (5%) was used. When comparing this condition to a treatment where the MM was used instead of seawater, a drastic reduction in growth was evident. For instance, the biomass production levels in LAMA 677 and LAMA 685 were approximately 90% and 70% lower, respectively, when using seawater. The seawater used in the medium culture was collected on the Brazilian continental shelf, where the currents have low nutrient concentrations, as reported by Yoneda [40]. These low nutrient concentrations may have led to less growth when compared to the MM. On the other hand, when considering the PHA content in the biomass, the differences were much smaller.

For example, LAMA 677 produced approximately 21% less P(3HB) in the seawater cultures. In fact, LAMA 685 exhibited a slight increase when using seawater, reaching 48.26% P(3HB), or a gain of 10%. It is possible that this difference occurred because the collected seawater was poor in nutrients, and this limiting condition favored the accumulation of P(3HB) in LAMA 685.

A low-cost medium composed of seawater as a source of nutrients and minerals, and residual biodiesel glycerol as the carbon source was also tested as a medium for bacterial PHA production. In this case, the medium was supplied with two concentrations of the carbon source, 5% and 10%. The first assay resulted in a biomass and P(3HB) content of 0.06 ± 0.01 g· L^{-1} and 31.7% for the LAMA 677 isolate, and 0.32 ± 0.01 g· L^{-1} and 53.6% for LAMA 685. When using the 10% residual glycerol culture media, LAMA 677 showed no growth, and LAMA 685 presented lower values in relation to the other treatment (5% glycerol). This result was also observed by Zhu et al. [35] in *Burkholderia cepacia* ATCC 17759 cultures with residual glycerol in concentrations ranging from 3% to 9%. The authors reported that the increase in the carbon source concentration resulted in a gradual reduction of biomass and PHA content in the cells. Shrivastav et al. [9] evaluated the growth potential of organisms isolated from terrestrial and marine environments, using residual glycerol at concentrations of 1%, 2%, 5%, and 10% as the carbon sources, and reported high growth in the experiments with 1% and 2% carbon, and reduced growth in the experiments with 5% and 10%. This difference may have occurred as a result of the impurities (salts, esters, and alcohol) in the residual glycerol, which may have affected the metabolic processes of the microorganism. The use of residual glycerol is being investigated as an opportunity to reduce the costs of bacterial PHA production. For example, the new *Pannonibacter phragmitetus* ERC8 isolate was found to be capable of producing PHA (0.43 g· L^{-1}) from residual glycerol as the sole carbon source. The maximum PHA production was 1.36 g· L^{-1} when a low concentration (0.80%) of residual glycerol was applied [41]. Naranjo et al. [42] demonstrated the valorization of glycerol for P(3HB) production when working with *Bacillus megaterium*. The study successfully produced 4.8 g· L^{-1} of P(3HB) using 2% (w/v) purified glycerol under controlled conditions. Similarly, Jincy et al. [43] performed statistical optimization for P(3HB) production (0.60 g· L^{-1}) using 2% (v/v) residual glycerol by *Bacillus firmus* NII 0830. A study by Hermann-Krauss et al. [6] compared residual glycerol to pure glycerol and did not reveal any negative effects in terms of productivity or polyester properties in *Haloferax mediterranei*. This finding demonstrated that expensive carbon sources for archaic PHA production can be replaced with low residual glycerol phase surplus products from the biodiesel production process.

Conversely, in work carried out by Rodríguez-Contreras et al. [44], the authors obtained high cell dry mass and growth rates when glycerol was used together with glucose in a fermentation with *Cupriavidus necator*. When analyzing the biomass and the growth of *Burkholderia sacchari* using glycerol as a carbon source, the strain properly synthesized P(3HB); however, the biopolymers obtained from both fermentations with glycerol showed low molecular masses.

The impurities present in the residual glycerol vary according to the raw material and the biodiesel production process, as seen from the characterizations. Onwudili and Williams [45] described the composition of a residual glycerol containing 20.8% methanol, 33.1% esters, 1.52% moisture and 2.28% ash. Mothes et al. [34] analyzed residual glycerol from several companies and reported compositions of 5.3% to 14.2% moisture, 0.01% to 1.7% methanol, and 1% to 6% salts. Thompson and He [46] reported extremely variable methanol concentrations, ranging from 23.4% to 37.5%. According to the results obtained in the test with the MM and 5% pure or residual glycerol, the best biomass production was identified with the use of residual glycerol. Under this condition, LAMA 677 reached 1.39 ± 0.05 g· L^{-1}, representing an increase of 0.28 g· L^{-1} or 25.2%, compared to the use of pure glycerol. LAMA 685 reached a value of 2.71 ± 0.96 g· L^{-1}, with the biomass 2.85 times greater than that recorded in the test with pure glycerol. Thus, the best growth rates occurred when residual glycerol was added to the culture medium. It is possible that this result was obtained as a function of the nutrients contained in the carbon source, as Thompson and He [46] found varying contents of calcium (11.0 to 163.3 mg·L^{-1}), potassium (216.7 mg·L^{-1}), magnesium (0.4 to 126.7 mg·L^{-1}), phosphorus (12.0 to 134.7 mg·L^{-1}), sulfur (14.0 to 128.0 mg·L^{-1}), and sodium (1.07 to 1.4 mg·L^{-1}) when analyzing residual glycerol from several companies.

LAMA 677 had a lower biomass P(3HB) value of 11.34% (absolute value) when grown in residual glycerol (5%) than when grown in a medium with pure glycerol. Other studies have also shown the negative effect of residual glycerol on the accumulation of PHA compared to pure glycerol when used in the same concentrations. This difference also occurred in a study by Kawata and Aiba [13], which used residual and pure glycerol at a concentration of 5% in cultures of *Halomonas* sp. KM-1. The study reported the lowest PHA content in the medium with added biodiesel by-product, compared to purified glycerol. When *Cupriavidus necator* bacteria was cultured in pure and residual glycerol, Posada et al. [47] measured 57.1 g· L^{-1} of PHA in the purified substrate and 27.8 g· L^{-1} in the crude substrate. Kangsadan et al. [36] reported a PHA yield of 83.23% of the biomass with pure glycerol, and 78.26% in the crude form. AndreeBen et al. [48] reported contents of 11.85% of polymer with the use of pure glycerol and 5.24% of PHA with the addition of crude glycerol in cultures of recombinant *Escherichia coli*. This fact can be explained by the influence of impurities in the residual glycerol on PHA synthesis.

In a recent study by de Paula et al. [49], *Pandoraea* sp. MA03 showed strong potential to produce P(3HB) from crude glycerol. Experiments were performed for a 10–50 g· L^{-1} carbon source, and the best values for P(3HB) production were shown in crude glycerol cultivations, compared to pure glycerol, with a polymer accumulation ranging from 49.0% to 63.6% cell dry weight. Based on the P(3HB) production parameters of the evaluated organisms, it is possible to conclude that LAMA 685 has a higher capacity to tolerate impurities in crude glycerol, even though lower growth rates occurred in the medium formulated with seawater and residual glycerol (10%), especially considering that LAMA 677 did not show any growth under the same conditions. Additionally, the cultivation of LAMA 685 in the MM with residual glycerol (5%) resulted in superior biomass, 1.76 g· L^{-1}, when compared to the MM with pure glycerol (5%). Moreover, LAMA 677 showed an increase of 0.28 g· L^{-1} in total biomass when using residual glycerol, although the biomass results obtained from both organisms in the MM supplemented with 5% pure glycerol were similar. In addition, LAMA 685 was found to be very similar in P(3HB) content in the assays of the MM with 5% pure or residual glycerol, whereas LAMA 677 achieved a lower P(3HB) content in the residual glycerol cultivation.

The isolates cultured in low-cost media (5% seawater and residual glycerol) expressed a P(3HB) content of 0.02 g· L^{-1} in LAMA 677 and 0.17 g· L^{-1} in LAMA 685. When compared to the results of Pandian et al. [14], who evaluated the PHA production of *Bacillus megaterium* SRKP-3 (isolated organism from the marine environment) in a medium formulated with low-value inputs including seawater, rice bran and dairy residues, the low PHA content is verified, as they obtained results ranging from 0.196 to 6.376 g· L^{-1} of PHA. However, in the study by Pandian et al. [14], the culture medium used for the polymer synthesis could be considered more nutritionally rich, as, according to Silva [50], the effluents of the dairy industry are characterized by high amounts of organic matter, vitamins and minerals. Therefore, the medium culture used by Pandian et al. [14] was composed of higher concentrations of nutrients compared to the medium used in this study, thus possibly explaining the better polymer contents reported.

4. Conclusions

It is possible to conclude that marine bacteria have great potential for PHA production. Specifically, the use of marine bacteria from the epipelagic zone, which are exposed to a myriad of substrates, increases the opportunity to accumulate the biopolymer as an energy reserve. This ability was verified in the laboratory when isolates were able to use various carbon sources simulating agro-industrial residues (starch, CMC, glycerol, etc.). Two isolates that were efficient in producing PHA in a high concentration from pure glycerol were further investigated. In addition, the potential of those selected bacteria to synthesize P(3HB) using seawater and residual glycerol from biodiesel as the culture media was revealed. These bacteria will be further characterized in order to optimize their production and evaluate their performance in other low-cost substrates, as well to conduct their molecular identification. These results open a new avenue to explore marine bacteria as efficient converters of by-products into biomass rich in PHA, thus reducing the production costs.

Acknowledgments: ICGEB/CNPq (Brazil, Process 577915/2008-8) and CNPq/INCT-Mar COI (Brazil, Process 565062/2010-7) supported this work. We are also thankful to Jose Angel Alvarez Perez, coordinator of South Atlantic MAR-ECO Patterns and Processes of the Ecosystems, and Andre Silva Barreto for providing samples. We also thank CNPq for the scholarship provided to Andre Oliveira De Souza Lima (Process 311010/2015-6) and Maria Cecilia Miotto, as well to Santa Catarina State Govern for Nathalia Aparecida Santos Castilho's scholarship.

Author Contributions: Rodrigo Yoji Uwamori Takahashi and Andre Oliveira De Souza Lima conceived and designed the study and experiments. Rodrigo Yoji Uwamori Takahashi, Nathalia Aparecida Santos Castilho and Marcus Adonai Castro Da Silva performed the experiments under Andre Oliveira De Souza Lima's supervision. Rodrigo Yoji Uwamori Takahashi and Maria Cecilia Miotto contributed to the analysis. Rodrigo Yoji Uwamori Takahashi and Maria Cecilia Miotto wrote the paper with the suggestions/corrections of Andre Oliveira De Souza Lima.

Conflicts of Interest: The authors declare no conflict of interest.

References

1. Rehm, B.H.A.; Steinbüche, A. Biochemical and genetic analysis of PHA synthases and other proteins required for PHA synthesis. *Int. J. Biol. Macromol.* **1999**, *25*, 3–19. [CrossRef]

2. Sudesh, K.; Abe, H.; Doi, Y. Synthesis, structure and properties of polyhydroxyalkanoates: Biological polyesters. *Prog. Polym. Sci.* **2000**, *25*, 1503–1555. [CrossRef]

3. Sheu, D.S.; Chen, W.M.; Yang, J.Y.; Chang, R.C. Thermophilic bacterium *Caldimonas taiwanensis* produces poly(3-hydroxybutyrate-co-3-hydroxyvalerate) from starch and valerate as carbon sources. *Enzym. Microb. Technol.* **2009**, *44*, 289–294. [CrossRef]

4. Tay, B.Y.; Lokesh, B.E.; Lee, C.Y.; Sudesh, K. Polyhydroxyalkanoate (PHA) accumulating bacteria from the gut of higher termite *Macrotermes carbonarius* (Blattodea: Termitidae). *World J. Microbiol. Biotechnol.* **2010**, *26*, 1015–1024. [CrossRef]

5. Valentin, H.E.; Broyles, D.L.; Casagrande, L.A.; Colburn, S.M.; Creely, W.L.; Delaquil, P.A.; Felton, H.M.; Gon-zalez, K.A.; Houmiel, K.L.; Lutke, K.; et al. PHA production, from bacteria to plants. *Int. J. Biol. Macromol.* **1999**, *25*, 303–306. [CrossRef]

6. Hermann-Krauss, C.; Koller, M.; Muhr, A.; Fasl, H.; Stelzer, F.; Braunergg, G. Archaeal production of polyhydroxyalkanoate (PHA) co-and terpolyesters from biodiesel industry-derived by-products. *Archaea* **2013**. [CrossRef] [PubMed]

7. Koller, M.; Maršálek, L.; Dias, M.M.S.; Braunegg, G. Producing microbial polyhydroxyalkanoate (PHA) biopolyesters in a sustainable manner. *New Biotechnol.* **2017**, *37*, 24–38. [CrossRef] [PubMed]

8. Quillaguamán, J.; Guzmán, H.; Van-Thuoc, D.; Hatti-Kaul, R. Synthesis and production of polyhydroxyalkanoates by halophiles: Current potential and future prospects. *Appl. Microbiol. Biotechnol.* **2010**, *85*, 1687–1696. [CrossRef] [PubMed]

9. Shrivastav, A.; Mishra, S.K.; Shethia, B.; Pancha, I.; Jain, D.; Mishra, S. Isolation of promising bacterial strains from soil and marine environment for polyhydroxyalkanoates (PHAs) production utilizing *Jatropha* biodiesel byproduct. *Int. J. Biol. Macromol.* **2010**, *47*, 283–287. [CrossRef] [PubMed]

10. Tan, D.; Xue, Y.S.; Aibaidula, G.; Chen, G.Q. Unsterile and continuous production of polyhydroxybutyrate by *Halomonas* TD01. *Bioresour. Technol.* **2011**, *102*, 8130–8136. [CrossRef] [PubMed]

11. Yue, H.; Ling, C.; Yang, T.; Chen, X.; Chen, Y.; Deng, H.; Wu, Q.; Chen, J.; Chen, G.-Q. A seawater-based open and continuous process for polyhydroxyalkanoates production by recombinant *Halomonas campaniensis* LS21 grown in mixed substrates. *Biotechnol. Biofuels* **2014**, *7*, 108. [CrossRef]

12. Margesin, R.; Schinner, F. Potential of halotolerant and halophilic microorganisms for biotechnology. *Extremophiles* **2001**, *5*, 73–83. [CrossRef] [PubMed]

13. Kawata, Y.; Aiba, S. Poly(3-hydroxybutyrate) production by isolated *Halomonas* sp. KM-1 using waste glycerol. *Biosci. Biotechnol. Biochem.* **2010**, *74*, 175–177. [CrossRef] [PubMed]

14. Pandian, S.R.; Venkatraman, D.; Kalishwaralal, K.; Rameshkumar, N.; Jeraraj, M.; Gurunathan, S. Optimization and fed-batch production of PHB utilizing dairy waste and sea water as nutrient sources by *Bacillus megaterium* SRKP-3. *Bioresour. Technol.* **2010**, *101*, 705–711. [CrossRef] [PubMed]

15. Yin, J.; Chen, J.C.; Wu, Q.; Chen, G.Q. Halophiles, coming stars for industrial biotechnology. *Biotechnol. Adv.* **2015**, *33*, 1433–1442. [CrossRef] [PubMed]

16. Spiekermann, P.; Rehm, B.H.A.; Kalscheuer, R.; Baumeister, D.; Steinbüchel, A. A sensitive, viable-colony staining method using Nile red for direct screening of bacteria that accumulate polyhydroxyalkanoic acids and other lipid storage compounds. *Arch. Microbiol.* **1999**, *171*, 73–80. [CrossRef] [PubMed]

17. Baumann, P.; Baumann, L.; Mandel, M. Taxonomy of Marine Bacteria: The Genus *Beneckea. J. Bacteriol.* **1971**, *107*, 268–294. [PubMed]

18. Alves, L.P.; Almeida, A.T.; Cruz, L.M.; Pedrosa, F.O.; De Souza, E.M.; Chubatsu, L.S.; Müller-Santos, M.; Valdameri, G. A simple and efficient method for poly-3-hydroxybutyrate quantification in diazotrophic bacteria within 5 minutes using flow cytometry. *Braz. J. Med. Biol. Res.* **2017**, *50*, e5492. [CrossRef] [PubMed]

19. Zuriani, R.; Vigneswari, S.; Azizan, M.N.M.; Majid, M.I.A.; Amirul, A.A.A. High throughput Nile red fluorescence method for rapid quantification of intracellular bacterial polyhydroxyalkanoates. *Biotechnol. Bioprocess Eng.* **2013**, *18*, 472–478. [CrossRef]

20. Arikawa, H.; Sato, S.; Fujiki, T.; Matsumoto, K. Simple and rapid method for isolation and quantitation of polyhydroxyalkanoate by SDS-sonication treatment. *J. Biosci. Bioeng.* **2017**, *S1389–S1723*, 30664–30668. [CrossRef] [PubMed]

21. Degelau, A.; Scheper, T.; Bailey, J.E.; Guske, C. Fluorometric measurement of poly-β hydroxybutyrate in *Alcaligeneseutrophus* by flow cytometry and spectrofluorometry. *Appl. Microbiol. Biotechnol.* **1995**, *42*, 653–657. [CrossRef]

22. Reddy, C.S.K.; Ghai, R.; Rashmi; Kalia, V.C. Polyhydroxyalkanoates: An overview. *Bioresour. Technol.* **2003**, *87*, 137–146. [CrossRef]

23. Hedgpeth, J. *Classification of Marine Environments*; Reseck, J., Jr., Ed.; Marine Biology: Englewood Cliffs, NJ, USA, 1957; pp. 18–27.

24. Longhurst, R.A.; Harrison, G.W. Vertical nitrogen flux from the oceanic photic zone by diel migrant zooplankton and nekton. *Deep Sea Res. Part A Oceanogr. Res. Pap.* **1988**, *35*, 881–889. [CrossRef]

25. Chien, C.C.; Chen, C.C.; Choi, M.H.; Kung, S.S.; Wei, Y.H. Production of poly-ß-hydroxybutyrate (PHB) by *Vibrio* spp. isolated from marine environment. *J. Biotechnol.* **2007**, *132*, 259–263. [CrossRef] [PubMed]

26. Mahishi, L.H.; Tripathi, G.; Rawal, S.K. Poly(3-hydroxybutyrate) (PHB) synthesis by recombinant *Escherichia coli* harbouring *Streptomyces aureofaciens* PHB biosynthesis genes: Effect of various carbon and nitrogen sources. *Microbiol. Res.* **2003**, *158*, 19–27. [CrossRef] [PubMed]

27. Mohandas, S.P.; Balan, L.; Lekshmi, N.; Cubelio, S.S.; Philip, R.; Sing, I.S.B. Production and characterization of Polyhydroxybutyrate from *Vibrio Harveyi* MCCB 284 utilizing glycerol as carbon source. *J. Appl. Microbiol.* **2016**, *122*, 698–707. [CrossRef] [PubMed]

28. Van-Thuoc, D.; Quillaguamán, J.; Mamo, G.; Matiason, B. Utilization of agricultural residues for poly(3-hydroxybutyrate) production by *Halomonasboliviensis* LC1. *J. Appl. Microbiol.* **2008**, *104*, 420–428. [PubMed]

29. Bertrand, J.L.; Ramsay, B.A.; Ramsay, J.A.; Chavarie, C. Biosynthesis of poly-β-hydroxyalkanoates from pentoses by *Pseudomonas pseudoflava. Appl. Environ. Microbiol.* **1990**, *56*, 3133–3138. [PubMed]

30. Silva, L.F.; Taciro, M.K.; Ramos, M.E.M.; Carter, J.M.; Pradella, J.G.C.; Gomez, J.G.C. Poly-3-hydroxybutyrate (P3HB) production by bacteria from xylose, glucose and sugarcane bagasse hydrolysate. *J. Ind. Microbiol. Biotechnol.* **2004**, *31*, 245–254. [CrossRef] [PubMed]

31. Fernández, D.; Rodríguez, E.; Bassas, M.; Viñas Solanas, A.M.; Liorens, J.; Marquéz, A.M.; Manresa, A. Agro-industrial oily wastes as substrates for PHA production by the new strain *Pseudomonas aeruginosa* NCIB 40045: Effect of culture conditions. *Biochem. Eng. J.* **2005**, *26*, 159–167. [CrossRef]

32. He, W.; Tian, W.; Zhang, G.; Chen, G.-Q.; Zhang, Z. Production of novel polyhydroxyalkanoates by *Pseudomonas stutzeri* 1317 from glucose and soybean oil. *FEMS Microbiol. Lett.* **1998**, *169*, 45–49. [CrossRef]

33. Cavalheiro, J.M.B.T.; Almeida, M.C.M.D.; Grandfils, C.; Fonseca, M.M.R. Poly(3-hydroxybutyrate) production by *Cupriavidus necator* using waste glycerol. *Process Biochem.* **2009**, *44*, 509–515. [CrossRef]

34. Mothes, G.; Schnorpfeil, C.; Ackermann, J.U. Production of PHB from crude glycerol. *Eng. Life Sci.* **2007**, *7*, 475–479. [CrossRef]

35. Zhu, C.; Nomura, C.T.; Perrotta, J.A.; Stipanovic, A.J.; Nakas, J.P. Production and characterization of poly-3-hydroxybutyrate from biodiesel-glycerol by *Burkholderia cepacia* ATCC 17759. *Biotechnol. Prog.* **2010**, *26*, 424–430. [PubMed]

36. Kangsadan, T.; Swadchaipon, N.; Kongruang, S. Value-added utilization of crude glycerol from biodiesel production by microbial synthesis of polyhydroxybutyrate-valerate. *Curr. Opin. Biotechnol.* **2011**, *22*, S1–S35. [CrossRef]

37. Piccoli, R.A.M.; Silva, E.S.; Taciro, M.K.; Maiorano, A.E.; Ribeiro, C.M.S.; Rodrigues, M.F.A. Produção de polihidroxibutirato a partir de glicerol resíduo da produção de biodiesel. In *Simpósio Nacional de Bioprocessos*; 17, 2011, Caxias do Sul. Anais...; Associação Brasileira de Engenharia Química: Caxias do Sul, Brazil, 2011; pp. 1–6.

38. Rodrigues, R.C. Condições de Cultura Para a Produção de Poli(3-hidroxibutirato) por *Ralstoniaeu tropha* a partir de Resíduos de Indústrias de Alimento. Master's Thesis, Universidade Federal de Santa Catarina, Trindade, Florianópolis, Brasil, 2005.

39. Nascimento, V.M.; Silva, L.F.; Gomez, J.G.C.; Fonseca, G.G. Growth of *Burkholderia sacchari* LFM 101 cultivated in glucose, sucrose and glycerol at different temperatures. *Sci. Agricola.* **2016**, *73*, 429–433. [CrossRef]

40. Yoneda, N.T. Área Temática: Plâncton, 1999. Centro de Estudos do Mar, Universidade Federal do Paraná. Available online: http://www.brasil-rounds.gov.br/round7/arquivos_r7/PERFURACAO_R7/refere/pl% E2ncton.pdf (accessed on 23 December 2011).

41. Ray, S.; Prajapati, V.; Patel, K; Triedi, U. Optimization and characterization of PHA from isolate *Pannonibacter phragmitetus* ERC8 using glycerol waste. *Int. J. Biol. Macromol.* **2016**, *86*, 741–749. [CrossRef] [PubMed]

42. Naranjo, J.M.; Posada, J.A.; Higuita, J.C.; Cardona, C.A. Valorization of glycerol through the production of biopolymers: The PHB case using *Bacillus megaterium*. *Bioresour. Technol.* **2013**, *133*, 38–44. [CrossRef] [PubMed]

43. Jincy, M.; Sindhu, R.; Pandey, A.; Binod, P. Bioprocess development for utilizing biodiesel industry generated crude glycerol for production of poly-3-hydroxybutyrate. *J. Sci. Ind. Res.* **2013**, *72*, 596–602.

44. Rodríguez-Contreras, A.; Koller, M.; Dias, M.M.S.; Calaffel-Monfort, M.; Braunegg, G.; Marqués-Calvo, M.S. Influence of glycerol on poly(3-hydroxybutyrate) production by *Cupriavidus necator* and *Burkholderia sacchari*. *Biochem. Eng. J.* **2015**, *94*, 50–57. [CrossRef]

45. Onwudili, J.A.; Williams, P.T. Hydrothermal reforming of bio-diesel plant waste: Products distribution and characterization. *Fuel* **2010**, *89*, 501–509. [CrossRef]

46. Thompson, J.C.; He, B.B. Characterization of crude glycerol from biodiesel production from multiple feedstocks. *Appl. Eng. Agric.* **2006**, *22*, 261–265. [CrossRef]

47. Posada, J.A.; Naranjo, J.M.; López, J.A.; Higuita, J.C.; Cardona, C.A. Design and analysis of poly-3-hydroxybutyrate production processes from crude glycerol. *Process Biochem.* **2011**, *46*, 310–317. [CrossRef]

48. Andreeßen, B.; Lange, A.B.; Robenek, H.; Steinbüchel, A. Conversion of glycerol to poly(3-Hydroxypropionate) in recombinant *Escherichia coli*. *Appl. Environ. Microbiol.* **2010**, *76*, 622–626. [CrossRef] [PubMed]

49. De Paula, F.C.; Kakazu, S.; de Paula, C.B.C.; Contiero, J. Polyhydroxyalkanoate production from crude glycerol by newly isolated *Pandoraea* sp. *J. King Saud Univ. Sci.* **2017**, *29*, 166–173. [CrossRef]

50. Silva, A.M.X.P. Degradação de Efluentes Lácteo sem Reactores UASB Com Recirculação. Master's Thesis, Universidade de Aveiro, Aveiro, Portugal, 2008.

bioengineering

MDPI

Article

Biodegradable Polymeric Substances Produced by a Marine Bacterium from a Surplus Stream of the Biodiesel Industry

Sourish Bhattacharya [1,†], Sonam Dubey [2,†], Priyanka Singh [3], Anupama Shrivastava [4] and Sandhya Mishra [2,*]

1   Process Design and Engineering Cell, CSIR-Central Salt and Marine Chemicals Research Institute, Bhavnagar 364002, India; sourishb@csmcri.org
2   Salt and Marine Chemicals, CSIR-Central Salt and Marine Chemicals Research Institute, Bhavnagar 364002, India; sonamdubey20@gmail.com
3   DTU BIOSUSTAIN, Novo Nordisk Foundation Center for Biosustainability, Technical University of Denmark, Lyngby 2800, Denmark; prnksingh254@gmail.com
4   Research & Product Development, Algallio Biotech Private Limited, Vadodara 390020, India; anupamashrivastav@gmail.com
*   Correspondence: smishra@csmcri.org
†   These authors contributed equally to this work.

Academic Editor: Martin Koller
Received: 13 September 2016; Accepted: 24 November 2016; Published: 30 November 2016

Abstract: Crude glycerol is generated as a by-product during transesterification process and during hydrolysis of fat in the soap-manufacturing process, and poses a problem for waste management. In the present approach, an efficient process was designed for simultaneous production of 0.2 g/L extracellular ε-polylysine and 64.6% (w/w) intracellular polyhydroxyalkanoate (PHA) in the same fermentation broth (1 L shake flask) utilizing *Jatropha* biodiesel waste residues as carbon rich source by marine bacterial strain (*Bacillus licheniformis* PL26), isolated from west coast of India. The synthesized ε-polylysine and polyhydroxyalkanoate PHA by *Bacillus licheniformis* PL26 was characterized by thermogravimetric analysis (TGA), differential scanning colorimetry (DSC), Fourier transform infrared spectroscopy (FTIR), and ^{1}H Nuclear magnetic resonance spectroscopy (NMR). The PHA produced by *Bacillus licheniformis* was found to be poly-3-hydroxybutyrate-co-3-hydroxyvalerate (P3HB-co-3HV). The developed process needs to be statistically optimized further for gaining still better yield of both the products in an efficient manner.

Keywords: crude glycerol; polyhydroxyalkanoate; ε-polylysine; *Bacillus licheniformis*; fermentation

1. Introduction

Most of the global economy is driven by petroleum fuels as the main source of energy. However, due to market fluctuation, it is moving towards a sustainable bio based economy as fossil reserves are projected to decline completely by 2050 [1–3]. In addition to over exploitation of petroleum deposits, climate change and other negative environmental effects from exhaust gases lead researchers for the search of renewable alternatives such as biodiesel [4]. Biodiesel is an appealing alternative [5,6], which is clean burning, non-toxic and biodegradable [7]. It is a fatty acid methyl ester compound produced by a transesterification process of animal or plant oils with methanol in the presence of a catalyst [8–10]. Generally, glycerol is obtained in huge amount as a by-product in production of biodiesel [11,12]. With every 100 lbs of biodiesel produced by transesterification of vegetable oils or animal fats, 10 lbs of crude glycerol is generated [13,14]. However, the tremendous growth of the biodiesel industry has created a glycerol surplus that resulted in a dramatic 10-fold decrease in crude glycerol prices over the last few years. This decrease in prices resulted a problem for the glycerol producing and refining industries and also the economic viability of the biodiesel industry has also been greatly affected [15,16].

For sustainable development and commercialization of biodiesel production, effective utilization of crude glycerol into value added products are desired. However, conversion of crude glycerol will also promote the accretion of integrated biorefineries. 1,3-propanediol, citric acid, PHA's, ε-polylysine, butanol, hydrogen, ethanol, phytase, lipase, succinic acid, docosahexaenoic acid, eicosapentanoic acid, monoglycerides, lipids and syngas can be produced from crude glycerol through microbial strains [17–31]. However, many of the available technologies need further optimization and development in the form of efficient and sustainable form for its incorporation in bio-refineries.

The utilization of low quality glycerol obtained as by-product of biodiesel production is a big challenge as this glycerol cannot be used for direct food and cosmetic uses. An effective usage for conversion of crude glycerol to specific products may cut down the biodiesel production costs. The process for biodiesel preparation and generating value added products simultaneously through the use of waste generated during the biodiesel production is a very effective approach [32]. Clearly, the development of processes to convert crude glycerol into higher-value products is an urgent need.

Various studies on tuning the material properties of the polyhydroxyalkanoate (PHA) polymer are carried out for its higher applicability in diverse areas [33]. Nerve 2010 reported production of ε-polylysine from *Streptomyces albulus* (CCRC 11814) utilizing crude glycerol as carbon source through aerobic fermentation yielding 0.2 g/L ε-polylysine. However, growth rate of *Streptomyces albulus* (CCRC 11814) was slow due to other impurities such as methanol and other salts present in crude glycerol.

ε-polylysine and polyhydroxyalkanoates are important biopolymers which can be conjugated with other biopolymers for its various applications e.g., ε-polylysine may be utilized as water absorbable hydrogels, drug carriers and anticancer agents. Simultaneously, PHA may be used in drug delivery systems, atrial septal defect repair and cardiovascular stents. ε-polylysine and polyhydroxyalkanoates are important biopolymers which are synthesized through microbe in an efficient and eco-friendly manner. However, these biopolymers may be conjugated with other biopolymers for its application as water absorbable hydrogels, drug carriers and anticancer agents. The hydrogels prepared from these complex biopolymers may be used for its application in quick peritoneal repair and prevention of post-surgical intraabdominal adhesions.

In the present study, for sustainable development of biodiesel production, efforts have been made for effective utilization of crude glycerol as the carbon source for simultaneous production of ε-polylysine and PHA. The present bioconversion route will effectively convert the waste stream of biodiesel production into value added products. In addition, metabolic engineering may used in future for improving product yield of such strains.

2. Materials and Methods

2.1. Materials

2.1.1. Chemicals

Peptone, yeast extract, iron(III) citrate, NaCl, MgCl$_2$, Na$_2$SO$_4$, CaCl$_2$, KCl, NaHCO$_3$, KBr, SrCl$_2$, H$_3$BO$_3$, Na$_2$O$_3$Si, NaF, (NH$_4$)(NO$_3$) and Na$_2$HPO$_4$ were purchased from M/S Hi-Media Limited Mumbai, and were of the highest purity available. ε-PL was procured from Handary SA, Brussels, Belgium, Polyhydroxybutyrate and 3-hydroxyvalerate from Sigma, Bangalore, India.

2.1.2. Growth and Maintenance of *Bacillus licheniformis*

Bacillus licheniformis was isolated from sea brine of experimental salt farm, Bhavnagar, India. It was maintained on Zobell marine agar plates containing (g/L) peptone 5.0; yeast extract 1.0; iron(III) citrate 0.1; NaCl 19.45; MgCl$_2$ 8.8; Na$_2$SO$_4$ 3.24; CaCl$_2$ 1.8; KCl 0.55; NaHCO$_3$ 0.16; KBr 0.08;

SrCl$_2$ 0.034; H$_3$BO$_3$ 0.022; Na$_2$O$_3$Si 0.004; NaF 0.0024; (NH$_4$)(NO$_3$) 0.0016; Na$_2$HPO$_4$ 0.008; agar, 1.5, at pH 7.6 $\pm$ 0.2. 5 mL of glycerol was added to the above medium. The slants were incubated at 37 °C for 4 days and then stored at 4 °C.

Experiments were carried out in 250 mL Erlenmeyer flasks with 100 mL of production medium with following components (g/L): yeast extract, 10; glucose, 50; (NH$_4$)$_2$SO$_4$, 15; MgSO$_4$, 0.5; K$_2$HPO$_4$, 0.8; KH$_2$PO$_4$, 1.4; FeSO$_4$, 0.04; ZnSO$_4$, 0.04. The pH of the medium was adjusted to 6.8 with 1 N NaOH before sterilization [34]. 10% (v/v) of a 48-h-old culture (approximately 8.9 $\times$ 10^8 cells/mL) was used as inoculum. Shake flask cultures of the organism were incubated at temperature 37 $\pm$ 2 °C with continuous agitation at 150 rpm for 96 h. These fermentation parameters were kept uniform for all the studies conducted. All experiments were carried out in triplicates.

2.2. Fermentation

2.2.1. Culture Media

The strain *Bacillus licheniformis* PL26 was cultivated in Zobell marine broth, Himedia, Mumbai, India. The media was adjusted to pH 7.6 $\pm$ 0.2. The plates were incubated at 37 °C temperature for 48 h.

2.2.2. Inoculum Development

The seed culture inoculated with loopful of *Bacillus licheniformis* was incubated overnight at 30 °C in an incubator shaker at 120 rpm. The inoculum for the production batch was prepared by using a single colony of *B. licheniformis* PL26 having 100 mL working volume.

2.2.3. Simultaneous Production of ε-Polylysine and PHA

The marine bacteria was cultured in Zobell marine broth to obtain seed culture having O.D.$_{600}$ of 2.3. Zobell marine medium comprising (g/L) peptone 5.0; yeast extract 1.0; iron(III) citrate 0.1; NaCl 19.45; MgCl$_2$ 8.8; Na$_2$SO$_4$ 3.24; CaCl$_2$ 1.8; KCl 0.55; NaHCO$_3$ 0.16; KBr 0.08; SrCl$_2$ 0.034; H$_3$BO$_3$ 0.022; Na$_2$O$_3$Si 0.004; NaF 0.0024; (NH$_4$)(NO$_3$) 0.0016; Na$_2$HPO$_4$ 0.008 in one liter of the medium maintained at pH 7.6 $\pm$ 0.2. 20% seed culture was inoculated in the production medium which contained 20 g crude glycerol, yeast extract 5 g, (NH$_4$)$_2$SO$_4$ 10 g, K$_2$HPO$_4$ 0.8 g, KH$_2$PO$_4$ 1.36 g, MgSO$_4$ 0.5 g, ZnSO$_4$ 0.04 g, FeSO$_4$ 0.03 g in one litre of the medium maintained at pH 8.9 $\pm$ 0.2.

2.2.4. Analysis of ε-Polylysine

The culture broth was harvested after fermentation and cells were separated by centrifugation at 15,296$\times$ *g* rcf for 10 min in refrigerated centrifuge. 1 mL of supernatant was added to 1 mL of 1 mM methyl orange, mixed thoroughly under shaking condition along with incubating it at 37 °C for 60 min [35]. Further, the solution was centrifuged at 15,296$\times$ *g* rcf for 10 min in refrigerated centrifuge and absorbance of the supernatant was measured at 465 nm on UV-vis spectrophotometer (Varian, Palo Alto, CA, USA). A standard curve was derived from measurements with known amounts (0.1–2 mg/mL) of standard ε-PL procured from Handary S.A. [36].

2.2.5. Percentage Carbon Utilization of *Bacillus licheniformis* PL26

Percentage carbon utilized by *Bacillus licheniformis* PL26 was calculated a

$$\text{Carbon utilization } (\%) = \frac{\text{Total utilized carbon by bacteria}}{\text{Total carbon present in the medium}} \times 100 \qquad (1)$$

Glycerol estimation was carried out by using "Waters Alliance" high performance chromatographic system equipped with RI detector (Waters 2414 model, Waters India Ltd., Bangalore, India) and separation module (Waters 2695 model, Waters India Ltd., Bangalore, India). Chromatographic separations were performed on an "Aminex HPX-87H" column (300 $\times$ 7.8 mm)

(Bio-Rad Laboratories, Richmond, CA, USA) with a precolumn (30 × 4.6 mm) of the same stationary phase (DVB-S, hydrogen form, Richmond, CA, USA). Isocratic elution at a flow rate of 0.6 mL/min was carried out using a mixture of 5 mM sulfuric acid. Peak detection was made by keeping the cells of the RI detector at 30 °C. The samples were appropriately degassed, twice diluted with double-distilled water, filtered through a "Whatman" 0.45-µm filter membrane (GE Healthcare Life Sciences, Little Chalfont, Buckinghamshire, UK), and then injected (50-µL loop volume). Data were obtained and processed by using "Waters EMPOWER" software (waters Corporation India, Bangalore, India). Peak identification was carried out by spiking the sample with pure standards and comparing the retention times with those of pure compounds.

2.2.6. Extraction and Purification of PHA

After completion of 96 h fermentation, culture broth was centrifuged at 15,296× *g* rcf for 10 min in refrigerated centrifuge. The cell pellets were oven dried overnight at 60 °C. Cellular digestion of dried cell pellet was carried out by re-suspending it in 6% (v/v) sodium hypochlorite solution followed by centrifugation at 10,000 rpm for 5 min. Further, the digested cell pellets were washed twice with methanol followed by distilled water to remove the traces of impurities resulting in a purified product, which was further dissolved in chloroform and weighed after air drying [37].

2.3. Purification of ε-Polylysine

2.3.1. Precipitation of Polycationic ε-PL with TPB⁻ Anion from the Supernatant

After removal of cells through centrifugation, the supernatant obtained was treated with sodium tetraphenylborate for precipitating ε-PL as a polyelectrolyte salt with the TPB⁻ anion. The polycationic ε-PL salt with the TPB⁻ anion was further purified by washing the mixed precipitate with acetone to remove triphenylborate and benzene. Thereafter, the precipitate reacted with 1 M HCl for obtaining ε-PL hydrochloride.

2.3.2. Analytical Methods

FT-IR spectra of obtained PHA were recorded on a Perkin-Elmer Spectrum GX (FT-IR System, Waltham, MA, USA) instrument. ^{1}H Nuclear magnetic resonance spectroscopy of PHA was determined on Bruker Avance-II 500 (Ultra shield) spectrometer, Bangalore, India, at 500 MHz, in CDCl$_3$. Proton ^{1}H NMR spectroscopy was also used to determine copolymer composition through running standards of 3HB and 3HV. Differential scanning calorimetry (DSC) of PHA was carried out using a DSC 204 F1 phoenix instrument with Netzsch software (NETZSCH Technologies India Pvt. Ltd., Chennai, India). The PHA samples were scanned from −20 °C to 500 °C with the heating rate of 10 °C/min. Glass transition temperature and onset melting points were determined in the scan between −20 °C and 500 °C in DSC analysis. Thermo-gravimetric analysis (TGA) of PHA was carried out in temperature range of 27–500 °C using TG 209 F1 instrument (NETZSCH Technologies India Pvt. Ltd., Chennai, India).

3. Results and Discussions

3.1. Simultaneous ε-Polylysine and PHA Production by B. licheniformis PL26

After complete submerged fermentation of 96 h at an agitation of 220 rpm, fermentation broth was centrifuged to obtain supernatant containing ε-polylysine and biomass for PHA extraction.

Simultaneous production of ε-polylysine and PHA by *B. licheniformis* PL26 was obtained utilizing crude glycerol as the carbon source. However, *B. licheniformis* is able to produce 0.2 g/L ε-polylysine extracellularly in the fermentation broth along with 64.59% PHA with respect to dry cell weight i.e., 1.1 g/L P(3HB-co-3HV) having 96 h production age at 37 °C (Figures S1–S3).

In the present case, *B. licheniformis* PL26 is able to produce both ε-polylysine and PHA in the same fermentation broth, which is not reported till date. However, *Streptomyces albulus* (CCRC 11814) is reported to produce ε-polylysine utilizing crude glycerol [3]. *S. albulus* being an Actinomycetes, it possesses a relatively slower growth rate as compared to *Bacillus* sp. As previously reported, *S. albulus* produces ε-polylysine after 120 h in M3G medium containing glucose as the carbon source, 0.2 g/L of ε-polylysine was produced by *Streptomyces albulus* (CCRC 11814) after 168 h [3], but in present study, ε-polylysine was produced after 96 h by *B. licheniformis* at a concentration of 0.2 g/L. In addition, fast growth rate and less production age of *B. licheniformis* which is producing 64.59% PHA with respect to dry cell weight i.e., 1.1 g/L P(3HB-co-3HV), which may be considered as the additional advantage of the process. Simultaneously, crude glycerol was replaced with analytical grade pure glycerol with similar concentration of the carbon source and it was found that pure glycerol yielded 0.06 g/L ε-polylysine and 0.4 g/L P(3HB-co-3HV).

0.2 g/L of ε-Poly-L-lysine produced from *Bacillus licheniformis* PL26 is in similar range with respect to its production from a wild strain *Streptomyces albulus* CCRC 11814 as reported in the literature [3]. However, similar sort of system developed by Moralejo-Ga'rate 2013, wherein using microbial community, simultaneous production of PHA and polyglucose was done [37]. Similarly, using halobacterium *Haloferax mediterranei*, simultaneously poly(3-hydroxybutyrate-*co*-3-hydroxyvalerate and extracellular polysaccharide (EPS) was produced [38]. Few microbes or microbial consortium have potential to produce two different polymers simultaneously using single carbon and nitrogen source.

3.2. Percentage Carbon Utilization of Bacillus licheniformis PL26

Carbon utilization percentage of isolated *Bacillus licheniformis* PL26 is shown in tab:bioengineering-03-00034-t001. *Bacillus licheniformis* isolated from salt pan has a 30% total carbon utilization percentage as it utilizes 0.21% total carbon from 0.7% total carbon present in the production medium.

Table 1. Percentage carbon utilization by *Bacillus licheniformis*.

Parameter	Concentration
Total Carbon content in fermentation medium	0.7%
Total Carbon left in the supernatant after complete fermentation (96 h production age)	0.41%
Carbon present in the biomass	0.21%
Percentage carbon utilized	30%

3.3. Characterization of Purified PHA

The polymer extracted from *B. licheniformis* PL26 grown in production media was characterized through TGA, DSC, NMR and *Fourier transform infrared spectroscopy* (FTIR).

Figure 1 indicates TGA analysis to analyze the thermal decomposition of the extracted polymer through the thermogravimetric analyzer. The extracted polymer showed 0.28 g weight loss out of 0.35 g till 500 °C temperature. Mass change of 0.21 g PHA out of 0.35 g PHA was found in the temperature range of 225–325 °C.

The DSC analysis shown in Figure 2 indicates the melting temperature of the standard sample and the extracted polymer through sodium hypochlorite treatment. The thermal degradation was obtained at 268 °C for the obtained polymer and 250 °C of standard PHA from Sigma Aldrich.

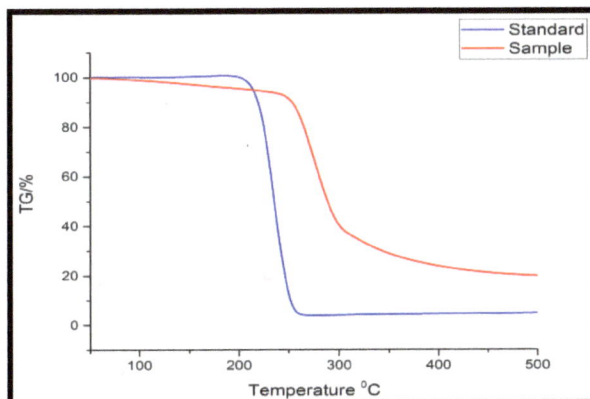

Figure 1. Thermogravimetric analysis (TGA) of purified PHA obtained from *Bacillus licheniformis.*

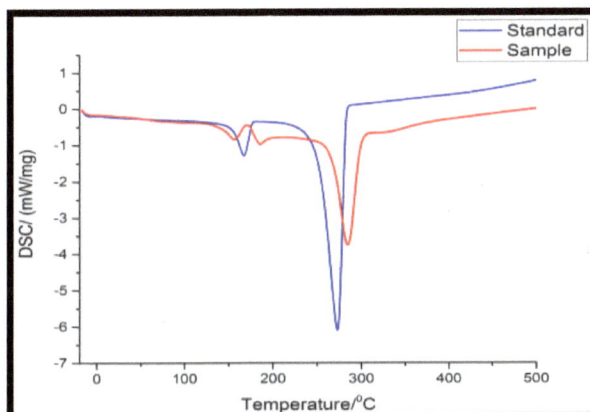

Figure 2. Differential scanning calorimetry (DSC) of purified PHA from *Bacillus licheniformis.*

^{1}H NMR spectra of the extracted polymer, standard PHB and standard 3 hydroxy valerate were found to be comparable with respect to each other. Prominent peaks were observed at δ = 1.6 ppm for CH_3, δ = 2.4 ppm for CH_2 and δ = 5.2 ppm for CH group (Figure 3).

Infra red (IR) spectra (Figure 4) showed intense peaks at 1724 and 1283 cm^{-1} corresponding to –C=O and –CH group which are in correlation of peaks of standard PHA. Peaks at 1380, 1456 and 2932 correspond to –CH_3, –CH_2 and –CH group which are in correlation of peaks of standard PHA as shown in Figure 5.

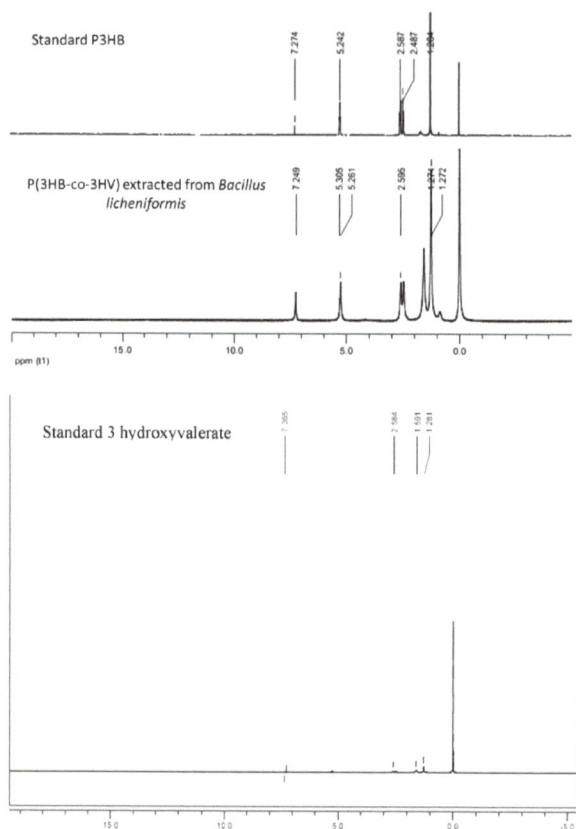

Figure 3. Nuclear magnetic resonance (NMR) spectra of purified product along with standard PHB and standard 3 hydroxy valerate.

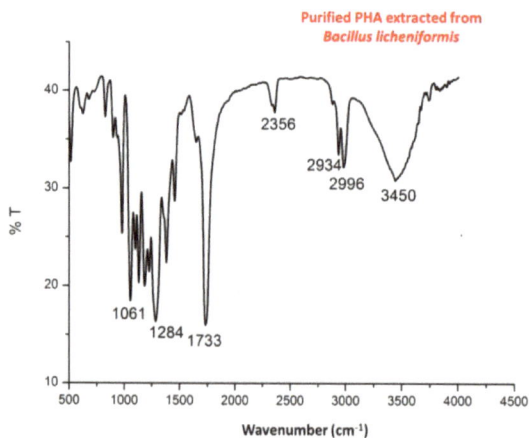

Figure 4. Fourier transform infrared spectroscopy (FTIR) spectra of purified PHA recovered from *Bacillus licheniformis*.

Standard PHA procured from Sigma Aldrich

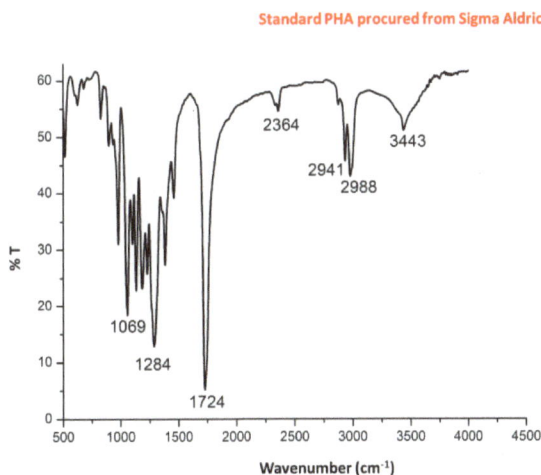

Figure 5. Fourier transform infrared spectroscopy (FTIR) spectra of standard PHA procured from Sigma Aldrich.

3.4. Characterization of ε-Polylysine

^{1}H NMR of ε-Polylysine in D_2O Isolated from *B. licheniformis* PL26

Protons (H^a, H^c) attached to α-amino groups arrived together as broad singlet at δ 3.76 ppm and protons (H^b, H^d) attached to ε-amino groups arrived together as broad singlet at δ 3.14 ppm. Protons attached to β and β' carbons come at δ 1.75 ppm as broad singlet. While the other protons attached to carbons come at δ 1.47 ppm (4H, ε and ε') and δ 1.30 ppm (4H, γ and γ' respectively (Figure 6). Overall, the peaks showing peptide linkage between α-carboxyl group and the ε-amino group, confirming the structure as ε-polylysine.

As per Jia et al., 2010, chemical shift of ε-H in the ε-polylysine units $\delta_{\varepsilon H}$ and at the N-terminal $\delta'_{\varepsilon H}$ are 3.097 and 2.863 of 5 KDa ε-polylysine protein [39]. However, similar results were obtained in the present case, wherein chemical shift of ε-H in the ε-polylysine units, $\delta_{\varepsilon H}$ is at δ 3.14 ppm, which are in similar range with respect to the mentioned reports.

The described process showed the potential of utilizing *Bacillus licheniformis* for the production of PHA (64.59% w/w w.r.t cell dry mass i.e., 1.1 g/L P(3HB-co-3HV)) as well as ε-polylysine (0.2 g/L) using crude glycerol as the carbon source. In addition, there are no such reports as per our knowledge wherein two polymers are produced at a time in the same fermentation broth. Previously, Bera et al., 2014 reported microbial synthesis of such polymer (P(3HB-co-3HV)) by *Halomonas hydrothermalis* (MTCC accession no. 5445) from seaweed derived levulinic acid at a concentration of 57.5% PHA/dry cell weight and Ghosh et al., 2011 reported microbial synthesis of PHA from biodiesel by-products i.e., from crude glycerol and *Jatropha* deoiled cake hydrolysate at a concentration of 75% PHA/dry cell weight. However, further optimization will be required for efficient production of PHA and ε-polylysine at larger scale.

The most important parameter responsible for cost of the production in PHA is substrate for carbon and energy source. The economic feasibility of PHA would depend on few important factors like growth rate of microbe for generation of biomass, substrate cost and recovery process including the solvent involved. However, production of other biopolymer along with PHA in the fermentation medium will have additional advantage in further reducing the production cost. PHA production from biodiesel waste stream can reduce production cost and at the same time solves the problem of waste disposal. In order to increase the PHA yield, the concentration of carbon source and other nutrient source desired for the microbial production may be optimized.

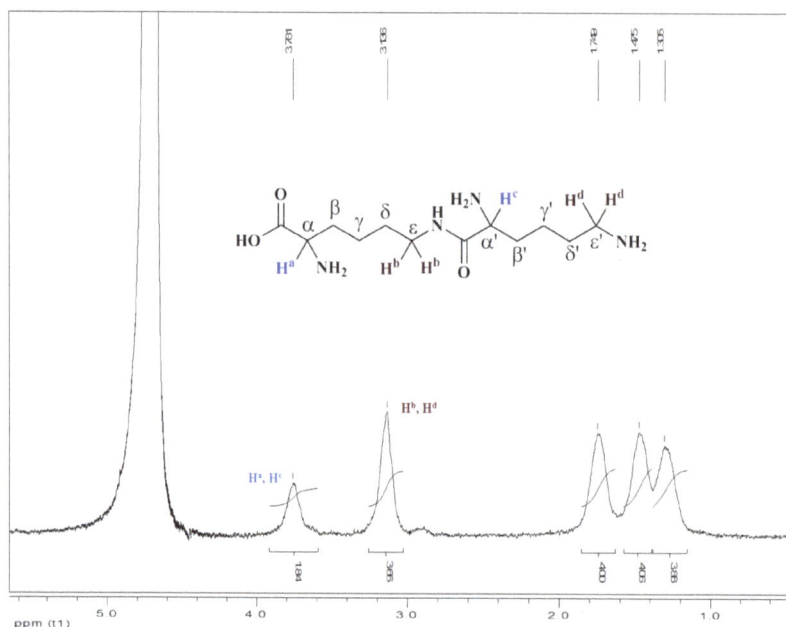

Figure 6. ^{1}H NMR of ε-PL in D_2O isolated from *Bacillus licheniformis*.

4. Conclusions

In the present study, an integrated process for simultaneous production of extracellular ε-polylysine and intracellular P(3HB-co-3HV) developed through marine bacterial strain (*Bacillus licheniformis*) isolated from west coast of India utilizing Jatropha biodiesel waste residues as carbon rich source. A maximum of 0.2 g/L ε-polylysine content and 64.6% (w/w) P(3HB-co-3HV) production with respect to dry biomass was obtained in the fermentation broth using *Bacillus licheniformis*. ε-polylysine and PHA are important class of biopolymers which have various applications in food, agriculture, medicine, pharmacy, controlled drug release, tissue engineering, etc. Although, the present approach provides a solution for the effective utilization of biodiesel by-product, still, the developed process needs to be optimized further for gaining still better yield of both the products for its acclamation as cost effective and sustainable process.

Supplementary Materials: The following are available online at http://www.mdpi.com/2306-5354/3/4/34/s1, Figure S1: effect of production age (h.) on ε-polylysine yield at 35 °C, Figure S2: effect of production age (h) on ε-polylysine yield at 37 °C, Figure S3: effect of production age (h) on ε-polylysine yield at 40 °C.

Acknowledgments: Financial assistance from CSIR as a part of EMPOWER scheme is gratefully acknowledged. We also acknowledge analytical science division of CSIR-CSMCRI for their timely support and cooperation. Authors are also grateful to CSC 0105, CSC 0203, OLP 0077 for providing financial support. We extend our gratitude to BDIM for providing PRIS no. CSIR-CSMCRI—043/2016 registration number.

Author Contributions: Sourish Bhattacharya developed the protocol for isolation and screening of potential ε-polylysine halophilic bacteria, followed by optimizing medium composition for simultaneous production of ε-polylysine and polyhydroxyalkanoate in the medium. Further, the extraction, purification and characterization of ε-polylysine was also conducted by him. Priyanka Singh assisted in isolation of bacteria from west coast of India and screening for identifying potential ε-polylysine halophilic bacterial isolate. Anupama Shrivastava and Sonam Dubey optimized the process for extraction and purification of polyhydroxyalkanoate. Sandhya Mishra monitored the overall experiments being performed and helped in the interpretation of results and polishing the manuscript.

Conflicts of Interest: The authors declare no conflict of interest.

References and Notes

1. Campbell, C.J.; Laherrère, J.H. The end of cheap oil. *Sci. Am.* **1998**, *3*, 78–83. [CrossRef]
2. Sheehan, J.; Camobreco, V.; Duffield, J.; Graboski, M.; Shapouri, H. Life Cycle Inventory of Biodiesel and Petroleum Diesel for Use in an Urban Bus. Available online: http://www.nrel.gov/docs/legosti/fy98/24089.pdf (accessed on 17 January 2016).
3. Nerve, Z. Upgrading of Biodiesel-Derived Glycerol in the Biosynthesis of ε-Poly-L-Lysine: An Integrated Biorefinery Approach. Available online: http://wiredspace.wits.ac.za/handle/10539/9253?show=full (accessed on 17 January 2016).
4. Vasudevan, P.T.; Briggs, M. Biodiesel production—Current state of the art and challenges. *J. Ind. Microbiol. Biotechnol.* **2008**, *35*, 421–430. [CrossRef] [PubMed]
5. Adhikari, S.; Fernando, S.D.; To, S.D.F.; Bricka, R.M.; Steele, P.H.; Haryanto, A. Conversion of glycerol to hydrogen via a steam reforming process over nickel catalysts. *Energy Fuels* **2008**, *22*, 1220–1226. [CrossRef]
6. Wirawan, S.S.; Tambunan, A.H. The current status and prospects of biodiesel development in Indonesia: A review. In Proceedings of the Third Asia Biomass Workshop, Tsukuba, Japan, 16 November 2006.
7. Marchetti, J.M.; Miguel, V.U.; Errazu, A.F. Possible methods for biodiesel production. *Renew. Sustain. Energy Rev.* **2007**, *11*, 1300–1311. [CrossRef]
8. Johnson, D.T.; Taconi, K.A. The glycerin glut: Options for the value-added conversion of crude glycerol resulting from biodiesel production. *Environ. Prog.* **2007**, *26*, 338–348. [CrossRef]
9. Thompson, J.C.; He, B.B. Characterization of crude glycerol from biodiesel production from multiple feedstock. *Appl. Eng. Agric.* **2006**, *22*, 261–265. [CrossRef]
10. Dasari, M.A.; Kiatsimkul, P.P.; Sutterlin, W.R.; Suppes, G.J. Low-pressure hydrogenolysis of glycerol to propylene glycol. *Appl. Catal. A Gen.* **2005**, *281*, 225–231. [CrossRef]
11. Lemke, D. Volumes of Versatility. Auri Ag Innovation News.
12. Gallan, M.; Bonet, J.; Sire, R.; Reneaume, J.; Plesu, A.E. From residual to useful oil: Revalorization of glycerine from the biodiesel synthesis. *Bioresour. Technol.* **2009**, *100*, 3775–3778. [CrossRef] [PubMed]
13. McCoy, M. Glycerin surplus. *Chem. Eng. News* **2006**, *84*, 7–8. [CrossRef]
14. Yazdani, S.S.; Gonzalez, R. Anaerobic fermentation of glycerol: A path to economic viability for the biofuels industry. *Curr. Opin. Biotechnol.* **2007**, *18*, 213–219. [CrossRef] [PubMed]
15. Ghosh, P.K.; Mishra, S.C.P.; Gandhi, M.R.; Upadhyay, S.C.; Paul, P.; Anand, P.S.; Popat, K.M.; Shrivastav, A.V.; Mishra, S.K.; Ondhiya, N.; et al. Integrated Process for the Production of *Jatropha* Methyl Ester and by Products. EU Patent 2,475,754, 18 July 2012.
16. Yang, F.; Hanna, M.A.; Sun, R. Value-added uses for crude glycerol—A byproduct of biodiesel production. *Biotechnol. Biofuels* **2012**, *5*, 1–10. [CrossRef] [PubMed]
17. Ito, T.; Nakashimada, Y.; Senba, K.; Matsui, T.; Nishio, N. Hydrogen and ethanol production from glycerol-containing wastes discharged after biodiesel manufacturing process. *J. Biosci. Bioeng.* **2005**, *100*, 260–265. [CrossRef] [PubMed]
18. Shrivastav, A.; Mishra, S.K.; Shethia, B.; Pancha, I.; Jain, D.; Mishra, S. Isolation of promising bacterial strains from soil and marine environment for polyhydroxyalkanoates (PHAs) production utilizing *Jatropha* biodiesel byproduct. *Int. J. Biol. Macromol.* **2010**, *47*, 283–287. [CrossRef] [PubMed]
19. Rymowicz, W.; Rywińska, A.; Marcinkiewicz, M. High-yield production of erythritol from raw glycerol in fed-batch cultures of *Yarrowia lipolytica*. *Biotechnol. Lett.* **2009**, *31*, 377–380. [CrossRef] [PubMed]
20. Tang, S.; Boehme, L.; Lam, H.; Zhang, Z. *Pichia pastoris* fermentation for phytase production using crude glycerol from biodiesel production as the sole carbon source. *Biochem. Eng. J.* **2009**, *43*, 157–162. [CrossRef]
21. Volpato, G.; Rodrigues, R.C.; Heck, J.X.; Ayub, M.A.Z. Production of organic solvent tolerant lipase by *Staphylococcus caseolyticus* EX17 using raw glycerol as substrate. *J. Chem. Technol. Biotechnol.* **2008**, *83*, 821–828. [CrossRef]

22. Scholten, E.; Renz, T.; Thomas, J. Continuous Cultivation Approach for Fermentative Succinic Acid Production from Crude Glycerol by *Basfia succiniciproducen* DD1. *Biotechnol. Lett.* **2009**, *31*, 1947–1951. [CrossRef] [PubMed]

23. Ethier, S.; Woisard, K.; Vaughan, D.; Wen, Z.Y. Continuous culture of the microalgae *Schizochytrium limacinum* on biodiesel-derived crude glycerol for producing docosahexaenoic acid. *Bioresour. Technol.* **2011**, *102*, 88–93. [CrossRef] [PubMed]

24. Athalye, S.K.; Garcia, R.A.; Wen, Z.Y. Use of biodiesel-derived crude glycerol for producing eicosapentaenoic acid (EPA) by the fungus *Pythium Irregular*. *J. Agric. Food. Chem.* **2009**, *57*, 2739–2744. [CrossRef] [PubMed]

25. Choi, W.J.; Hartono, M.R.; Chan, W.H.; Yeo, S.S. Ethanol Production from Biodiesel-Derived Crude Glycerol by Newly Isolated *Kluyvera Cryocrescen*. *Appl. Microbiol. Biotechnol.* **2011**, *89*, 1255–1264. [CrossRef] [PubMed]

26. Oh, B.R.; Seo, J.W.; Heo, S.Y.; Hong, W.K.; Luo, L.H.; Joe, M.; Park, D.H.; Kim, C.H. Enhancement of ethanol production from glycerol in a *Klebsiella pneumoniae* mutant strain by the inactivation of lactate dehydrogenase. *Bioresour. Technol.* **2011**, *102*, 3918–3922. [CrossRef] [PubMed]

27. Taconi, K.A.; Venkataramanan, K.P.; Johnson, D.T. Growth and solvent production by *Clostridium pasteurianum* ATCC® 6013™ utilizing biodiesel-derived crude glycerol as the sole carbon source. *Environ. Prog. Sustain. Energy* **2009**, *28*, 100–110. [CrossRef]

28. Poblete-Castro, I.; Binger, D.; Oehlert, R.; Rohde, M. Comparison of mcl-Poly (3-hydroxyalkanoates) synthesis by different *Pseudomonas putida* strains from crude glycerol: Citrate accumulates at high titer under PHA-producing conditions. *BMC Biotechnol.* **2014**, *14*, 962. [CrossRef] [PubMed]

29. Moita, R.; Freches, A.; Lemos, P.C. Crude glycerol as feedstock for polyhydroxyalkanoates production by mixed microbial cultures. *Water Res.* **2014**, *58*, 9–20. [CrossRef] [PubMed]

30. Hermann-Krauss, C.; Koller, M.; Muhr, A.; Fasl, H.; Stelzer, F.; Braunegg, G. Archaeal production of polyhydroxyalkanoate (PHA) co-and terpolyesters from biodiesel industry-derived by-products. *Archaea* **2013**, *2013*, 129268. [CrossRef] [PubMed]

31. González-Pajuelo, M.; Andrade, J.C.; Vasconcelos, I. Production of 1,3-propanediol by *Clostridium butyricum* VPI 3266 using a synthetic medium and raw glycerol. *J. Ind. Microbiol. Biotechnol.* **2004**, *31*, 442–446. [CrossRef] [PubMed]

32. Bera, A.; Dubey, S.; Bhayani, K.; Mondal, D.; Mishra, S.; Ghosh, P.K. Microbial synthesis of polyhydroxyalkanoate using seaweed-derived crude levulinic acid as co-nutrient. *Int. J. Biol. Macromol.* **2015**, *72*, 487–494. [CrossRef] [PubMed]

33. Chheda, A.H.; Vernekar, M.R. Improved production of natural food preservative ε-poly-L-lysine using a novel producer *Bacillus cereus*. *Food Biosci.* **2014**, *30*, 56–63. [CrossRef]

34. Itzhaki, R.F. Colorimetric method for estimating polylysine and polyarginine. *Anal. Biochem.* **1972**, *50*, 569–574. [CrossRef]

35. Dhangdhariya, J.H.; Dubey, S.; Trivedi, H.B.; Pancha, I.; Bhatt, J.K.; Dave, B.P.; Mishra, S. Polyhydroxyalkanoate from marine *Bacillus megaterium* using CSMCRI's Dry Sea Mix as a novel growth medium. *Int. J. Biol. Macromol.* **2015**, *76*, 254–261. [CrossRef] [PubMed]

36. Chheda, A.H.; Vernekar, M.R. Enhancement of ε-poly-l-lysine (ε-PL) production by a novel producer *Bacillus cereus* using metabolic precursors and glucose feeding. *3 Biotech* **2015**, *5*, 839–846. [CrossRef]

37. Koller, M.; Chiellini, E.; Braunegg, G. Study on the Production and Re-use of Poly (3-hydroxybutyrate-co-3-hydroxyvalerate) and Extracellular Polysaccharide by the Archaeon *Haloferax mediterranei* Strain DSM 1411. *Chem. Biochem. Eng. Q.* **2015**, *29*, 87–98. [CrossRef]

38. Moralejo-Gárate, H.; Palmeiro-Sánchez, T.; Kleerebezem, R.; Mosquera-Corral, A.; Campos, J.L.; van Loosdrecht, M. Influence of the cycle length on the production of PHA and polyglucose from glycerol by bacterial enrichments in sequencing batch reactors. *Biotechnol. Bioeng.* **2013**, *110*, 3148–3155. [CrossRef] [PubMed]

39. Jia, S.; Fan, B.; Dai, Y.; Wang, G.; Peng, P.; Jia, Y. Fractionation and Characterization of ε-poly-L-lysine from *Streptomyces albulus* CGMCC 1986. *Food Sci. Biotechnol.* **2010**, *19*, 361–366. [CrossRef]

bioengineering

MDPI

Article

Utilization of Sugarcane Bagasse by *Halogeometricum borinquense* Strain E3 for Biosynthesis of Poly(3-hydroxybutyrate-*co*-3-hydroxyvalerate)

Bhakti B. Salgaonkar * and Judith M. Bragança

Department of Biological Sciences, Birla Institute of Technology and Science Pilani, K K Birla, Goa Campus, NH-17B, Zuarinagar, Goa 403 726, India; judith@goa.bits-pilani.ac.in
* Correspondence: salgaonkarbhakti@gmail.com; Tel.: +91-832-2580-305

Academic Editor: Martin Koller
Received: 29 March 2017; Accepted: 22 May 2017; Published: 25 May 2017

Abstract: Sugarcane bagasse (SCB), one of the major lignocellulosic agro-industrial waste products, was used as a substrate for biosynthesis of polyhydroxyalkanoates (PHA) by halophilic archaea. Among the various wild-type halophilic archaeal strains screened, *Halogeometricum borinquense* strain E3 showed better growth and PHA accumulation as compared to *Haloferaxvolcanii* strain BBK2, *Haloarcula japonica* strain BS2, and *Halococcus salifodinae* strain BK6. Growth kinetics and bioprocess parameters revealed the maximum PHA accumulated by strain E3 to be 50.4 ± 0.1 and 45.7 ± 0.19 (%) with specific productivity (qp) of 3.0 and 2.7 (mg/g/h) using NaCl synthetic medium supplemented with 25% and 50% SCB hydrolysate, respectively. PHAs synthesized by strain E3 were recovered in chloroform using a Soxhlet apparatus. Characterization of the polymer using crotonic acid assay, X-ray diffraction (XRD), differential scanning calorimeter (DSC), Fourier transform infrared (FT-IR), and proton nuclear magnetic resonance (^{1}H-NMR) spectroscopy analysis revealed the polymer obtained from SCB hydrolysate to be a co-polymer of poly(3-hydroxybutyrate-*co*-3-hydroxyvalerate) [P(3HB-*co*-3HV)] comprising of 13.29 mol % 3HV units.

Keywords: archaea; halophiles; sugarcane bagasse; polyhydroxyalkanoates; *Halogeometricum borinquense* strain E3; Soxhlet extractor

1. Introduction

Conventional plastics obtained from non-renewable petrochemical resources are creating environmental havoc due to their non-degradable nature. To solve this problem, various bio-based materials derived from renewable resources have been explored as a replacement for conventional plastics. These materials could be (i) directly extracted from biomass as polysaccharides, lignocelluloses, proteins, and lipids; (ii) chemically synthesized, e.g., by in vitro polymerization of bio-derived monomers such as lactate to produce poly(lactic acid) (PLA); or (iii) biologically synthesized by microorganisms, in vivo polymerization of hydroxylalkanoic acid (HA) units to polyhydroxyalkanoates (PHAs) [1]. PHAs are synthesized and accumulated as inclusions by microorganisms when the available nitrogen/phosphorus gets depleted while carbon is in excess. PHAs are synthesized either in the inner membrane, on a central scaffold, or in the cytoplasm of cells and aggregated in the form of globular, water-insoluble granules [2,3].

High production costs, downstream processing, and low yields are the major hurdles for the commercial production and application of PHA, making microbially synthesized PHA 5–10 times more expensive than petroleum-derived polymers [4]. Carbon sources/substrates represent half of the PHA fermentation cost [5–7]. Various strategies such as replacing commercial substrates with inexpensive renewable agro-industrial waste, finding novel high PHA-accumulating microorganisms or microbial strain improvements, and reducing the cost of PHA recovery/downstream process can make the overall fermentation process more cost-effective [4].

Sugarcane (*Saccharum officinarum*) is the world's largest cash crop, with Brazil being the leading producer, followed by India, and is exploited for the production of sugar, jaggery, ethanol, molasses, alcoholic beverages (rum), soda, etc. [8]. Sugarcane bagasse (SCB) is the leftover, fibrous residue of sugarcane stalk after the extraction of juice and is a major lignocellulosic, inexpensive byproduct of the sugarcane industry [9]. SCB needs special attention for its management, and is primarily used as a source of energy for electricity/biogas production. It also serves as a raw material for fermentation processes in the production of various products such as enzymes (cellulose, lipase, xylanase, inulinase, and amylase), animal feed (single cell protein), amino acids, organic acids, bioethanol, bioplastics, etc. [10].

SCB is comprised of cellulose (46%), hemicelluloses (27%), lignin (23%), and ash (4%) [9]. Cellulose is a homopolysaccharide comprised of a linear chain of $\beta(1{\rightarrow}4)$ linked D-glucose units, whereas hemicellulose is a heteropolysaccharide consisting of many different sugar monomers such as xylan (consisting units of pentose sugar, xylose), glucuronoxylan (consisting units of glucuronic acid and xylose), arabinoxylan (consisting of copolymers of two pentose sugars, arabinose and xylose), glucomannan (consisting of D-mannose and D-glucose), and xyloglucan (consisting of units of glucose and xylose). Degradation of these polymers would yield various sugars that could serve as a substrate for PHA production. The fibrous nature of SCB makes its microbial degradation slower and more difficult and hence limits its utilization. To overcome this, pre-treatment of SCB is usually carried out for improved substrate availability and to speed up the fermentation process. Acid hydrolysis is a fast and simple method, mostly performed to release sugars from the SCB that can be readily used by microorganisms, rather than feeding solid bagasse, which is time-consuming and interferes with the downstream processing [11].

Much insight has been gained into PHA synthesis by using non-halophilic microorganisms as compared to their halophilic counterparts. PHA production by halophiles has been studied among members of the families *Halobacteriaceae* (Domain *Archaea*) and *Halomonadaceae* (Domain *Bacteria*). Kirk and Ginzburg (in 1972) first documented the occurrence of PHA granules in halophilic archaea [12]. Subsequent research on members of the family *Halobacteriaceae* revealed significant PHA synthesis by the following genera: *Haloferax* (*Hfx. mediterranei*), *Haloarcula* (*Har. marismortui*), *Halobacterium* (*Hbt. noricense*), *Haloterrigena* (*Htg. hispanica*), *Halogeometricum* (*Hgm. borinquense*), *Halococcus* (*Hcc. dombrowskii*), *Natrinema* (*Natrinema pallidum*); *Halobiforma* (*Hbf. haloterrestris*), and *Halopiger* (*Hpg. aswanensis*) [13–20]. Among the halophilic bacteria, members of the genus *Halomonas* (*H. boliviensis*) are known to synthesize a large amount of PHA from a variety of substrates such as glucose, maltose, starch hydrolysate, etc. [21,22].

Overall, there are few reports on PHA production using agro-industrial waste/cheap substrates by employing halophilic microorganisms. *Haloferax mediterranei* is the most widely studied and is reported to produce a copolymer of poly(3-hydroxybutyrate-*co*-3-hydroxyvalerate) [P(3HB-*co*-3HV)] using various renewable agro-industrial waste products like extruded corn starch, rice bran, wheat bran, hydrolyzed whey, waste stillage from the rice-based ethanol industry, vinasse, etc. [23–26]. Pramanik et al. and Taran reported the synthesis of a homopolymer of hydroxybutyrate (PHB) by *Haloarcula marismortui* MTCC 1596 and *Haloarcula* sp. IRU1 by utilization of vinasse waste from the ethanol industry and petrochemical wastewater, respectively [27,28].

To date, to the best of our knowledge, there has been no report on the utilization of sugarcane bagasse for PHA production by members of the domain Archaea, especially the genus *Halogeometricum*. Therefore, in the present study, the ability of halophilic archaeal isolates to accumulate PHA from SCB hydrolysate was examined. The growth kinetics and bioprocess parameters during growth and PHA production by *Hgm. borinquense* strain E3 were examined and the polymer synthesized was characterized using physico-chemical analysis techniques.

2. Materials and Methods

2.1. HalophilicArchaeal Strains and Media Used

Four halophilic archaeal isolates (GenBank/DDBJ database accession number), *Halococcus salifodinae* strain BK6 (AB588757), *Haloferax volcanii* strain BBK2 (AB588756), and *Haloarcula japonica* strain BS2 (HQ455798), isolated from solar salterns of Ribandar in Goa, India, and *Hgm. borinquense* strain E3 (AB904833), obtained from Marakkanam in Tamil Nadu, India, were used in the present study [29,30]. The cultures were maintained on complex media (Table 1), i.e., NTYE (<u>N</u>aCl <u>T</u>ryptone <u>Y</u>east <u>E</u>xtract), NT (<u>N</u>aCl <u>T</u>ri-sodium citrate), EHM (<u>E</u>xtremely <u>H</u>alophilic <u>M</u>edium). The NSM (<u>N</u>aCl <u>S</u>ynthetic <u>M</u>edium) with a varying concentration of SCB hydrolysate as per the requirements was used as a production medium (Table 1).

Table 1. Composition of maintenance and production media used in the study.

Ingredients (g/L)	Maintenance Media			Production Medium
	NTYE	NT	EHM	NSM
NaCl	250.0	250.0	250.0	200.0
$MgSO_4 \cdot 7H_2O$	20.0	20.0	20.0	-
$MgCl_2 \cdot 6H_2O$	-	-	-	13.0
KCl	5.0	2.0	2.0	4.0
Tryptone	5.0	-	-	-
Yeast Extract	3.0	10.0	10.0	1.0
Tri-Sodium citrate	-	3.0	-	-
$CaCl_2 \cdot 2H_2O$	-	-	0.36	1.0
NaBr			0.23	-
$NaHCO_3$	-	-	0.06	0.2
NH_4Cl	-	-	-	0.2
KH_2PO_4	-	-	-	0.5
Peptone	-	-	5.0	-
$FeCl_3 \cdot 6H_2O$	-	-	Trace	0.005

NTYE: <u>N</u>aCl <u>T</u>ryptone <u>Y</u>east <u>E</u>xtract; NT: <u>N</u>aCl <u>T</u>ri-sodium citrate; EHM: <u>E</u>xtremely <u>H</u>alophilic <u>M</u>edium; NSM: <u>N</u>aCl <u>S</u>ynthetic <u>M</u>edium. NSM with various concentration of SCB was used as production media. Agar 1.8% (*w/v*) was use as solidifying agent. The pH of the medium was adjusted to 7.0–7.2 using 1M NaOH.

2.2. Procurement, Processing, and Hydrolysis of Sugarcane Bagasse

Sugarcane bagasse (SCB) was collected from a local sugarcane juice extractor, from Vasco-da-Gama, Goa, India. It was dried under sunlight for 3–5 days, cut into small ~5–10 cm pieces, followed by pulverization to a fine powder using a blender. The powdered form of the waste was subjected to dilute acid hydrolysis. Briefly, 5 gm of the SCB powder was added to 100 mL of 0.75% (*v/v*) sulfuric acid in water. The mixture was heated at 100 °C for 1 h under reflux using an allihn condenser. The hydrolysate was filtered using non-absorbent cotton to separate the solid residue from the liquid hydrolysate. The liquid hydrolysate was neutralized to pH 7.0–7.4 using NaOH, followed by sterilization at 121 °C for 10 min and storage at 4 °C [11].

2.3. Characterization of the SCB

The SCB powder was characterized for the following physical and chemical parameters. Total solids (TS) and volatile solids (VS) were estimated according to the American Public Health Association (APHA) [31]. The chemical oxygen demand (COD) was determined as described by Raposo et al. [32]. The carbon (C), hydrogen (H), nitrogen (N), and sulfur (S) contents of the SCB were determined using a CHNS Analyzer (Elementar, Rhine Main area near Frankfurt, Germany). The total carbohydrate content of the SCB hydrolysate was estimated by the phenol sulphuric acid method, as described by Dubois et al. [33]. Total Kjeldahl nitrogen (TKN) was determined as described by

Labconco [34]. All the physical and chemical characterization was performed in triplicate to determine means and standard deviations.

2.4. Screening of the Halophilic Archaeal Isolates for PHA Accumulation using SCB Hydrolysate

Halophilic archaeal isolates were screened for the production of PHA using NSM (Table 1) and Nile Red stain. Briefly, the NSM plates were prepared by adding 1.8% agar (w/v) to the medium followed by autoclaving; while still molten, the medium was supplemented with varying concentrations (0.5%–30% v/v) of SCB hydrolysate along with 50 μL of Nile Red stain [stock of 0.01% (w/v) Nile red in DMSO] such that the final concentration was0.5 μg/mL medium. Twenty microliters of log phase (three-day-old) halophilic archaeal cultures were spot inoculated on the agar plates and incubated at 37 °C for 6–7 days. The plates were exposed to ultraviolet (UV) light using gel documentation system (BIO-RAD Laboratories, Hercules, CA, USA) andthe emitted fluorescence from the culture was quantified using TotalLab Quant software [16,35].

2.5. Selection and Further Study of the Best PHA Producer Strain

Based on it having the best growth and fluorescence on NSM supplemented with SCB hydrolysate, the halophilic archaeon *Hgm. borinquense* strain E3 was selected for further study. Preliminary screening indicted the strain E3 to grow up to 30% (v/v) of SCB hydrolysate. Therefore, the concentration of SCB hydrolysate that inhibited the growth of strain E3 was determined by growing the culture on NSM agar plates containing higher concentrations of the SCB hydrolysate, i.e., 50%, 75%, and 100%. NSM with 100% SCB hydrolysate was prepared by dissolving the medium ingredients in the directly SCB hydrolysate. Based on the growth observed on NSM agar plates, growth of the strain E3 was further recorded in NSM broth with that particular concentration of SCB hydrolysate.

2.6. The Growth Kinetics and PHA Quantification

Growth and intracellular PHA content for *Hgm. borinquense* strain E3 were determined as follows. An actively growing mid-log phase culture (3–4 days) of *Hgm. borinquense* strain E3 grown in NGSM (NaCl Glucose Synthetic Medium) containing 0.2% glucose was used as a starter culture. One percent of the starter culture was inoculated in NSM (NaCl Synthetic Medium) containing 25% and 50% (v/v) SCB hydrolysate. The flasks were maintained at 37 °C, 110 rpm on a rotary shaker (Skylab Instruments, Mumbai, India). At regular intervals of 24 h, aliquots of the culture broth wereaseptically withdrawn and the following parameters were monitored: (i) culture growth was monitored by recording the absorbance at 600 nm using a UV-visible spectrophotometer (UV-2450, Shimadzu, Tokyo, Japan) with the respective medium as blank, (ii) Cell Dry Mass (CDM) was determined by centrifuging 2 mL of the culture broth at 12,000 g for 15 min, washing the pellet with distilled water, and recentrifuging it at 12,000 g for 15 min, followed by drying at 60 °C until a constant weight was obtained. Since the SCB hydrolysate had some particle participation, the dry mass of the plain medium (without culture) was taken and subtracted from the culture CDM so as to avoid error. (iii) Total carbohydrates were determined colorimetrically according to Dubios et al. and compared with the standard curve [33]; (iv) the pH of the medium was monitored using a pH meter; and (v) polymer quantification was done by converting PHA to crotonic acid using concentrated sulfuric acid. The absorbance was recorded at 235 nm using a UV-visible spectrophotometer (UV-2450, Shimadzu, Tokyo, Japan) and compared with the standard curve for PHB [36]. All the experiments were performed in triplicate to determine means and standard deviations.

2.7. Extraction of the PHA

The PHA extraction from the biomass was done as described by Sánchez et al. with slight modifications [37]. Briefly, *Hgm. borinquense* strain E3 was grown in NSM containing 25% SCB hydrolysate for six days. The cells were harvested by centrifuging at 12,000 g for 10 min using Eppendorf centrifuge 5810R (Hamburg, Germany). The cell pellet was dried for 12 h at 60 °C in an oven (Bio Technics, Mumbai, India). The dried cells were ground using mortar and pestle and extracted for 8–10 h at 60 °C in a soxhlet extractor using chloroform. Up to 95% of the chloroform was recollected using rotary evaporator (Rotavapor R-210, Büchi, Switzerland) and the remaining 5% of the chloroform-containing polymer was poured into a clean glass Petri dish and kept undisturbed during the total evaporation of the chloroform to give a uniform polymer film.

2.8. Characterization of the PHA

Characterization of the polymer was done using UV-visible spectrophotometry (crotonic acid assay), X-ray diffraction (XRD) analysis, differential scanning calorimeter (DSC) analysis, Fourier transform infrared (FT-IR) spectroscopy, and proton nuclear magnetic resonance (^{1}H-NMR) spectroscopy, as described in detail by Salgaonkar and Bragança [30].

3. Results and Discussion

3.1. Sugarcane Bagasse (SCB)

The SCB used in the present study appeared greenish-brown with a sweet odor and upon pulverization had total solids (TS) of 94.3 ± 0.14%, volatile solids (VS) of 92.7 ± 0.14%, and COD of 1.18 ± 0.05 g/Kg. The carbon (C), hydrogen (H), nitrogen (N), and sulfur (S) content of the SCB was found to be C = 43.45 ± 0.12%, H = 6.0 ± 0.27%, N = 0.26 ± 0.02%, and S = 0.32 ± 0.12%. Hydrolysis of SCB using dilute sulfuric acid yielded total carbohydrates of 12.64 ± 0.7 g/L and total Kjeldahl nitrogen (TKN) of 0.7 g/L (Table S1).

3.2. Screening for PHA using SCB Hydrolysate

All four halophilic archaeal isolates were able to grow on NSM plates with Nile Red dye supplemented with SCB hydrolysate as substrate. Upon exposure of the plates to UV light, only three cultures, *Hfx. volcanii* strain BBK2, *Har. japonica* strain BS2, and *Hgm. borinquense* strain E3, showed bright orange fluorescence, indicating the accumulation of PHA (Figure S1). *Hcc. salifodinae* strain BK6 showed weak growth but failed to show any fluorescence. The intensity of fluorescence exhibited by the cultures varied, in the order *Hgm. borinquense* strain E3 > *Har. japonica* strain BS2 > *Hfx. volcanii* strain BBK2. *Hgm. borinquense* strain E3 grew faster and showed better fluorescence, which directly correlates withthe amount of polymer accumulated over a range of SCB concentrations (Figure S1). Preliminary work on *Hgm. borinquense* strain E3 proved it to be the best accumulator of PHA in an NGM medium supplemented with 2% glucose [30].

3.3. Optimization of SCB Hydrolysate Concentration

Figure 1 represents the growth of *Hgm. borinquense* strain E3 on NSM agar plates and broth containing various concentrations of SCB hydrolysate. The SCB hydrolysate optimization studies revealed that the culture could tolerate and grew up to 75% (v/v) SCB hydrolysate on NSM agar plates (Figure 1A). Interestingly, when grown in NSM broth (Figure 1B), the culture grew only up to 50% (v/v) SCB hydrolysate and failed to grow at higher concentrations.

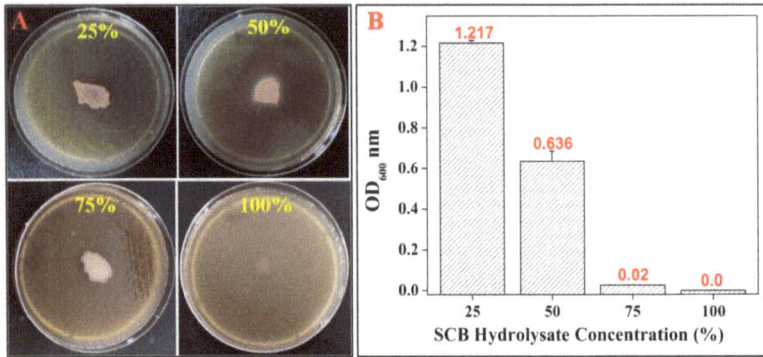

Figure 1. Growth of *Hgm. borinquense* strain E3 on (**A**) NSM agar plates and (**B**) NSM broth, containing various concentrations of SCB hydrolysate.

3.4. Growth Profile of Hgm. borinquense Strain E3 and Polymer Quantification Study

The time course of growth of *Hgm. borinquense* strain E3 in NSM containing 25% and 50% SCB hydrolysate is presented in Figure 2. Table 2 gives a comparison of the various kinetics and bioprocess parameters used to determine growth and PHA production by *Hgm. borinquense* strain E3 using SCB hydrolysate. In the presence of 25% SCB hydrolysate, isolate E3 showed a 48-h lag and reached 3.17 ± 0.19 g/L of maximum cell dry mass (CDM), containing 1.6 ± 0.09 g/L of PHA. In the presence of 50% SCB hydrolysate, isolate E3 exhibited a longer lag phase of 96 h and reached 4.15 ± 0.7 g/L of maximum CDM, containing 1.9 ± 0.3 g/L of PHA. The lag phase of the culture depends on the definite environmental conditions. This prolonged lag phase could be reduced by increasing the inoculum size and decreasing the effect of culture conditions on the growth, which can be achieved by acclimatizing the starter culture to the ingredients of production medium by pre-culturing the isolate in the presence of the respective substrates concentration. The substantial quantity of carbohydrates, i.e., 12.64 ± 0.7 g/L in the SCB hydrolysate, served as the basic essential carbon source required for the growth and synthesis of PHA by *Hgm. borinquense* strain E3. The rapid consumption of the total carbohydrates by the isolate was observed as the growth progressed and a steady drop in the pH of the production medium from 7.2 to 5.0 was also noted.

To the best of our knowledge, haloarchaea have not been explored for their potential to utilize SCB hydrolysate for the synthesis of PHA. Bioprocess parameters (Table 2) revealed the maximum PHA accumulation by *Hgm. borinquense* strain E3 to be 50.4 ± 0.1 and 45.7 ± 0.19 (%) with specific productivity (qp) of 3.0 and 2.7 (mg/g/h) using NaCl synthetic medium supplemented with 25% and 50% SCB hydrolysate, respectively. A recent investigation by Pramanik et al. reported the potential of haloarchaeon *Har. marismortui* MTCC 1596 to produce 23 ± 1.0 and 30 ± 0.3 (%) P(3HB) with specific productivity (qp) of 1.21 and 1.39 (mg/g/h) using a nutrient-deficient medium (NDM) supplemented with 10% and 100% raw and treated vinasse, respectively [27]. The specific productivity (qp) attained by *Hgm. borinquense* strain E3 using SCB hydrolysate was higher compared to *Har. marismortui* MTCC 1596 grown in the presence of vinasse. Silva et al. (2004) investigated the synthesis of P(3HB) by *Burkholderia sacchari* IPT 101 and *Burkholderia cepacia* IPT 048 by feeding the cultures with SCB hydrolysate as a carbon source. It was noted that both the cultures reached 4.4 g/L of dry biomass, containing 62% and 53% of P(3HB) in the case of *B. sacchari* IPT 101 and *B. cepacia* IPT 048, respectively. The specific production rate and yield coefficient of the PHA were 0.11 (g/L/h) and 0.39 (g/g) for *B. sacchari* IPT 101, whereas it was 0.09 (g/L/h) and 0.29 (g/g) for *B. cepacia* IPT 048 [7].

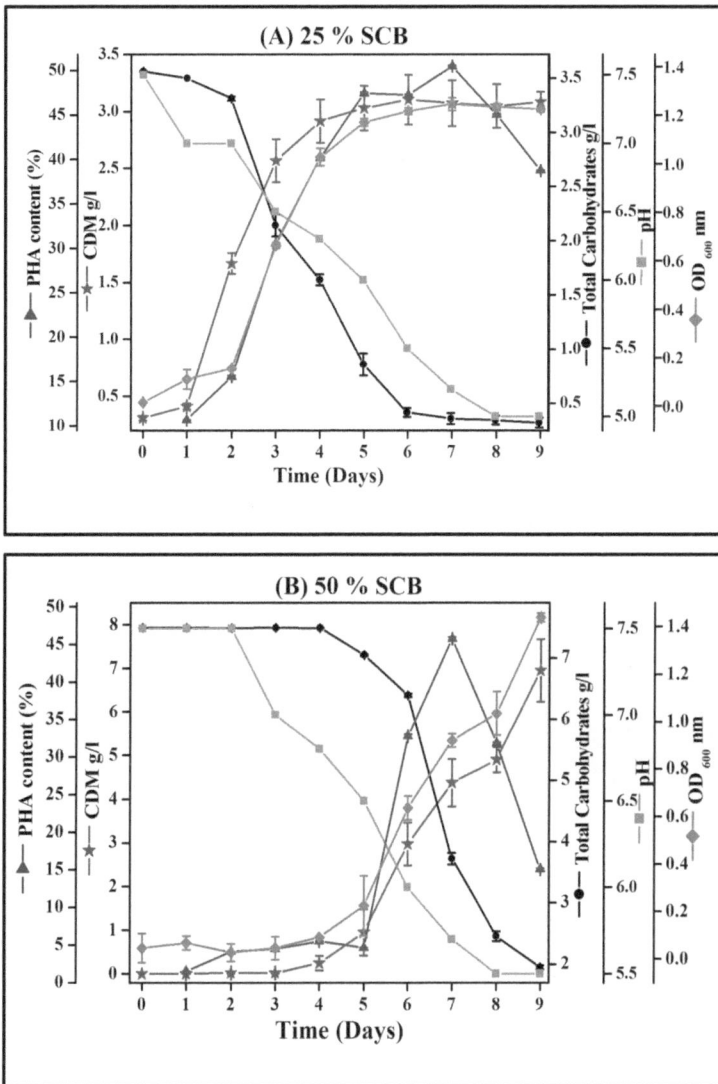

Figure 2. Growth profile and PHA production by *Hgm. borinquense* strain E3 in NSM containing (**A**) 25% and (**B**) 50% SCB hydrolysate.

Attempts have been made to reduce the fermentation cost of PHA by employing various haloarchaeal strains and examining their ability to utilize inexpensive substrates. Danis et al. showed the ability of *Natrinema pallidum*1KYS1 to produce 0.075, 0.055, 0.091, 0.039, 0.077, and 0.464 g/L of polymer by utilizing various waste products such as corn starch, sucrose, whey, melon, apple, and tomato as carbon substrates [18]. Pramanik et al. studied the ability of *Har. marismortui* to utilize 10% raw vinasse and 100% pre-treated vinasse to produce 2.8 g/L and 4.5 g/L of PHB [27]. Similarly, Bhattacharyya et al. employed *Hfx. mediterranei* to produce 19.7 g/L and 17.4 g/L from 25% and 50% pre-treated vinasse, respectively [26]. Also, 24.2 g/L PHBV biosynthesis was observed in *Hfx. mediterranei* with extruded cornstarch [23].

Table 2. Kinetic and bioprocess parameters during growth and PHA production by *Hgm. borinquense* strain E3 (present study) in comparison with *Har. marismortui* MTCC 1596 (literature).

Halophilic Archaeal Strain	Production Medium	Lag (h)	CDM (g/L)	PHA (g/L)	PHA Content (%)	μmax (1/h)	qp [a] (mg/g/h)	Y$_{P/S}$ [b]	Vol. Productivity [c] (g/L/h)	Reference
Hgm. borinquense strain E3	NSM25% SCB	48	3.17 ± 0.19	1.6 ± 0.09	50.4 ± 0.1	0.017	3.0	0.448	0.0095	Present study
	NSM50% SCB	96	4.15 ± 0.7	1.9 ± 0.3	45.7 ± 0.19	0.023	2.7	0.253	0.0113	
	NGSM2% Glucose	-	5.78 ± 0.4	4.25 ± 0.045	73.5 ± 0.045	ND	4.3	0.212	0.0252	[30] *
Har. marismortui MTCC 1596	NDM10% Raw vinasse	96 ± 12	12.0 ± 0.20	2.8 ± 0.2	23 ± 1.0	0.086	1.21	2.17	0.015	[27]
	NDM100% treated vinasse	144 ± 12	15.0 ± 0.35	4.5 ± 0.2	30 ± 0.3	0.128	1.39	0.77	0.020	

NSM: NaCl Synthetic Medium; NGSM: NaCl Glucose Synthetic Medium; NDM: Nutrient Deficient Medium; ND: not determined; μmax: maximum specific growth rate; CDM: Cell Dry Mass; PHA: Polyhydroxyalkanoate; *Hgm*: *Halogeometricum*; *Har*: *Haloarcula*; MTCC: Microbial Type Culture Collection; [a] Specific production rate of PHA (qp) = PHA (g/L)/time (h) × CDM (g/L) [38]; [b] Yield coefficient of PHA (Y$_{P/S}$) = PHA (g/L)/ total organic carbon (g/L). The Y$_{P/S}$ was calculated based on total carbohydrate in NSM with 25% (3.5642 g/L) and 50% SCB hydrolysate (7.494 g/L) [26]; [c] Volume productivity of PHA = PHA (g/L)/ time (h) [39]. Time of growth for *Hgm. borinquense* strain E3 using NSM supplemented with glucose and SCB hydrolysate was 168 h (7 days), whereas that for *Har. marismortui* MTCC 1596 using NDM and 10% raw and 100% treated vinasse was 192 h (8 days) and 216 h (9 days), respectively [27]; * Some of the bioprocess parameters are not mentioned in the reference.

3.5. Bench Scale Polymer Production and Extraction by Hgm. borinquense Strain E3

The polymer was extracted from the cell biomass as shown in Figure S2. The dried cells of *Hgm. borinquense* strain E3 (before Soxhlet extraction), when subjected to concentrated H_2SO_4 hydrolysis, showed a clear peak at 235 nm, which is indicative of crotonic acid, indicating the presence of PHA (Figure S3A) [36]. After Soxhlet extraction, no peak at 235 nm was observed, thus confirming the complete extraction of the polymer from the cell mass (Figure S3B).

3.6. Polymer Characterization

The polymer obtained using SCB hydrolysate (Figure S2J) appeared orange due to the co-extraction of carotenoid pigment along with some cellular lipids from the cells of *Hgm. borinquense* strain E3. These impurities were taken care of by treating the polymer with acetone for 10 min [30]. This polymer was characterized using a UV-visible spectrophotometer, XRD, DSC, FT-IR, and NMR analysis.

3.6.1. UV-Visible Spectrophotometric Analysis

Concentrated H_2SO_4 hydrolysis of the polymer obtained from SCB gave a characteristic peak at 235 nm of crotonic acid, which corresponded with the standard PHB (Sigma-Aldrich, St. Louis, MO, USA) (inset of Figure S3) and also with the copolymer P(3HB-*co*-3HV) synthesized by *Hgm. borinquense* strain E3 and *Hfx. mediterranei* "DSM 1411"when fed with substrates such as glucose and raw vinasse, respectively [26,30].

3.6.2. XRD Analysis

Figure 3 represents the X-ray diffraction (XRD) patterns of the polymer obtained from SCB hydrolysate in comparison with standard PHB (Sigma-Aldrich, St. Louis, MO, USA). The profile of polymer from SCB exhibited prominent peaks at $2\theta = 13.8°$, $17.4°$, $21.8°$, $25.8°$, and $30.7°$, corresponding to (020), (110), (101), (121), and (002) reflections of the orthorhombic crystalline lattice. Overall, the diffraction pattern was similar to that of the PHB. However, peak shifts as well as a decrease in peak intensity were observed in the case of the SCB polymer when compared with standard PHB. It was clearly observed that the diffraction peaks between $2\theta = 0–25°$ were broadened and drastically decreased in intensity for the SCB polymer. A broadening of the peaks indicates a decrease in crystallinity, i.e., the amorphous nature of the polymer [40]. The crystallite size L (nm) was determined for the highest peaks using the Scherrer equation, which is defined as: $L(nm) = 0.94 \lambda / B Cos\theta$, where λ is the wavelength of the X-ray radiation, which is 1.542 Å(wavelength of the Cu); B is the full width at half maximum (FWHM) in radians; and θ is the Bragg angle [41,42]. The crystallite size for the highest peak (020) in the case of standard PHB was found to be 22.3 nm, which decreased drastically to 10.4 nm for peak (110) of the SCB polymer. For the diffraction peak of the (110) plane of SCB polymer, an increase in the FWHM was observed as compared to the standard PHB. This clearly indicates a decrease in the crystallite size, given that the peak width is inversely proportional to the crystallite size. The crystallite size matched more or less to the one reported for the copolymer P(3HB-*co*-3HV) synthesized by *Hgm. borinquense* strain E3 by utilizing glucose, which was found to be 12.17 nm for the (110) peak [30].

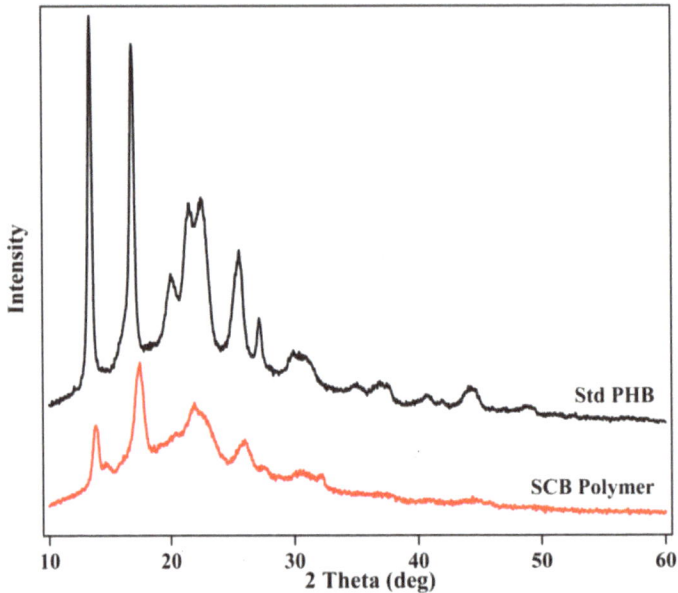

Figure 3. Comparison of X-ray diffraction patterns of standard PHB and polymer obtained from *Hgm. borinquense* strain E3 grown in NSM containing SCB hydrolysate.

3.6.3. DSC Analysis

The thermograms derived from differential scanning calorimetry (DSC) analysis for the polymer obtained using SCB hydrolysate and standard PHB (Sigma Aldrich, St. Louis, MO, USA) are represented in Figure 4. The polymer obtained from SCB hydrolysate exhibited two melting endotherms at Tm_1 = 136.5 °C and Tm_2 = 149.4 °C, whereas standard PHB displayed a single melting endotherm at Tm = 169.2 °C. The degradation temperature (Td) peaks for the SCB polymer and PHB were at 275.4 °C and 273.2 °C, respectively (Table 3). A recent study by Buzarovska et al. (2009) reported two melting endotherms for pure copolymer PHBV containing 13 mol% 3-hydroxyvalerate (3HV) [43]. The lower melting peak (Tm_1) at ~138 °C could be due to the melting of the primary formed crystallites, whereas the upper one (Tm_2) at 152 °C is mostly due to the recrystallization of species during the scan [43]. The existence of multiple melting peaks in a polymer indicates that the polymers have varying monomer units such as 3HB and 3HV units [44]. *Haloferax mediterranei* is known to produce PHA with multiple melting endotherms by utilizing various carbon substrates [45]. Chen et al. (2006) and Koller et al. (2007) showed the ability of *Haloferax mediterranei* ATCC 33500/DSM 1411 to utilize extruded cornstarch/whey sugars as carbon substrates for the production of copolymer P(3HB-*co*-3HV) containing 10.4 mol% and 6 mol% of 3-hydroxyvalerate (3HV), respectively. The P(3HB-*co*-3HV) produced by strain DSM 1411 showed two melting peaks at 150.8 °C (Tm_1) and 158.9 °C (Tm_2), whereas the melting endotherms for strain ATCC 33500 were at 129.1 °C (Tm_1) and 144.0 °C (Tm_2) [13,46].

Table 3. Comparison of the DSC data of a polymer synthesized by *Hgm. borinquense* strain E3 using SCB waste with data from the literature.

PHA from Various Substrates	Haloarchaeal Isolate	DSC Characterization (°C)			Reference
		T*m*1	T*m*2	T*d*	
SCB	*Hgm. borinquense* strain E3	136.59	149.4	275.4	Present study
Glucose		138.15	154.74	231.08	[30]
Cornstarch	*Hfx. mediterraneic* ATCC 33500	129.1	144.0	NR	[13]
Whey	*Hfx. mediterranei* DSM 1411	150.8	158.9	241	[46]

T*m*—melting temperature; T*d*—degradation temperature NR-not reported

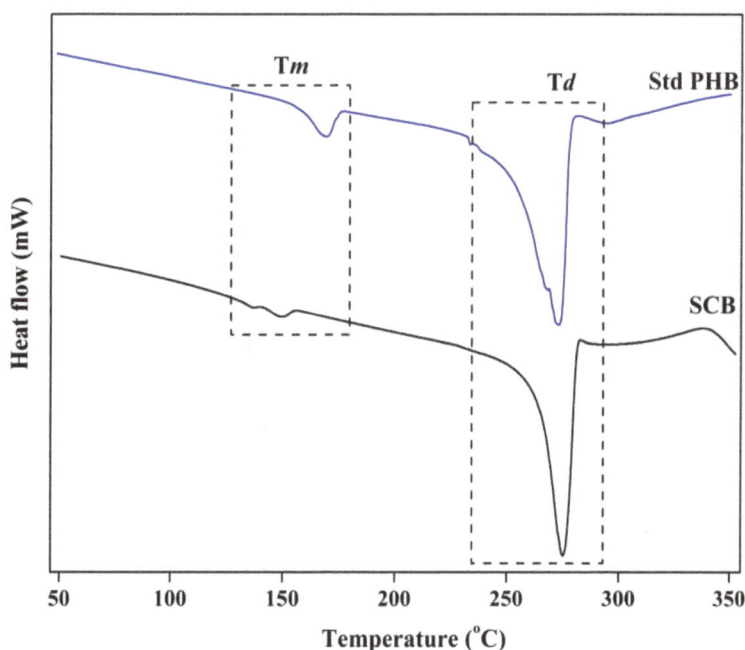

Figure 4. Comparison of DSC curves of standard PHB and polymer obtained from *Hgm. borinquense* strain E3 grown in NSM containing SCB hydrolysate.

3.6.4. FT-IR Analysis

The FT-IR spectra of polymer obtained using SCB hydrolysate were compared with those of the standard PHB (Sigma Aldrich, St. Louis, MO, USA) (Figure 5). The IR spectra of polymer obtained from SCB and standard PHB exhibited one intense absorption band at 1724 cm^{-1} and 1731 cm^{-1}, respectively, characteristic of ester carbonyl group (C=O) stretching. A band at 1281 cm^{-1} represents C–O–C stretching, whereas one in the region 3100–2800 cm^{-1}, i.e., 2983 cm^{-1} and 2981 cm^{-1}, represents C–H stretching (Figure 5). Apart from these, other prominent bands were also observed, which may be due to interactions between the OH and C=O groups resulting in a shift of the stretching [40]. The peaks obtained for a polymer obtained from SCB hydrolysate matched well with those of the standard polymer.

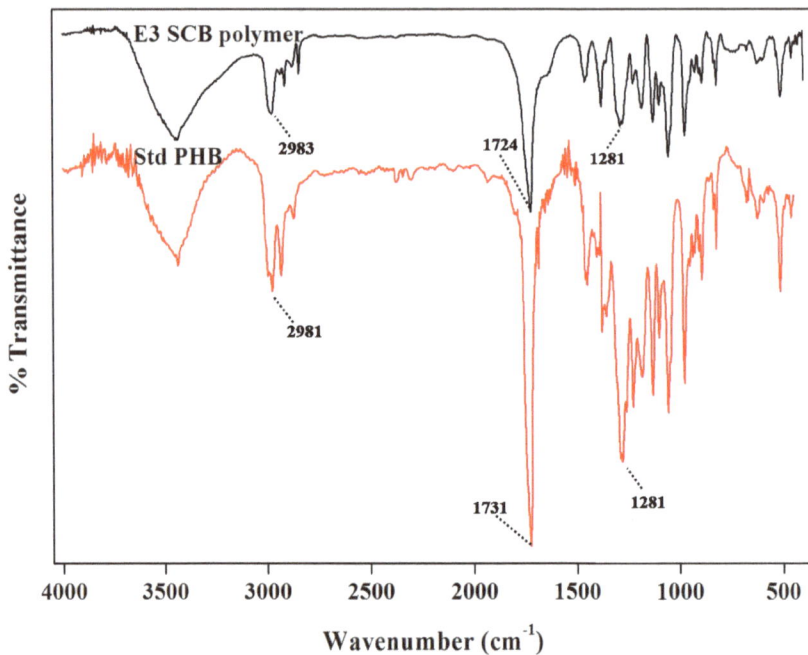

Figure 5. FT-IR spectra of standard PHB and polymer obtained from *Hgm. borinquense* strain E3 grown in NSM containing SCB hydrolysate.

3.6.5. ^{1}H-NMR Analysis

^{1}H-NMR scans of the polymer obtained from *Hgm. borinquense* strain E3 using SCB hydrolysate are represented in Figure 6. The chemical shift of the peaks and their chemical structure are represented in Table 4. Characteristic peaks at 0.889 ppm and 1.26 ppm are of methyl (CH$_3$) from 3-hydroxyvalerate (3HV) and 3-hydroxybutyrate (3HB) unit, respectively. Therefore, it can be confirmed that the polymer obtained using SCB hydrolysate is a co-polymer of Poly(3-hydroxybutyrate-*co*-3-hydroxyvalerate) [P(3HB-*co*-3HV)]. The signals obtained from ^{1}H-NMR correlated with those reported by Bhattacharyya et al. (2012) and Chen et al. (2006) for a co-polymer P(3HB-*co*-3HV)] obtained from *Hfx. mediterranei* strain DSM 1411 and strain ATCC 33500 by utilization of molasses spent wash (vinasse) and cornstarch, respectively (Figure 6, Table 4) [13,26]. Moreover, the ^{1}H NMR spectrum of homopolymer of 3HB (P3HB) showed only one prominent peak at 1.25 ppm of methyl (CH$_3$) from HB unit [35]. The co-polymer of P(3HB-*co*-3HV) comprised of 13.29% 3HV units, which was calculated as described by Salgaonkar and Bragança [30]. Interestingly, copolymer P(3HB-*co*-3HV)] containing a higher amount of 3HV (21.47% 3HV) was synthesized by the same E3 strain in NSM media with glucose as the substrate. The drastic reduction in 3HV units from 21.47% (glucose) to 13.29% (SCB) could be due to the inhibition of propionyl-coenzyme A synthesis, an important precursor of 3HV monomer by various byproducts of the SCB hydrolysate.

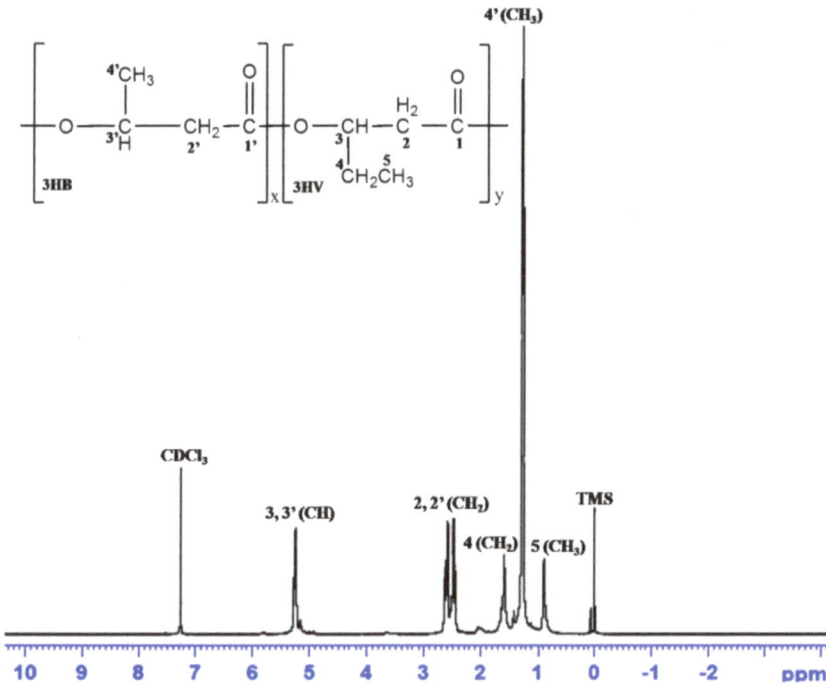

Figure 6. ^{1}H-NMR spectra of polymer from *Hgm. borinquense* strain E3 grown in NSM containing SCB hydrolysate.

Table 4. Comparison of the chemical shift of the peaks and their chemical structure, obtained from ^{1}H-NMR data of polymer synthesized by *Hgm. borinquense* strain E3 using SCB waste with data from the literature.

PHA from Various Substrates	Haloarchaeal Isolate	Relative Chemical Structure					Reference
		CH$_3$ (3HB)	CH$_2$ (3HV/3HB)	CH (3HV/3HB)	CH$_3$ (3HV)	CH$_2$ (3HV)	
		Chemical Shifts of Each Peak (ppm)					
SCB	*Hgm. borinquense* strain E3	1.26–1.27	2.44–2.63	5.22–5.27	0.889	1.618–1.635	Present study
Glucose		1.26–1.28	2.44–2.63	5.26	0.85–0.91	1.6	[30]
Cornstarch	*Hfx. mediterranei* ATCC 33500	1.2	2.5	5.2	0.9	1.6	[13]
Vinasse	*Hfx. mediterranei* DSM 1411	1.26–1.28	2.43-2.645	5.22-5.28	0.86–0.95	1.586	[26]

HB: hydroxybutyrate; HV: hydroxyvalerate; CH$_3$: methyl; CH$_2$: methylene; CH: methane.

PHA accumulation by extremely halophilic archaea and moderately halophilic and/or halotolerant bacteria, inhabiting hypersaline and marine regions of countries such as China, Turkey, Bolivia, Vietnam, India, etc., has been documented [6,16,18,21,22]. Moderately halophilic bacteria belonging to the genus *Halomonas* such as *H. boliviensis* LC1, *H. nitroreducens*, and *H. salina* have been reported to accumulate 56.0%, 33.0%, and 55.0% (w/w) CDM of homopolymer of 3-hydroxybutyrate (3HB), i.e., P(3HB) by utilizing versatile substrates such as starch hydrolysate, glucose, and glycerol, respectively [22,47]. Similarly, Van-Thuoc et al. (2012) reported the ability of halophilic and halotolerant bacteria *Bacillus* sp. ND153 and *Yangia pacifica* QN271 to accumulate P(3HB) (65.0 and 48.0% w/w

CDM)/PHBV (71.0 and 31.0% w/w of CDM) when glucose with or without propionate was provided as the carbon source [48]. However, there are very few reports on halophilic bacteria such as *H. campisalis* MCM B-1027 and *Yangia pacifica* ND199/ND218 synthesizing copolymer PHBV, irrespective of precursors like propionic/valeric acid in the culture medium [49,50]. Shrivastav et al.(2010) reported the utilization of *Jatropha* biodiesel byproduct as a substrate by *Bacillus sonorensis* strain SM-P-1S and *Halomonas hydrothermalis* strain SM-P-3M for the production of 71.8 and 75.0% (w/w) CDM of P(3HB), respectively [50].

Various members of halophilic archaea, belonging to the family *Halobacteriaceae*, such as *Halopiger aswanensis* strain 56 and *Hgm. borinquense* strain TN9 have been reported to accumulate 34.0% and 14.0% (w/w) CDM of homopolymer of P(3HB) by utilizing versatile substrates such as glucose, yeast extract, butyric acid, and sodium acetate [16,20]. Interestingly, *Hfx.* mediterranei is known to accumulate 23.0% (w/w) CDM of copolymer PHBV from glucose naturally, without any addition of precursor [23]. There are limited reports on the utilization of SCB hydrolysates as substrates for PHA production by microorganisms. Silva et al. reported the ability of two Gram-negative soil bacteria, *Burkholderia sacchari* IPT 101 and *Burkholderia cepacia* IPT 048, to accumulate poly-3-hydroxybutyrate (P3HB) when cultivated in SCB hydrolysate by submerged fermentation (SMF) [7].

Yu and Stahl reported the ability of the Gram-negative bacterium *Ralstonia eutropha* to synthesize both P(3HB) and P(3HB-*co*-3HV) when grown on SCB hydrolysate along with glucose as a carbon substrate. However, the bacterium failed to synthesize the polymer when grown in a hydrolysate solution devoid of glucose and was also unable to utilize pentose sugars like xylose and arabinose as a sole source of carbon [51]. Interestingly, in the present study, isolate *Hgm. borinquense* strain E3 was able to grow and synthesize PHA [P(3HB-*co*-3HV)] from crude SCB hydrolysate without any supplementation of carbon substrate (glucose) or prior treatment of the SCB hydrolysate for the removal of inhibitors. The strain E3 was also able to utilize arabinose and xylose when supplied as the sole source of carbon. Further studies should be done to investigate the cell and polymer yield after pre-treatment of the SCB hydrolysate to remove toxic substances. Also, the effect of glucose or other carbon substrates as supplements to SCB hydrolysate could be investigated with respect to increasing the PHA yield.

4. Conclusions

In the present study, NSM with crude SCB hydrolysate was used as a carbon substrate for the production of PHA by the extremely halophilic archaeon, *Hgm. borinquense* strain E3. The maximum PHA accumulation was observed on the seventh day, reaching a total dry biomass of 3.17 and 4.15 g/L, containing 50.4% and 45.7% PHA. The polymer exhibited two melting endotherms and was identified to be a co-polymer of P(3HB-*co*-3HV) comprising of 13.29% 3HV. Strain E3 accumulated a substantial quantity of PHA using crude SCB hydrolysate without any prior treatment or additional carbon substrate. Investigation into enhancing the quality and yield of P(3HB-*co*-3HV) from SCB hydrolysate could be achieved by: (i) standardization of the production medium, by additional supplement of carbon substrate such as glucose/xylose, and (ii) detoxification/pre-treatment of SCB hydrolysate for removal of inhibitors. Reducing the long lag phase of the culture by increasing the inoculum concentrations and optimization of other cultivation parameters such as pH, temperature, aeration, or salt can further increase the biomass and polymer yield.

The potential application of *Hgm. borinquense* strain E3 for the utilization of agro-industrial waste such as SCB has been clearly demonstrated in the present study. Since India's economy is dominated by agriculture and agro-based industries, large amounts of agro-industrial waste are being generated. Various agro-industrial waste products that can be degraded by such halophilic microbes should be explored for the production of biopolymers as this may help with both managing waste and cutting down the costs of commercial substrates. PHAs from halophilic archaea can be looked upon as a promising prospect for exploring novel bioplastics. This polymer can be further studied for various medical applications like tissue engineering and as a scaffold in organ culture.

Supplementary Materials: The supplementary materials are available online at http://www.mdpi.com/2306-5354/4/2/50/s1.

Acknowledgments: Bhakti B. Salgaonkar thank the Council of Scientific and Industrial Research (CSIR) India for a Senior Research Fellowship (SRF) (09/919(0016)/2012-EMR-I. The authors thank the Sophisticated Instrumentation Facility (SIF), Chemistry Division, VIT University, Vellore for FT-IR and ^{1}H-NMR analysis. The authors are grateful to Narendra Nath Ghosh, Department of Chemistry, BITS Pilani Goa Campus, for the DSC analysis.

Author Contributions: Bhakti B. Salgaonkar and Judith M. Bragança conceived the idea, designed the experiments, and analyzed the data. Bhakti B. Salgaonkar performed the experiments and drafted the manuscript, which was reviewed and edited by Judith M. Bragança.

Conflicts of Interest: The authors declare no conflict of interest.

Abbreviations

APHA	American Public Health Association
CDM	cell dry mass
COD	chemical oxygen demand
DSC	Differential scanning calorimetry
EHM	Extremely Halophilic Medium
FT-IR	Fourier transform infrared
NMR	nuclear magnetic resonance
Har.	*Haloarcula*
Hbf.	*Halobiforma*
Hbt.	*Halobacterium*
Hcc.	*Halococcus*
Hfx.	*Haloferax*
Hgm.	*Halogeometricum*
Hpg.	*Halopiger*
Htg.	*Haloterrigena*
Nnm.	*Natrinema*
NSM	NaCl synthetic medium
NTYE	NaCl Tryptone Yeast extract
P(3HB-*co*-3HV)	poly(3-hydroxybutyrate-*co*-3-hydroxyvalerate)
P3HB	Poly-3-hydroxybutyrate
PHA	Polyhydroxyalkanoates
SCB	sugarcane bagasse
TKN	Total Kjeldahl nitrogen
TS	total solids
VS	volatile solids
XRD	X-ray diffraction

References

1. Chen, G.Q.; Patel, M.K. Plastics derived from biological sources: Present and future: A technical and environmental review. *Chem. Rev.* **2011**, *112*, 2082–2099. [CrossRef] [PubMed]
2. Jendrossek, D.; Pfeiffer, D. New insights in the formation of polyhydroxyalkanoate granules (carbonosomes) and novel functions of poly(3-hydroxybutyrate). *Environ. Microbiol.* **2014**, *16*, 2357–2373. [CrossRef] [PubMed]
3. Vadlja, D.; Koller, M.; Novak, M.; Braunegg, G.; Horvat, P. Footprint area analysis of binary imaged *Cupriavidus necator* cells to study PHB production at balanced, transient, and limited growth conditions in a cascade process. *Appl. Microb. Biotechnol.* **2016**, *100*, 10065–10080. [CrossRef] [PubMed]
4. Koller, M.; Maršálek, L.; de Sousa Dias, M.M.; Braunegg, G. Producing microbial polyhydroxyalkanoate (PHA) biopolyesters in a sustainable manner. *New Biotechnol.* **2017**, *37*, 24–38. [CrossRef] [PubMed]
5. Valappil, S.P.; Boccaccini, A.R.; Bucke, C.; Roy, I. Polyhydroxyalkanoates in Gram-positive bacteria: Insights from the genera *Bacillus* and *Streptomyces*. *Antonie Leeuwenhoek.* **2007**, *91*, 1–17. [CrossRef] [PubMed]

6. Han, J.; Hou, J.; Liu, H.; Cai, S.; Feng, B.; Zhou, J.; Xiang, H. Wide distribution among halophilic archaea of a novel polyhydroxyalkanoate synthase subtype with homology to bacterial type III synthases. *Appl. Environ. Microbiol.* **2010**, *76*, 7811–7819. [CrossRef] [PubMed]
7. Silva, L.F.; Taciro, M.K.; Ramos, M.M.; Carter, J.M.; Pradella, J.G.C.; Gomez, J.G.C. Poly-3-hydroxybutyrate (P3HB) production by bacteria from xylose, glucose and sugarcane bagasse hydrolysate. *J. Ind. Microbiol. Biotechnol.* **2004**, *31*, 245–254. [CrossRef] [PubMed]
8. Parameswaran, B. Sugarcane bagasse. In *Biotechnology for Agro-Industrial Residues Utilisation*; Springer: Dordrecht, The Netherlands, 2009; pp. 239–252.
9. Pippo, W.A.; Luengo, C.A. Sugarcane energy use: Accounting of feedstock energy considering current agro-industrial trends and their feasibility. *Int. J. Energy Environ. Eng.* **2013**, *4*. [CrossRef]
10. Obruca, S.; Benesova, P.; Marsalek, L.; Marova, I. Use of lignocellulosic materials for PHA production. *Chem. Biochem. Eng. Q.* **2015**, *29*, 135–144. [CrossRef] [PubMed]
11. Lavarack, B.P.; Griffin, G.J.; Rodman, D. The acid hydrolysis of sugarcane bagasse hemicellulose to produce xylose, arabinose, glucose and other products. *Biomass Bioenergy* **2002**, *23*, 367–380. [CrossRef]
12. Kirk, R.G.; Ginzburg, M. Ultrastructure of two species of halobacterium. *J. Ultrastruct. Res.* **1972**, *41*, 80–94. [CrossRef]
13. Chen, C.W.; Don, T.M.; Yen, H.F. Enzymatic extruded starch as a carbon source for the production of poly(3-hydroxybutyrate-co-3-hydroxyvalerate) by *Haloferax mediterranei*. *Process Biochem.* **2006**, *41*, 2289–2296. [CrossRef]
14. Han, J.; Lu, Q.; Zhou, L.; Zhou, J.; Xiang, H. Molecular characterization of the phaECHm genes, required for biosynthesis of poly(3-hydroxybutyrate) in the extremely halophilicarchaeon *Haloarculamarismortui*. *Appl. Environ. Microbiol.* **2007**, *73*, 6058–6065. [CrossRef] [PubMed]
15. Romano, I.; Poli, A.; Finore, I.; Huertas, F.J.; Gambacorta, A.; Pelliccione, S.; Nicolaus, G.; Lama, L.; Nicolaus, B. *Haloterrigena hispanica* sp. nov., an extremely halophilicarchaeon from Fuente de Piedra, Southern Spain. *Int. J. Syst. Evol. Microbiol.* **2007**, *57*, 1499–1503. [CrossRef] [PubMed]
16. Salgaonkar, B.B.; Mani, K.; Bragança, J.M. Accumulation of polyhydroxyalkanoates by halophilic archaea isolated from traditional solar salterns of India. *Extremophiles* **2013**, *17*, 787–795. [CrossRef] [PubMed]
17. Legat, A.; Gruber, C.; Zangger, K.; Wanner, G.; Stan-Lotter, H. Identification of polyhydroxyalkanoates in *Halococcus* and other haloarchaeal species. *Appl. Microbiol. Biotechnol.* **2010**, *87*, 1119–1127. [CrossRef] [PubMed]
18. Danis, O.; Ogan, A.; Tatlican, P.; Attar, A.; Cakmakci, E.; Mertoglu, B.; Birbir, M. Preparation of poly(3-hydroxybutyrate-co-hydroxyvalerate) films from halophilic archaea and their potential use in drug delivery. *Extremophiles* **2015**, *19*, 515–524. [CrossRef] [PubMed]
19. Hezayen, F.F.; Rehm, B.H.A.; Eberhardt, R.; Steinbuchel, A. Polymer production by two newly isolated extremely halophilic archaea: Application of a novel corrosion-resistant bioreactor. *Appl. Microbiol. Biotechnol.* **2000**, *54*, 319–325. [CrossRef] [PubMed]
20. Hezayen, F.F.; Gutierrez, M.C.; Steinbuchel, A.; Tindall, B.J.; Rehm, B.H.A. *Halopiger aswanensis* sp. nov., a polymerproducing and extremely halophilicarchaeon isolated from hypersaline soil. *Int. J. Syst. Evol. Microbiol.* **2010**, *60*, 633–637. [CrossRef] [PubMed]
21. Guzmán, H.; Van-Thuoc, D.; Martín, J.; Hatti-Kaul, R.; Quillaguamán, J. A process for the production of ectoine and poly(3-hydroxybutyrate) by *Halomonas boliviensis*. *Appl. Microbiol. Biotechnol.* **2009**, *84*, 1069–1077. [CrossRef] [PubMed]
22. Quillaguaman, J.; Hashim, S.; Bento, F.; Mattiasson, B.; Hatti-Kaul, R. Poly(β-hydroxybutyrate) production by a moderate halophile, *Halomonas boliviensis* LC1 using starch hydrolysate as substrate. *J. Appl. Microbiol.* **2005**, *99*, 151–157. [CrossRef] [PubMed]
23. Huang, T.Y.; Duan, K.J.; Huang, S.Y.; Chen, C.W. Production of polyhydroxyalkanoates from inexpensive extruded rice bran and starch by *Haloferax mediterranei*. *J. Ind. Microbiol. Biotechnol.* **2006**, *33*, 701–706. [CrossRef] [PubMed]
24. Koller, M.; Hesse, P.; Bona, R.; Kutschera, C.; Atlić, A.; Braunegg, G. Potential of various archae-and eubacterial strains as industrial polyhydroxyalkanoate producers from whey. *Macromol. Biosci.* **2007**, *7*, 218–226. [CrossRef] [PubMed]

25. Bhattacharyya, A.; Saha, J.; Haldar, S.; Bhowmic, A.; Mukhopadhyay, U.K.; Mukherjee, J. Production of poly-3-(hydroxybutyrate-*co*-hydroxyvalerate) by *Haloferax mediterranei* using rice-based ethanol stillage with simultaneous recovery and re-use of medium salts. *Extremophiles* **2014**, *18*, 463–470. [CrossRef] [PubMed]

26. Bhattacharyya, A.; Pramanik, A.; Maji, S.K.; Haldar, S.; Mukhopadhyay, U.K.; Mukherjee, J. Utilization of vinasse for production of poly-3-(hydroxybutyrate-*co*-hydroxyvalerate) by *Haloferax mediterranei*. *AMB Express* **2012**, *2*, 34. [CrossRef] [PubMed]

27. Pramanik, A.; Mitra, A.; Arumugam, M.; Bhattacharyya, A.; Sadhukhan, S.; Ray, A.; Mukherjee, J. Utilization of vinasse for the production of polyhydroxybutyrate by *Haloarcula marismortui*. *Folia Microbiol.* **2012**, *57*, 71–79. [CrossRef] [PubMed]

28. Taran, M. Utilization of petrochemical wastewater for the production of poly(3-hydroxybutyrate) by *Haloarcula* sp. IRU1. *J. Hazard. Mater.* **2011**, *188*, 26–28. [CrossRef] [PubMed]

29. Mani, K.; Salgaonkar, B.B.; Bragança, J.M. Culturable halophilic archaea at the initial and final stages of salt production in a natural solar saltern of Goa, India. *Aquat. Biosyst.* **2012**, *8*, 15. [CrossRef] [PubMed]

30. Salgaonkar, B.B.; Bragança, J.M. Biosynthesis of poly(3-hydroxybutyrate-*co*-3-hydroxyvalerate) by *Halogeometricumborinquense* strain E3. *Int. J. Biol. Macromol.* **2015**, *78*, 339–346. [CrossRef] [PubMed]

31. American Public Health Association; American Water Works Association. *Standard Methods for the Examination of Water and Wastewater: Selected Analytical Methods Approved and Cited by the United States Environmental Protection Agency*, 20th ed.; American Public Health Association: Washington, DC, USA, 1981.

32. Raposo, F.; De la Rubia, M.A.; Borja, R.; Alaiz, M. Assessment of a modified and optimised method for determining chemical oxygen demand of solid substrates and solutions with high suspended solid content. *Talanta* **2008**, *76*, 448–453. [CrossRef] [PubMed]

33. Dubois, M.; Gilles, K.A.; Hamilton, J.K.; Rebers, P.A.T.; Smith, F. Colorimetric method for determination of sugars and related substances. *Anal. Chem.* **1956**, *28*, 350–356. [CrossRef]

34. Labconco, C. *A Guide to Kjeldahl Nitrogen Determination Methods and Apparatus*; Labconco Corporation: Houston, TX, USA, 1998.

35. Salgaonkar, B.B.; Mani, K.; Braganca, J.M. Characterization of polyhydroxyalkanoates accumulated by a moderately halophilic salt pan isolate *Bacillus megaterium* strain H16. *J. Appl. Microbiol.* **2013**, *114*, 1347–1356. [CrossRef] [PubMed]

36. Law, J.H.; Slepecky, R.A. Assay of poly-β-hydroxybutyric acid. *J. Bacteriol.* **1961**, *82*, 33–36. [PubMed]

37. Sánchez, R.J.; Schripsema, J.; da Silva, L.F.; Taciro, M.K.; Pradella, J.G.; Gomez, J.G.C. Medium-chain-length polyhydroxyalkanoic acids (PHA mcl) produced by Pseudomonas putida IPT 046 from renewable sources. *Eur. Polym. J.* **2003**, *39*, 1385–1394. [CrossRef]

38. Follonier, S.; Panke, S.; Zinn, M. A reduction in growth rate of *Pseudomonas putida* KT2442 counteracts productivity advances in medium-chain-length polyhydroxyalkanoate production from gluconate. *Microb. Cell Factories* **2011**, *10*, 25. [CrossRef] [PubMed]

39. Castilho, L.R.; Mitchell, D.A.; Freire, D.M. Production of polyhydroxyalkanoates (PHAs) from waste materials and by-products by submerged and solid-state fermentation. *Bioresour. Technol.* **2009**, *100*, 5996–6009. [CrossRef] [PubMed]

40. Da Silva Pinto, C.E.; Arizaga, G.G.C.; Wypych, F.; Ramos, L.P.; Satyanarayana, K.G. Studies of the effect of molding pressure and incorporation of sugarcane bagasse fibers on the structure and properties of poly (hydroxy butyrate). *Compos. Part A Appl. Sci. Manuf.* **2009**, *40*, 573–582. [CrossRef]

41. Vidhate, S.; Innocentini-Mei, L.; D'Souza, N.A. Mechanical and electrical multifunctional poly(3-hydroxybutyrate-*co*-3-hydroxyvalerate)-multiwall carbon nanotube nanocomposites. *Polym. Eng. Sci.* **2012**, *52*, 1367–1374. [CrossRef]

42. Oliveira, L.M.; Araújo, E.S.; Guedes, S.M.L. Gamma irradiation effects on poly(hydroxybutyrate). *Polym. Degrad. Stab.* **2006**, *91*, 2157–2162. [CrossRef]

43. Buzarovska, A.; Grozdanov, A.; Avella, M.; Gentile, G.; Errico, M. Poly(hydroxybutyrate-*co*-hydroxyvalerate)/ titanium dioxide nanocomposites: A degradation study. *J. Appl. Polym. Sci.* **2009**, *114*, 3118–3124. [CrossRef]

44. Sudesh, K. *Polyhydroxyalkanoates from Palm Oil: Biodegradable Plastics*; Springer Science & Business Media: New York, NY, USA, 2012; Volume 76.

45. Hermann-Krauss, C.; Koller, M.; Muhr, A.; Fasl, H.; Stelzer, F.; Braunegg, G. Archaeal production of polyhydroxyalkanoate (PHA) co-and terpolyesters from biodiesel industry-derived by-products. *Archaea* **2013**, *2013*, 129268. [CrossRef] [PubMed]

46. Koller, M.; Hesse, P.; Bona, R.; Kutschera, C.; Atlić, A.; Braunegg, G. Biosynthesis of high quality polyhydroxyalkanoate co-and terpolyesters for potential medical application by the archaeon *Haloferax mediterranei*. *Macromol. Symp.* **2007**, *253*, 33–39. [CrossRef]

47. Cervantes-Uc, J.M.; Catzin, J.; Vargas, I.; Herrera-Kao, W.; Moguel, F.; Ramirez, E.; Lizama-Uc, G. Biosynthesis and characterization of polyhydroxyalkanoates produced by an extreme halophilic bacterium, *Halomonas nitroreducens*, isolated from hypersaline ponds. *J. Appl. Microbiol.* **2014**, *117*, 1056–1065. [CrossRef] [PubMed]

48. Van-Thuoc, D.; Huu-Phong, T.; Thi-Binh, N.; Thi-Tho, N.; Minh-Lam, D.; Quillaguaman, J. Polyester production by halophilic and halotolerant bacterial strains obtained from mangrove soil samples located in Northern Vietnam. *Microbiol. Open* **2012**, *1*, 395–406. [CrossRef] [PubMed]

49. Kulkarni, S.O.; Kanekar, P.P.; Nilegaonkar, S.S.; Sarnaik, S.S.; Jog, J.P. Production and characterization of a biodegradable poly(hydroxybutyrate-*co*-hydroxyvalerate) (PHB-*co*-PHV) copolymer by moderately haloalkalitolerant *Halomonas campisalis* MCM B-1027 isolated from Lonar Lake, India. *Bioresour. Technol.* **2010**, *101*, 9765–9771. [CrossRef] [PubMed]

50. Shrivastav, A.; Mishra, S.K.; Shethia, B.; Pancha, I.; Jain, D.; Mishra, S. Isolation of promising bacterial strains from soil and marine environment for polyhydroxyalkanoates (PHAs) production utilizing Jatropha biodiesel byproduct. *Int. J. Biol. Macromol.* **2010**, *47*, 283–287. [CrossRef] [PubMed]

51. Yu, J.; Stahl, H. Microbial utilization and biopolyester synthesis of bagasse hydrolysates. *Bioresour. Technol.* **2008**, *99*, 8042–8048. [CrossRef] [PubMed]

bioengineering

MDPI

Article

Production of Polyhydroxyalkanoates Using Hydrolyzates of Spruce Sawdust: Comparison of Hydrolyzates Detoxification by Application of Overliming, Active Carbon, and Lignite

Dan Kucera, Pavla Benesova, Peter Ladicky, Miloslav Pekar, Petr Sedlacek and Stanislav Obruca *

Faculty of Chemistry, Brno University of Technology, Purkynova 118, 612 00 Brno, Czech Republic;
Dan.Kucera@vut.cz (D.K.); pavla.benesova@vut.cz (P.B.); peter.ladicky@vut.cz (P.L.); pekar@fch.vut.cz (M.P.);
sedlacek-p@fch.vut.cz (P.S.)
* Correspondence: obruca@fch.vut.cz; Tel.: +420-541-149-486

Academic Editor: Martin Koller
Received: 28 April 2017; Accepted: 25 May 2017; Published: 28 May 2017

Abstract: Polyhydroxyalkanoates (PHAs) are bacterial polyesters which are considered biodegradable alternatives to petrochemical plastics. PHAs have a wide range of potential applications, however, the production cost of this bioplastic is several times higher. A major percentage of the final cost is represented by the price of the carbon source used in the fermentation. *Burkholderia cepacia* and *Burkholderia sacchari* are generally considered promising candidates for PHA production from lignocellulosic hydrolyzates. The wood waste biomass has been subjected to hydrolysis. The resulting hydrolyzate contained a sufficient amount of fermentable sugars. Growth experiments indicated a strong inhibition by the wood hydrolyzate. Over-liming and activated carbon as an adsorbent of inhibitors were employed for detoxification. All methods of detoxification had a positive influence on the growth of biomass and PHB production. Furthermore, lignite was identified as a promising alternative sorbent which can be used for detoxification of lignocellulose hydrolyzates. Detoxification using lignite instead of activated carbon had lower inhibitor removal efficiency, but greater positive impact on growth of the bacterial culture and overall PHA productivity. Moreover, lignite is a significantly less expensive adsorbent in comparison with activated charcoal and; moreover, used lignite can be simply utilized as a fuel to, at least partially, cover heat and energetic demands of fermentation, which should improve the economic feasibility of the process.

Keywords: polyhydroxyalkanoates; detoxification; lignite; *Burkholderia*

1. Introduction

Polyhydroxyalkanoates (PHAs) are polyesters which are synthesized by numerous naturally occurring microorganisms as energy and carbon storage materials. Moreover, due to their mechanical and technological properties resembling those of some petrochemical plastics, PHAs are generally considered a biodegradable alternative to petrochemical-based synthetic polymers [1]. PHAs have a wide range of potential applications, however, the production cost of these bioplastics are several times higher which complicates their production at an industrial scale [2].

A substantial percentage of the final cost is represented by price of carbon substrate [3]. This is the motivation for seeking alternative sources for PHAs production. Among numerous inexpensive or even waste substrates, lignocellulose materials—with sn annual generation of 80 billion tons—represent one of the most promising resources for biotechnological production of (not only) PHAs [4]. Nevertheless, utilization of lignocellulosic materials is accompanied by numerous obstacles stemming from the complex nature of these materials. To access fermentable sugars from cellulose and hemicellulose, a hydrolytic step is required. The hydrolysis of complex lignocellulose biomass is usually performed in two steps. Diluted mineral acid is used in the first step to hydrolyze hemicellulose and to disrupt the complex structure of lignocellulose, which enables subsequent enzymatic hydrolysis of cellulose [5]. Nevertheless, apart from utilizable sugars, also numerous microbial inhibitors such as organic acids (e.g., acetic, formic or levulinic acid), furfurals, and polyphenols are generated by the hydrolysis process. These substances usually reduce fermentability of the hydrolyzates and decrease yields of the biotechnological processes. This problem can be solved by introduction of detoxification. Generally, the aim of detoxification is to selectively remove or eliminate microbial inhibitors from the hydrolyzate prior to biotechnological conversion of the hydrolyzate into desired products [6].

Numerous detoxification methods are based on more or less selective removal of inhibitors by their adsorption on various sorbents. The most commonly used sorbent for this purpose is active carbon, nevertheless, this detoxification strategy suffers from high cost of the sorbent [6]. On the contrary, lignite represents a very promising, low-cost, and effective sorbent which has already been used for the treatment of wastewater to remove various organic and inorganic contaminants [7].

The woodworking industry generates a variety of solid waste materials, such as sawdust, shavings, or bark. It is true that many of these waste materials are already used in various applications. Wood waste is very often burned and used for heat and electricity generation. On the contrary, it could be a potentially inexpensive and renewable feedstock for biotechnological production of PHAs. For instance, Pan et al. employed *Burkholderia cepacia* for biotechnological production of PHAs from detoxified maple hemicellulosic hydrolyzate [8]. Further, Bowers et al. studied PHA production wood chips of *Pinus radiata* which were subjected to high-temperature mechanical pre-treatment or steam explosion in the presence of sulphur dioxide before being enzymatically hydrolyzed. *Novosphingobium nitrogenifigens* and *Sphingobium scionense* were used for PHA production on these hydrolyzates [9]. *Brevundimonas vesicularis* and *Sphingopyxis macrogoltabida* were employed by Silva et al. [10] to produce ter-polymer consisting of 3-hydroxybutyrate, 3-hydroxyvalerate, and lactic acid (3-hydroxypropionate) from acid hydrolyzed sawdust. Despite the fact that PHA production capabilities are exhibited by many bacterial strains, *Burkholderia cepacia* and *Burkholderia sacchari* are the most commonly used for PHA production from hydrolyzates of lignocellulosic materials [8,11,12].

In this study, wood hydrolyzate was utilized as a carbon source for production of polyhydroxyalkanoates. Moreover, since hydrolyzates contain substantial concentrations microbial inhibitors, various detoxification methods including the novel application of lignite as a sorbent are used to improve the fermentability of wood hydrolyzate based media and thus, the PHA yields obtained on this promising substrate.

2. Materials and Methods

2.1. Wood Hydrolyzate (WH) Preparation

Spruce sawdust was supplied by a wood processing company. The waste material was firstly dried to constant weight (80 °C for 24 h). Sawdust was then pretreated with diluted acid and thereafter subjected to enzymatic hydrolysis. To hydrolyze the hemicelluloses of raw material, 20% (w/v) pre-dried sawdust was treated by 4% H_2SO_4 for 60 min at 121 °C. Enzymatic hydrolysis, as a following step, was used for digestion of cellulose structure to release further fermentable saccharides. It was performed by adjusting the pH of the suspension to 5.0 by NaOH and cellulose was treated by 0.5% of Viscozyme L (Sigma-Aldrich, Deisenhofen, Germany) at 37 °C under permanent shaking for 24 h. Subsequently, solids were removed by filtration and the permeate, called wood hydrolyzate (WH), was used in the preparation of the cultivation medium and for PHA production.

2.2. Microorganisms and Cultivation

Burkholderia cepacia (CCM 2656) was purchased from Czech Collection of Microorganisms, Brno, Czech Republic. *Burkholderia sacchari* (DSM 17165) was purchased from Leibnitz Institute

DSMZ-German Collection of Microorganism and Cell Cultures, Braunschweig, Germany. The mineral salt medium for *B. cepacia* and *B. sacchari* cultivation was composed of: 1 g L^{-1} (NH$_4$)$_2$SO$_4$, 1.5 g L^{-1} KH$_2$PO$_4$, 9.02 g L^{-1} Na$_2$HPO$_4$·12H$_2$O, 0.1 g L^{-1} CaCl$_2$·2H$_2$O, 0.2 g L^{-1} MgSO$_4$·7H$_2$O, and 1 mL L^{-1} of microelement solution, the composition of which was as follows: 0.1 g L^{-1} ZnSO$_4$·7H$_2$O, 0.03 g L^{-1} MnCl$_2$·4H$_2$O, 0.3 g L^{-1} H$_3$BO$_3$, 0.2 g L^{-1} CoCl$_2$, 0.02 g L^{-1} CuSO$_4$·7H$_2$O, 0.02 g L^{-1} NiCl$_2$·6H$_2$O, 0.03 g L^{-1} Na$_2$MoO$_4$·2H$_2$O. The cultivations were performed in Erlenmeyer flasks (volume 100 mL) containing 50 mL of the cultivation medium. The temperature was set to 30 °C and the agitation to 180 rpm. The cells were harvested after 72 h of cultivation.

2.3. Detoxification of Hydrolyzates

Overliming was carried out as described by Ranatunga et al. [13], whereupon pH of the hydrolyzate was adjusted to approx. pH 10.0 using solid calcium hydroxide. The samples were then kept at 50 °C for 30 min, the pH was adjusted back to 7, and the sample was subsequently filtered through filter paper.

Detoxification with activated charcoal was performed as described by Pan et al. [8]. Charcoal was added to hydrolyzate in the ratio 1:20 (*w/v*) and stirred for 1 h at 60 °C. Solid particles were removed by filtration. Furthermore, detoxification with lignite was performed similarly, finely milled lignite power (grain size of under 0.2 mm) from South Moravian Coalfield (the northern part of the Vienna basin in the Czech Republic) was used.

2.4. Analytical Methods

All analyses of hydrocarbons and furfural were performed with a Thermo Scientific UHPLC system–UltiMate 3000. REZEX-ROA column (150 × 4.6 mm, 5 μm; City, Phenomenex, Torrance, California, USA) was used for separation. The mobile phase was 5 mN H$_2$SO$_4$ at a flow rate of 0.5 mL per min. Xylose and other saccharides were detected using a refractive index detector (ERC RefractoMax 520). Acetate, levulinic acid and furfural were detected with a Diode Array Detector (DAD-3000) at 284 nm.

Total phenolics were determinated as described by Li et al. [14] with the Folin–Ciocalteu reagent (Sigma-Aldrich). Gallic acid was used for calibration and total phenolics were expressed as milligrams of gallic acid equivalents per liter of wood hydrolyzate.

2.5. PHA Extraction and Content Analysis

To determine biomass concentration and PHA content in cells, samples (10 mL) were centrifuged and the cells were washed with distilled water. The biomass concentration expressed as cell dry weight (CDW) was analyzed as reported previously [15]. PHA content of dried cells was analyzed by gas chromatography (Trace GC Ultra, Thermo Scientific, Waltham, Massachusetts, USA) as reported by Brandl et al. [16]. Commercially available P(3HB-co-3HV) (Sigma Aldrich) composed of 88 mol. % 3HB and 12 mol. % 3HV was used as a standard; benzoic acid (LachNer, Neratovice, Czech Republic) was used as an internal standard.

3. Results and Discussion

The composite formed by cellulose, hemicellulose, and lignin is responsible for the remarkable resistance against hydrolysis and enzymatic attack [17]. Generally, proper pre-treatment of lignocellulose prior to its enzymatic hydrolysis by cellulases significantly improves fermentable sugar yields [18]. The combination of diluted acid hydrolysis (1% H$_2$SO$_4$) and enzymatic digestion of cellulose was used for hydrolysis of spruce sawdust. This approach yielded liquid hydrolyzate of wood (WH) and its composition is shown in Table 1.

Table 1. Composition of wood hydrolyzate (50 g of sawdust per 1 L of 4% H_2SO_4).

	Concentration
Glucose	4.5 g/L
Xylose	10.4 g/L
Ash	52,6 g/L
Polyphenols	1205 mg/L
Furfural	52.0 mg/L
Acetic acid	0.53 g/L
Levulinic acid	9.9 mg/L
5-HMF	*not detected*

Glucose is formed by the cleavage of cellulose. Hemicelluloses can be hydrolyzed to yield molecules such as xylose, arabinose, mannose, galactose, and uronic acid [19]. The total concentration of sugars in WH was determined to be 14.9 g L^{-1} (by the hydrolysis of 50 g L^{-1} of spruce sawdust). The only identified saccharides in WH are xylose (10.4 g L^{-1}) and glucose (4.5 g L^{-1}). Unfortunately, WH contains high concentrations of inhibiting substances such as polyphenols (1205 mg L^{-1}), furfural (52.0 mg L^{-1}), acetic acid (0.53 mg L^{-1}), and levulinic acid (9.9 mg L^{-1}). Polyphenols are likely to be released from waste wood biomass during the partial degradation of lignin by acid hydrolysis. Furfural is formed by degradation of reducing sugars at high pressure and low pH. Levulinic acid, the degradation product of furfural or 5-hydroxymethylfurfural, is formed in the same manner. Acetic acid is probably formed by deesterification of acetylated wood components. Moreover, the amount of ash is significant, which is a consequence of the application of sulfuric acid and subsequent neutralization by NaOH. High concentrations of salts may theoretically cause the inhibition of bacterial growth due to the induction of osmotic stress. On the contrary, mild osmotic up-shock was reported to support PHB accumulation in *Cupriavdius necator* H16 [20,21].

The hydrolyzate of waste wood biomass was used as the sole carbon source for PHA production employing *B. cepacia* and *B. sacchari*. Figure 1 demonstrates the negative impact of the presence of inhibitors on the intended biotechnological processes. In both cases, the WH was twice diluted prior to culturing and supplemented by mineral medium.

Yields of biomass were relatively low, approximately 1.0–1.5 g L^{-1}, and PHB content in CDW was about 10%. Total yield of PHB was around 0.1 g L^{-1}, which is very low.

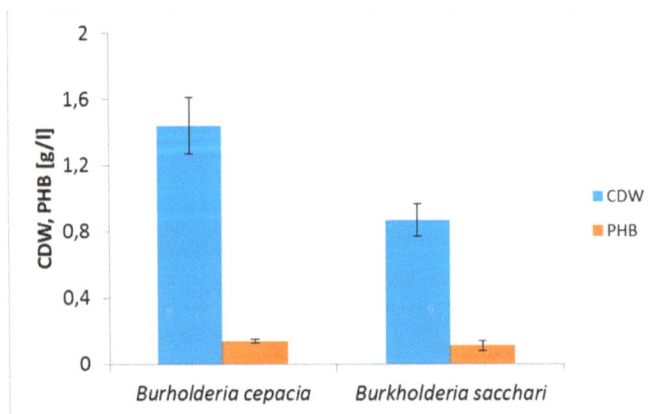

Figure 1. Cultivation of *B. cepacia* and *B. sacchari* on WH which composition is demonstrated in Table 1, WH was twice diluted and supplemented with mineral salts as described above. Cultivation conditions: 30 °C, 72 h, 180 rpm.

The effect of phenolic and other aromatic compounds, which may inhibit both microbial growth and product yield, are very variable, and can be related to specific functional groups. One possible mechanism is that phenolics interfere with the cell membrane by influencing its function and changing its protein-to-lipid ratio [22]. Undissociated acids enter the cell through diffusion over the cell membrane and then dissociate due to the neutral cytosolic pH. The dissociation of the acid leads to a decrease in the intracellular pH, which may cause cell death. This effect is promoted by furfural and 5-HMF which cause higher cell membrane permeation and disturb the proton gradient over the inner mitochondrial membrane which inhibits regeneration of ATP and eventually can lead to cellular death [23]. A different mechanism of action of growth inhibitors results in a stronger synergistic effect.

The presence of inhibitors, and especially polyphenols, in the wood hydrolyzate appears to be crucial for the intended biotechnological process. Therefore, we continued to focus on the elimination of microbial inhibitors. In the first phase, we compared two common detoxification procedures—separation by adsorption inhibitors on activated carbon and over-liming. Theoretically, overliming is effective due to precipitation or chemical destabilization of inhibitors [13] and activated charcoal could improve the fermentability of hydrolysate by absorbing phenolic compounds and other inhibitory substances [24].

The effect of various methods of detoxification on the concentration of the most important inhibitors present in hydrolysates and polyphenols is demonstrated in Figure 2.

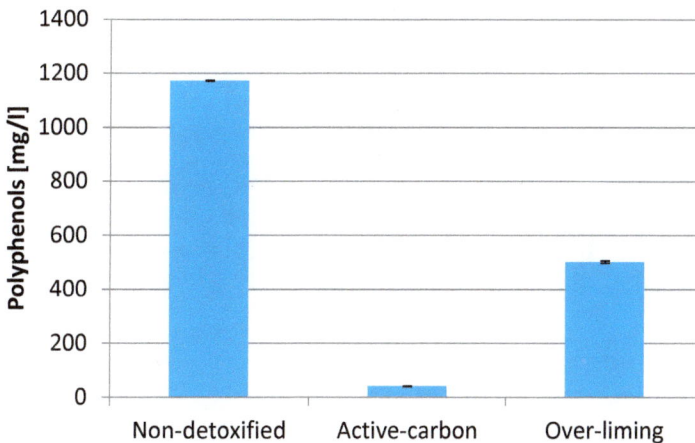

Figure 2. Detoxification employing over-liming and active carbon.

It is evident that both detoxification techniques significantly reduce the concentration of polyphenols in hydrolysates. More effective is the application of activated carbon, which can adsorb and thereby remove more than 90% of polyphenols. Table 2 demonstrates results of cultivation experiment with detoxified WH, employing the same PHB producers as in the previous test. Both methods of detoxification exhibited a positive influence on the growth of biomass, and this effect was more apparent with *B. cepacia*. More significantly, the effect of detoxification was reflected in the content of PHB in biomass. A positive effect on the biosynthesis of PHB occurred primarily in the strain of *B. sacchari*. PHB content reached nearly 90% of CDW. The yields were 8–12 times higher compared to the use of non-detoxified hydrolyzate. On the other hand, the detoxification process itself is time-consuming and particularly expensive, especially if activated carbon is used for detoxification of the hydrolyzates [25].

Table 2. Cultivation on detoxified hydrolyzates.

	Detoxification	Biomass (g/L)	PHB (%)	PHB (g/L)
Burkholderia sacchari	Non-detoxified	0.87	12.2	0.11
	Over-liming	1.57	88.7	1.39
	Activated carbon	1.01	87.6	0.89
Burkholderia cepacia	Non-detoxified	1.44	9.8	0.14
	Over-liming	2.86	30.0	0.86
	Activated carbon	1.40	74.7	1.05

A further aim of our experiments was to find an alternative sorbent, which would be comparable to activated carbon, but the cost of which would be significantly lower. After several pilot experiments, we focused on lignite. It is the youngest and the least carbonized brown coal, which consists of a macromolecular complex polyelectrolyte (e.g., humic acids), polysaccharides, polyaromatics, and carbon chains with sulfur, nitrogen, and oxygen-containing groups. Its cost is significantly lower than that of activated carbon. The price of activated carbon is currently around $1/kg [26] compared to a lignite price of $0.2/kg [26]. Moreover, the recovery of activated charcoal after its application as a sorbent in detoxification is practically impossible [27,28]. On the other side, lignite can be burned after absorbing the inhibitors and the energy released during the combustion process could provide energy which can at least partially cover energetic demands of the intended process of PHA production from waste wood biomass.

The adsorption capacity of lignite and its application as a sorbent is often a subject of interest. It is the price of conventional sorbents that leads to finding low-cost alternatives [29]. Over the last decade, there has been an increase in publications dealing with low-cost adsorbents for wastewater treatment [30]. For instance, lignite was used as a sorbent for removal of organic substances such as phenol [31] or inorganic components, especially heavy metals [32], from contaminated water solutions. Nevertheless, to our best of our knowledge despite its high sorption capacity and low cost, lignite has not been used as a sorbent for detoxification of complex lignocellulose hydrolyzates to increase their fermentability and yield of biotechnological products.

According to our results, lignite has a lower sorption capacity than activated charcoal. On the other hand, lignite is also able to eliminate a substantial amount of inhibitors, and thus potentially increase the fermentability of WH. Comparison of lignite and activated charcoal as a sorbent for microbial inhibitors is displayed in Table 3.

Table 3. Detoxification using active carbon and lignite

	Glucose (g/L)	Xylose (g/L)	Polyphenols (mg/L)	Furfural (mg/L)	Levulinic Acid (mg/L)	Acetic Acid (g/L)
Non-detoxified	4.4	10.0	998.8	41.4	10.0	0.5
Lignite	4.6	10.3	772.4	35.7	7.9	0.4
Activated carbon	4.5	10.1	23.8	3.8	3.6	0.4

The sorption properties of lignite depend on the number of sorption sites or functionalities [33]. Understandably, WH detoxified with lignite were also tested for cultivation. Detoxified hydrolyzates using lignite and activated carbon were used for the biotechnological production of PHB employing *B. cepacia* and *B. sacchari*. Figure 3 shows the results.

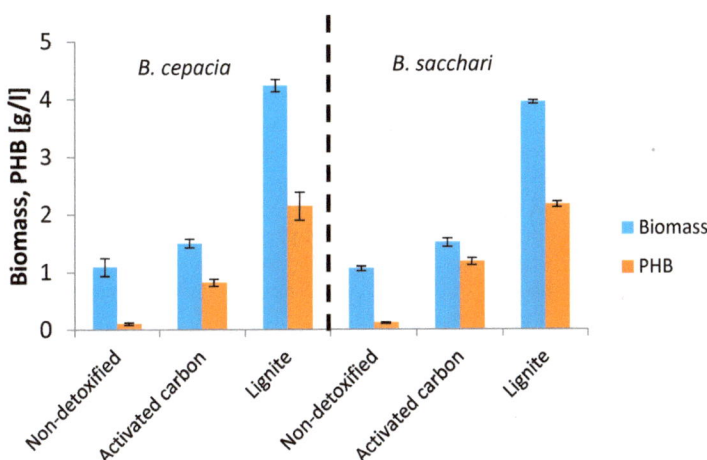

Figure 3. Cultivation on detoxified hydrolyzates.

It is interesting that the sorption capacity of lignite as detoxification strategy does not provide as good a removal of monitored inhibitors as in the case of activated carbon. However, lignite is comparable with activated charcoal, considering the overall yields of PHB. *B. Sacchari* reached markedly higher yields. These surprisingly higher yields obtained by replacing active carbon with lignite should be explained and our further experiments will be focused in this direction. We assume that, during the detoxification, lignite released substances that had a positive effect on the growth of soil bacteria. Lignite is a complex material compared to activated carbon which contains only carbon. Therefore, use of lignite could have an enriching effect on the composition of the production medium. This interesting and surprising feature could also be used in other biotechnological processes in which soil originating microorganisms are employed.

4. Conclusions

This article demonstrates the possibilities of utilization of lignite in biotechnology. Lignite can be used as a sorbent to detoxify wood hydrolyzate and its efficiency was comparable with commonly used activated carbon. This detoxification method was evaluated directly using the hydrolyzates to produce PHAs employing *Burkholderia cepacia* and *Burkholderia sacchari*. The results showed that the use of lignite considerably improved fermentability of wood hydrolyzates and enhanced PHA yields. Therefore, lignite can cope with significantly more expensive activated carbon.

Acknowledgments: This work was supported by the project Materials Research Centre at FCH BUT–Sustainability and Development No. LO1211 and by national COST project LD15031 of the Ministry of Education, Youth, and Sports of the Czech Republic and by the project GA15-20645S of the Czech Science Foundation (GACR).

Author Contributions: Dan Kucera, Peter Ladicky and Stanislav Obruca conceived and designed the experiments; Dan Kucera, Peter Ladicky, Stanislav Obruca, Petr Sedlacek, Miloslav Pekar and Pavla Benesova performed the experiments; Dan Kucera, Peter Ladicky, Pavla Benesova and Stanislav Obruca analyzed the data; Miloslav Pekar, Petr Sedlacek and Stanislav Obruca contributed reagents/materials/analysis tools; Dan Kucera and Stanislav Obruca wrote the paper.

Conflicts of Interest: The authors declare no conflict of interest. The founding sponsors had no role in the design of the study; in the collection, analyses, or interpretation of data; in the writing of the manuscript, or in the decision to publish the results.

References

1. Steinbüchel, A. Perspectives for Biotechnological Production and Utilization of Biopolymers: Metabolic Engineering of Polyhydroxyalkanoate Biosynthesis Pathways as a Successful Example. *Macromol. Biosci.* **2001**, *1*, 1–24. [CrossRef]

2. Ivanov, V.; Stabnikov, V.; Ahmed, Z.; Dobrenko, S.; Saliuk, A. Production and applications of crude polyhydroxyalkanoate-containing bioplastic from the organic fraction of municipal solid waste. *Int. J. Environ. Sci. Technol.* **2015**, *12*, 725–738. [CrossRef]

3. Choi, J.; Lee, S.Y. Factors affecting the economics of polyhydroxyalkanoate production by bacterial fermentation. *Appl. Microbiol. Biotechnol.* **1999**, *51*, 13–21. [CrossRef]

4. Obruca, S.; Benesova, P.; Kucera, D.; Petrik, S.; Marova, I. Biotechnological conversion of spent coffee grounds into polyhydroxyalkanoates and carotenoids. *New Biotechnol.* **2015**, *32*, 569–574. [CrossRef] [PubMed]

5. Amidon, T.E.; Wood, C.D.; Shupe, A.M.; Wang, Y.; Graves, M.; Liu, S. Biorefinery: Conversion of Woody Biomass to Chemicals, Energy and Materials. *J. Biobased Mater. Bioenergy* **2008**, *2*, 100–120. [CrossRef]

6. Canilha, L.; Kumar Chandel, A.; dos Santos Milessi, T.S.; Fernandes Antunes, F.A.; da Costa Freitas, W.L.; das Graças Almeida Fellipe, M.; da Silva, S.S. Bioconversion of Sugarcane Biomass into Ethanol: An Overview about Composition, Pretreatment Methods, Detoxification of Hydrolysates, Enzymatic Saccharification, and Ethanol Fermentation. *J. Biomed. Biotechnol.* **2012**, *2012*, 1–15. [CrossRef] [PubMed]

7. Doskočil, L.; Grasset, L.; Enev, V.; Kalina, L.; Pekař, M. Study of water-extractable fractions from South Moravian lignite. *Environ. Earth Sci.* **2015**, *73*, 3873–3885. [CrossRef]

8. Pan, W.; Perrotta, J.A.; Stipanovic, A.J.; Nomura, C.T.; Nakas, J.P. Production of polyhydroxyalkanoates by *Burkholderia cepacia* ATCC 17759 using a detoxified sugar maple hemicellulosic hydrolysate. *J. Ind. Microbiol. Biotechnol.* **2012**, *39*, 459–469. [CrossRef] [PubMed]

9. Bowers, T.; Vaidya, A.; Smith, D.A.; Lloyd-Jones, G. Softwood hydrolysate as a carbon source for polyhydroxyalkanoate production. *J. Chem. Technol. Biotechnol.* **2014**, *89*, 1030–1037. [CrossRef]

10. Silva, J.A.; Tobella, L.M.; Becerra, J.; Godoy, F.; Martínez, M.A. Biosynthesis of poly-β-hydroxyalkanoate by *Brevundimonas vesicularis* LMG P-23615 and *Sphingopyxis macrogoltabida* LMG 17324 using acid-hydrolyzed sawdust as carbon source. *J. Biosci. Bioeng.* **2007**, *103*, 542–546. [CrossRef] [PubMed]

11. Keenan, T.M.; Tanenbaum, S.W.; Stipanovic, A.J.; Nakas, J.P. Production and Characterization of Poly-β-hydroxyalkanoate Copolymers from *Burkholderia cepacia* Utilizing Xylose and Levulinic Acid. *Biotechnol. Prog.* **2004**, *20*, 1697–1704. [CrossRef] [PubMed]

12. Wang, Y.; Liu, S. Production of (R)-3-hydroxybutyric acid by *Burkholderia cepacia* from wood extract hydrolysates. *AMB Express* **2014**, *4*. [CrossRef] [PubMed]

13. Ranatunga, T.D.; Jervis, J.; Helm, R.F.; McMillan, J.D.; Wooley, R.J. The effect of overliming on the toxicity of dilute acid pretreated lignocellulosics: The role of inorganics, uronic acids and ether-soluble organics. *Enzyme Microb. Technol.* **2000**, *27*, 240–247. [CrossRef]

14. Li, H.-B.; Wong, C.-C.; Cheng, K.-W.; Chen, F. Antioxidant properties in vitro and total phenolic contents in methanol extracts from medicinal plants. *Food Sci. Technol.* **2008**, *41*, 385–390. [CrossRef]

15. Obruca, S.; Marova, I.; Melusova, S.; Mravcova, L. Production of polyhydroxyalkanoates from cheese whey employing *Bacillus megaterium* CCM 2037. *Ann. Microbiol.* **2011**, *61*, 947–953. [CrossRef]

16. Brandl, H.; Gross, R.A.; Lenz, R.W.; Fuller, R.C. *Pseudomonas oleovorans* as a source of poly(beta-hydroxyalkanoates) for potential application as a biodegradable polyester. *Appl. Environ. Microbiol.* **1988**, *54*, 1977–1982. [PubMed]

17. Peters, D. Raw Materials. In *White Biotechnology*; Ulber, R., Sell, D., Eds.; Springer: Berlin, Germany, 2007; pp. 1–30.

18. Obruca, S.; Benešová, P.; Maršálek, L.; Márová, I. Use of Lignocellulosic Materials for PHA Production. *Chem. Biochem. Eng. Q.* **2015**, *29*, 135–144. [CrossRef]

19. Wyman, C.E.; Decker, S.R.; Himmel, M.E.; Brady, J.W.; Skopec, C.E.; Viikari, L. *Hydrolysis of Cellulose and Hemicellulose*; Dumitriu, S., Ed.; CRC Press: New York, NY, USA, 2004; pp. 995–1034.

20. Obruca, S.; Marova, I.; Svoboda, Z.; Mikulikova, R. Use of controlled exogenous stress for improvement of poly (3-hydroxybutyrate) production in *Cupriavidus necator*. *Folia Microbiol.* **2010**, *55*, 17–22. [CrossRef] [PubMed]

21. Passanha, P.; Kedia, G.; Dinsdale, R.M.; Guwy, A.J.; Esteves, S.R. The use of NaCl addition for the improvement of polyhydroxyalkanoate production by *Cupriavidus necator*. *Bioresour. Technol.* **2014**, *163*, 287–294. [CrossRef] [PubMed]

22. Jönsson, L.J.; Alriksson, B.; Nilvebrant, N.-O. Bioconversion of lignocellulose: Inhibitors and detoxification. *Biotechnol. Biofuels* **2013**, *6*. [CrossRef] [PubMed]

23. Chandel, A.K.; da Silva, S.S.; Singh, O.V. Detoxification of Lignocellulose Hydrolysates: Biochemical and Metabolic Engineering Toward White Biotechnology. *Bioenergy Res.* **2013**, *6*, 388–401. [CrossRef]

24. Mussatto, S.I.; Roberto, I.C. Evaluation of nutrient supplementation to charcoal-treated and untreated rice straw hydrolysate for xylitol production by *Candida guilliermondii*. *Braz. Arch. Biol. Technol.* **2005**, *48*, 497–502. [CrossRef]

25. Babel, S.; Kurniawan, T.A. Low-cost adsorbents for heavy metals uptake from contaminated water: A review. *J. Hazard. Mater.* **2003**, *97*, 219–243. [CrossRef]

26. Lignite. Available online: https://www.alibaba.com (accessed on 26 April 2017).

27. Obruca, S.; Benesova, P.; Kucera, D.; Marova, I. Novel Inexpensive Feedstocks from Agriculture and Industry for Microbial Polyester Production. In *Recent Advances in Biotechnology; Microbial Biopolyester Production, Performance and Processing Microbiology, Feedstocks, and Metabolism*; Koller, M., Ed.; Bentham Science: Berlin, Germany, 2016; pp. 3–99.

28. Parawira, W.; Tekere, M. Biotechnological strategies to overcome inhibitors in lignocellulose hydrolysates for ethanol production: Review. *Crit. Rev. Biotechnol.* **2011**, *31*, 20–31. [CrossRef] [PubMed]

29. Gautam, R.K.; Mudhoo, A.; Lofrano, G.; Chattopadhyaya, M.C. Biomass-derived biosorbents for metal ions sequestration: Adsorbent modification and activation methods and adsorbent regeneration. *J. Environ. Chem. Eng.* **2014**, *2*, 239–259. [CrossRef]

30. De Gisi, S.; Lofrano, G.; Grassi, M.; Notarnicola, M. Characteristics and adsorption capacities of low-cost sorbents for wastewater treatment: A review. *Sustain. Mater. Technol.* **2016**, *9*, 10–40. [CrossRef]

31. Polat, H.; Molva, M.; Polat, M. Capacity and mechanism of phenol adsorption on lignite. *Int. J. Miner. Process.* **2006**, *79*, 264–273. [CrossRef]

32. Klucakova, M.; Pavlikova, M. Lignitic Humic Acids as Environmentally-Friendly Adsorbent for Heavy Metals. *J. Chem.* **2017**, *2017*, 1–5. [CrossRef]

33. Robles, I.; Bustos, E.; Lakatos, J. Adsorption study of mercury on lignite in the presence of different anions. *Sustain. Environ. Res.* **2016**, *26*, 136–141. [CrossRef]

bioengineering

MDPI

Article

Biosynthesis of Polyhydroxyalkanoate from Steamed Soybean Wastewater by a Recombinant Strain of *Pseudomonas* sp. 61-3

Ayaka Hokamura, Yuko Yunoue, Saki Goto and Hiromi Matsusaki *

Department of Food and Health Sciences, Faculty of Environmental and Symbiotic Sciences,
Prefectural University of Kumamoto, Kumamoto 862-8502, Japan; a_hokamura@pu-kumamoto.ac.jp (A.H.);
aocharirider@yahoo.co.jp (Y.Y.); g1675001@pu-kumamoto.ac.jp (S.G.)
* Correspondence: matusaki@pu-kumamoto.ac.jp; Tel.: +81-96-321-6697

Academic Editor: Martin Koller
Received: 12 May 2017; Accepted: 4 August 2017; Published: 8 August 2017

Abstract: *Pseudomonas* sp. 61-3 accumulates a blend of poly(3-hydroxybutyrate) [P(3HB)] homopolymer and a random copolymer, poly(3-hydroxybutyrate-*co*-3-hydroxyalkanoate) [P(3HB-*co*-3HA)], consisting of 3HA units of 4–12 carbon atoms. *Pseudomonas* sp. 61-3 possesses two types of PHA synthases, PHB synthase (PhbC) and PHA synthases (PhaC1 and PhaC2), encoded by the *phb* and *pha* loci, respectively. The P(94 mol% 3HB-*co*-6 mol% 3HA) copolymer synthesized by the recombinant strain of *Pseudomonas* sp. 61-3 (*phbC::tet*) harboring additional copies of *phaC1* gene is known to have desirable physical properties and to be a flexible material with moderate toughness, similar to low-density polyethylene. In this study, we focused on the production of the P(3HB-*co*-3HA) copolymer using steamed soybean wastewater, a by-product in brewing *miso*, which is a traditional Japanese seasoning. The steamed soybean wastewater was spray-dried to produce a powder (SWP) and used as the sole nitrogen source for the synthesis of P(3HB-*co*-3HA) by the *Pseudomonas* sp. 61-3 recombinant strain. Hydrolyzed SWP (HSWP) was also used as a carbon and nitrogen source. P(3HB-*co*-3HA)s with relatively high 3HB fractions could be synthesized by a recombinant strain of *Pseudomonas* sp. 61-3 (*phbC::tet*) harboring additional copies of the *phaC1* gene in the presence of 2% glucose and 10–20 g/L SWP as the sole nitrogen source, producing a PHA concentration of 1.0–1.4 g/L. When HSWP was added to a nitrogen- and carbon-free medium, the recombinant strain could synthesize PHA without glucose as a carbon source. The recombinant strain accumulated 32 wt% P(3HB-*co*-3HA) containing 80 mol% 3HB and 20 mol% medium-chain-length 3HA with a PHA concentration of 1.0 g/L when 50 g/L of HSWP was used. The PHA production yield was estimated as 20 mg-PHA/g-HSWP, which equates to approximately 1.0 g-PHA per liter of soybean wastewater.

Keywords: polyhydroxyalkanoate; PHA; copolymer; soybean wastewater

1. Introduction

Polyhydroxyalkanoates (PHAs) are accumulated in many bacteria as intracellular carbon and energy storage materials under nutrient-limited conditions in the presence of excess carbon [1–3]. PHAs are regarded as important environmentally compatible materials because of their potential use as biodegradable plastics with properties similar to petroleum-based plastics. PHAs can be divided into three groups based on their monomer structure. Short-chain-length PHAs (scl-PHAs) consisting of monomers with 3 to 5 carbon atoms, medium-chain-length PHAs (mcl-PHAs) consisting of monomers with 6 to 14 carbon atoms and scl-mcl-PHA copolymers consisting of both scl and mcl monomer units [4]. Poly(3-hydroxybutyrate) [P(3HB)], the principal member of the scl-PHAs, is both stiff and brittle. In contrast, mcl-PHA, consisting of mcl-3-hydroxyalkanoate (mcl-3HA) units with 6 to 14 carbon atoms, is generally amorphous because of its low crystallinity. PHA copolymers consisting of scl and mcl monomers synthesized by recombinant bacteria show various properties ranging from stiff to flexible depending on the monomer composition. Therefore, the composition ratio of scl and mcl monomers is crucial in PHA properties. In particular, the P(94% 3HB-*co*-6% 3HA) copolymer consisting of 3HA units of 6 to 12 carbon atoms possesses properties similar to low-density polyethylene (LDPE) [5]. However, the commercial development of PHAs has been limited because of their high production costs. Therefore, it is desirable that PHAs are economically produced from inexpensive carbon sources such as waste substrates.

In this regard, several studies have reported PHA production using waste substrates as carbon sources. P(3HB) homopolymer has been reported to be synthesized by *Cupriavidus necator* H16 (formerly *Ralstonia eutropha* H16) from plant oils as the sole carbon sources [6]. Furthermore, a random copolymer, P(3HB-*co*-3-hydroxyhexanoate), with a high PHA content has been synthesized from plant oils using a recombinant strain of *C. necator* PHB⁻4 (a PHA-negative mutant) harboring the PHA synthase gene from *Aeromonas cavie* [6]. Wong and Lee reported that P(3HB) could be synthesized from whey using the *Escherichia coli* strain GCSC 6576 harboring the PHA-biosynthetic operon from *C. necator* and the *ftsZ* gene from *E. coli* [7]. In a recent study, P(3HB) homopolymer and a P[3HB-*co*-3-hydroxyvalerate (3HV)] copolymer have been reported to be produced from waste such as oil extracted from spent coffee grounds [8] and waste from the olive oil industry [9]. *Cupriavidus* sp. KKU38 has been reported to synthesize P(3HB) from cassava starch hydrolysate [10]. A number of reviews on the topic have been published [11–13]. While there are numerous reports of the production of scl-PHA or mcl-PHA from waste, there are very few reports on the production of scl-mcl PHA from waste, although Wang et al. have reported scl-mcl PHA production from glycerol, a by-product of the biodiesel industry, using engineered *Escherichia coli* [13]. The production of scl-mcl PHAs consisting of 3HB and mcl-3HA from biomass sources is desirable for the dissemination of PHA as biodegradable plastics because the copolymer is expected to have various properties, ranging from stiff to flexible, depending on the monomer composition as described above.

Pseudomonas sp. 61-3 synthesizes two kinds of PHAs, a P(3HB) homopolymer and a random copolymer, P(3HB-*co*-3HA), consisting of 3-hydroxyalkanoate (3HA) units of 4–12 carbon atoms [14–16]. *Pseudomonas* sp. 61-3 possesses two types of PHA synthases, PHB synthase (PhbC) and PHA synthases (PhaC1 and PhaC2), encoded at the *phb* and *pha* loci, respectively [16]. PhbC shows substrate specificities for short-chain-length 3HA units, whereas PhaC1 and PhaC2 are able to incorporate a wide range of 3HA units of 4–12 carbon atoms into PHA. It has also been reported previously that PhaC1 is the major PHA providing enzyme in *Pseudomonas* sp. 61-3 [17].

Soybeans are used as raw materials in numerous Japanese foods such as *miso* (fermented soybean pastes), *shoyu* (soy sauce), *natto* (fermented soybean) and *tofu* (soybean protein curd), all of which produce wastewater during the manufacturing process. *Miso* is a traditional Japanese seasoning and many Japanese eat *miso* soup every day. However, the steamed soybean wastewater in produced in *miso* processing is a problem. The wastewater must be treated by a wastewater treatment facility as an activated sludge since the soybean wastewater still contains a large amount of organic compounds, resulting in an enormous cost. Following the production of one ton of *miso*, *shoyu*, or *tofu*, 740, 50, or 18 liters of wastewater is generated, respectively [18]. Their chemical oxygen demand is 32,000, 29,000 and 15,000 ppm, respectively, although they are more than 95% water. In Japan, over 100 million liters of wastewater is generated annually from soybean processed foods such as *miso*, *shoyu*, *natto* and *tofu*. Therefore, the utilization of this soybean wastewater is desirable. For example, there have been several reports describing the recovery of oligosaccharides from steamed soybean wastewater in *tofu* processing [19], the recovery of isoflavone aglycones from soy whey wastewater [20], and the use of the soybean-derived waste as biomass [21–24]. In this study, PHA production using steamed soybean wastewater as a nitrogen and/or carbon source was performed using a recombinant strain of *Pseudomonas* sp. 61-3. This is the first report describing scl-mcl-PHA production from steamed soybean wastewater.

2. Materials and Methods

2.1. Preparation and Hydrolysis of Steamed Soybean Wastewater and Starch

Steamed soybean wastewater was collected from barley *miso* (made from barley and soybean) brewery factory in Kumamoto prefecture, Japan and was spray dried to powder. The nutrient composition of the soybean wastewater powder (SWP) was analyzed by Japan Food Research Laboratories, one of the world's largest and most diversified testing services providers (Table 1). Since SWP contains a sufficient amount of protein, it was first used as a nitrogen source (1, 5, 10, 20 and 50 g/L) for PHA production without further treatment. According to the report by Kimura et al., the constituent sugars of the polysaccharides contained in soybean were arabinose (21.6 wt%), galactose (48.5 wt%), uronic acid (15.0 wt%), and xylose (or rhamnose) (14.9 wt%) [18]. Therefore, polysaccharides that can be used as carbon sources were also considered to be contained in the SWP. In order to investigate the use of a carbon source, an SWP hydrolysate was also prepared using the following two methods. One method involved hydrolysis using 0.6 N H_2SO_4 at 80 °C for 5 h. The other hydrolysis method used 5 N H_2SO_4 at 90 °C for 1 h. SWP or cornstarch (2.5 g, Kanto Chemical Co., Inc., Japan), as a carbon source control, was treated with 5 mL H_2SO_4 (0.6 N or 5 N) for 5 h or 1 h. After hydrolysis, the pH was adjusted to 7.0 using NaOH, and the hydrolysates were subsequently filter-sterilized or autoclaved. The hydrolyzed soybean wastewater powder (HSWP) was used for PHA production as the carbon and the nitrogen source at the concentrations of 10, 20, 30, 40, 50, 75, and 100 g/L. The cornstarch hydrolysate was used as the control of a carbon source.

Table 1. Nutrient composition of SWP (wt%).

Moisture	Crude Protein	Crude Fat	Fiber	Ash	Starch	Non-Fibrous Carbohydrate
12.37	10.91	0.17	0.13	12.83	1.75	61.84

2.2. Bacterial Strain, Plasmid, and Culture Conditions

Pseudomonas sp. 61-3 (*phbC::tet*), which is a *phbC*-negative mutant [16], and the recombinant strain were grown at 28 °C in a nutrient-broth (NB) medium consisting of 1% meat extract (Kyokuto Pharmaceutical Industrial Co., Ltd., Tokyo, Japan), 1% Bactopeptone (Difco Laboratories, Division of Becton Dickinson Company, Sparks, MD, USA) and 0.5% NaCl (pH 7.0). *Pseudomonas* sp. 61-3 (*phbC::tet*) harboring the pJKSc54-*phab*, carrying *phaC1* under the control of the *pha* promoter from *Pseudomonas* sp. 61-3 and *phbAB* under the control of the *phb* promoter from *C. necator*, was used to synthesize P(3HB-*co*-3HA) copolymer as described previously [5]. When needed, kanamycin (50 mg/L) and tetracycline (12.5 mg/L) were added to the medium for plasmid maintenance of this *Pseudomonas* sp. 61-3 recombinant strain.

2.3. Production and Analysis of PHA

Pseudomonas sp. 61-3 (*phbC::tet*) harboring pJKSc54-*phab* was grown on NB medium and transferred to 500 mL shaking flasks containing 100 mL of a nitrogen-free MS medium containing 0.9 g $Na_2HPO_4 \cdot 12H_2O$, 0.15 g of KH_2PO_4, 0.02 g of $MgSO_4 \cdot 7H_2O$ and 0.1 mL of trace element solution [15]. The culture with an initial absorbance at 600 nm of 0.05 was cultivated on a reciprocal shaker (130 strokes/min) at 28 °C for 48 h or 72 h. SWP as the sole nitrogen source was added to the medium and autoclaved. Filter-sterilized glucose (2 wt%) was aseptically added to the autoclaved medium as the sole carbon source. NH_4Cl (0.05%), as a control nitrogen source, was also added to the medium instead of SWP. HSWP and hydrolyzed cornstarch were filtered (ADVANTEC filter paper No.1, Toyo Roshi Kaisha, Ltd., Tokyo, Japan) to remove the residues and the filtrates were autoclaved, followed by they were aseptically added to the medium as both the nitrogen and the carbon sources, or the carbon source, respectively. Determination of cellular PHA composition by gas chromatography (GC) was performed as reported previously [13]. Approximately 30 mg of dry cells were subjected to methanolysis and the converted methyl esters were subjected to gas chromatography analysis on a Shimadzu GC-17A system equipped with an Inert Cap 1 capillary column (30 m × 0.25 mm, GL Sciences Inc., Tokyo, Japan) and a flame-ionization detector. Methyl caprylate was used as the internal standard. All cultivations were performed in triplicate.

3. Results

3.1. Nutrient Composition of SWP

SWP was obtained after spray drying of steamed soybean wastewater obtained from the processing of *miso*. The nutrient composition of SWP is shown in Table 1. Nutrients such as crude protein, crude fat, starch and non-fibrous carbohydrate present in the SWP could therefore be utilized for PHA production as nitrogen/carbon sources.

3.2. Utilization of SWP for PHA Production

Pseudomonas sp. 61-3 (*phbC::tet*) harboring the pJKSc54-*phab* plasmid was cultivated at 28 °C for either 48 h or 72 h in a nitrogen-free MS medium supplemented with 2% glucose and various concentrations (1, 5, 10, 20 and 50 g/L) of SWP as the nitrogen source. P(3HB-*co*-3HA) copolymers were synthesized by the recombinant strain using the SWP as the sole nitrogen source (Table 2). Compared with PHA production using 0.5 g/L of NH_4Cl as the sole nitrogen source, the PHA content (62 wt%) and the monomer composition of P(3HB-*co*-3HA) synthesized from 1 g/L of SWP were almost same but the dry cell weight (0.4–0.6 g/L) was much lower than the control (NH_4Cl). Because SWP was considered not to contain sufficient nitrogen source for cell growth, the concentration of SWP added to the medium was then increased. Increasing the concentration of SWP, caused an increase in the dry cell weight, whereas both the PHA content and the 3HB fraction in P(3HB-*co*-3HA) decreased. When the concentration of SWP was increased to 20 g/L or more, the SWP could not be completely dissolved in the medium since a precipitate was observed in the cell pellet after centrifugation of the culture broth. Overall, it was concluded that 10–20 g/L of SWP was suitable for PHA production, and that a PHA concentration of 1.0 to 1.4 g/L was produced. Thus, SWP could be used as the sole nitrogen source instead of NH_4Cl for P(3HB-*co*-3HA) production by this recombinant strain despite the fact that the PHA concentrations were lower than in the control experiment where 0.5 g/L of NH_4Cl was used as the sole nitrogen source.

Table 2. Biosynthesis of PHA *Pseudomonas* sp. 61-3 (*phbC::tet*) harboring pJKSc54-*phab* from SWP as the sole nitrogen source.

Nitrogen Source	Cultivation Time (h)	Dry Cell Weight (g/L)	PHA Content (wt%)	PHA Conc. (g/L)	PHA Composition (mol%)					
					3HB (C4)	3HHx (C6)	3HO (C8)	3HD (C10)	3HDD (C12)	3H5DD (C12′)
0.5 g/L NH₄Cl	48	3.2	62	2.0	87	trace	2	6	2	3
1 g/L SWP	48	0.4	60	0.2	87	trace	1	7	2	3
	72	0.6	66	0.4	87	1	1	6	2	3
5 g/L SWP	48	0.9	52	0.5	80	trace	2	9	4	5
	72	1.2	55	0.7	81	1	2	8	4	4
10 g/L SWP	48	1.4	42	0.6	80	trace	2	10	4	4
	72	1.8	58	1.0	76	1	3	11	4	5
20 g/L SWP	48	2.4	39	1.0	76	1	2	11	5	5
	72	3.0	46	1.4	73	1	3	12	5	6
50 g/L SWP	48	4.4	23	1.0	78	trace	2	10	6	4
	72	5.5	32	1.8	68	1	4	14	7	6

Cells were cultivated at 28 °C for 48 or 72 h in nitrogen-free MS medium containing glucose (2%) and SWP (0.1, 0.5, 1, 2 or 5%) as the sole carbon and nitrogen source, respectively. SWP, soybean waste powder; 3HB, 3-hydroxybutyrate; 3HHx, 3-hydroxyhexanoate; 3HO, 3-hydroxyoctanoate; 3HD, 3-hydroxydecanoate; 3HDD, 3hydroxydodecanoate; 3H5DD, 3-hydroxy-*cis*-5-dodecenoate.

We next investigated whether SWP could be utilized as the carbon source as well as the nitrogen source for the production of P(3HB-*co*-3HA). To achieve this, hydrolysis of SWP was carried out by two methods. First, hydrolysis of SWP was attempted at 80 °C for 5 h with 0.6 N H_2SO_4 and the hydrolysate (HSWP) was subsequently used for PHA production as both the carbon/nitrogen source. As a result, the recombinant strain accumulated P(3HB-*co*-3HA) using this HSWP (Table 3). The P(3HB-*co*-3HA) produced by the recombinant strain reached levels of 1.3 g/L using 75 g/L of HSWP. Similar to what was observed for SWP, the 3HB fraction of P(3HB-*co*-3HA) decreased with increasing concentration of HSWP. The composition ratio of scl- and mcl-monomers has a considerable influence on PHA qualities, in particular P(3HB-*co*-3HA) with too low a 3HB fraction is an amorphous polymer [15]. At a concentration of 20 to 75 g/L of HSWP, the recombinant strain accumulated 32–35 wt% PHA, which is a relatively high PHA content.

Table 3. Biosynthesis of PHA in *Pseudomonas* sp. 61-3 (*phbC::tet*) harboring pJKSc54-*phab* from HSWP (by 0.6 N H_2SO_4).

HSWP (g/L)	Dry Cell Weight (g/L)	PHA Content (wt%)	PHA Conc. (g/L)	PHA Composition (mol%)						
				3HB (C4)	3HV (C5)	3HHx (C6)	3HO (C8)	3HD (C10)	3HDD (C12)	3H5DD (C12')
10	0.6	26	0.2	88	1	trace	1	5	3	2
20	1.3	33	0.4	85	1	trace	1	6	4	3
30	2.0	35	0.7	84	1	trace	2	7	3	3
40	2.6	33	0.9	81	1	trace	2	7	5	4
50	3.1	32	1.0	80	1	trace	2	8	5	4
75	4.0	32	1.3	76	1	trace	2	10	6	5
100	4.2	21	0.9	78	1	1	3	8	6	3

Cells were cultivated at 28 °C for 48 h in carbon- and nitrogen-free MS medium containing HSWP (10, 20, 30, 40, 50, 75 or 100 g/L) as the nitrogen and the carbon sources. SWP was hydrolyzed by 0.6 N H_2SO_4 at 80 °C for 5 h and neutralized by 5 N NaOH. HSWP, Hydrolyzed soybean waste powder; 3HB, 3-hydroxybutyrate; 3HV, 3-hydroxyvalerate; 3HHx, 3-hydroxyhexanoate; 3HO, 3-hydroxyoctanoate; 3HD, 3-hydroxydecanoate; 3HDD, 3-hydroxydodecanoate; 3H5DD, 3-hydroxy-*cis*-5-dodecenoate.

Subsequently, SWP was hydrolyzed at 90 °C for 1 h with 5 N H_2SO_4 to hydrolyze it completely and the hydrolysate was used as the nitrogen and carbon source for PHA production in a similar manner as described above. The recombinant strain accumulated 26 wt% P(3HB-*co*-3HA) and produced 0.2–0.3 g/L of PHA from 10–20 g/L of HSWP (Table 4). However, PHA was barely produced at 40–50 g/L of HSWP, and the recombinant strain no longer grew at 75 g/L or more HSWP; this is likely due to the high concentration of salts formed by neutralization after hydrolysis. In conclusion, it was found that HSWP could be utilized as both the nitrogen and carbon source. In addition, the data suggested that hydrolysis of SWP using 0.6 N H_2SO_4 was better than using 5 N H_2SO_4 for the cell growth and PHA production.

Table 4. Biosynthesis of PHA in *Pseudomonas* sp. 61-3 (*phbC::tet*) harboring pJKSc54-*phab* from HSWP (by 5 N H_2SO_4).

HSWP (g/L)	Dry Cell Weight (g/L)	PHA Content (wt%)	PHA Conc. (g/L)	PHA Composition (mol%)						
				3HB (C4)	3HV (C5)	3HHx (C6)	3HO (C8)	3HD (C10)	3HDD (C12)	3H5DD (C12')
10	0.7	26	0.2	76	3	1	2	9	5	4
20	1.1	26	0.3	66	2	1	4	14	7	6
30	1.1	11	0.1	61	4	1	5	15	9	5
40	1.2	2	0.02	39	11	trace	5	21	19	5
50	0.9	2	0.02	42	16	trace	5	21	16	0
75	0	-	-	-	-	-	-	-	-	-
100	0	-	-	-	-	-	-	-	-	-

Cells were cultivated at 28 °C for 48 h in carbon- and nitrogen-free MS medium containing HSWP (10, 20, 30, 40, 50, 75 or 100 g/L) as the nitrogen and the carbon sources. SWP was hydrolyzed by 5 N H_2SO_4 at 90 °C for 1 h and neutralized by 5 N NaOH. HSWP, Hydrolyzed soybean waste powder; 3HB, 3-hydroxybutyrate; 3HV, 3-hydroxyvalerate; 3HHx, 3-hydroxyhexanoate; 3HO, 3-hydroxyoctanoate; 3HD, 3-hydroxydecanoate; 3HDD, 3-hydroxydodecanoate; 3H5DD, 3-hydroxy-*cis*-5-dodecenoate.

We attempted to produce PHA from hydrolyzed cornstarch as a carbon source to compare with SWP, using 0.05% NH$_4$Cl as the nitrogen source. First, hydrolysis of cornstarch (10 g/L) was attempted at 80 °C for 5 h with 0.6 N H$_2$SO$_4$. However, no cell growth was observed, which was attributed to the lack of time for hydrolysis or the lack of an amount of cornstarch added to the medium (Table 5). Therefore, hydrolysis of cornstarch (100 g/L) was subsequently performed for 8 h. As a result, the recombinant strain accumulated 71 wt% P(3HB-*co*-3HA) containing 86 mol% 3HB and 12 mol% 3HA (C$_6$-C$_{12}$) units from 100 g/L of hydrolyzed cornstarch (Table 5). Additionally, the recombinant strain grew well and the PHA produced reached levels of 2.9 g/L, which was the highest production level among the P(3HB-*co*-3HA)s reported so far. The recombinant strain also grew well and produced 54–63 wt% P(3HB-*co*-3HA) with high 3HB fraction (83–84 mol%) from cornstarch (14 and 20 g/L) hydrolyzed using 5 N H$_2$SO$_4$ (Table 5), whereas the cell growth and accumulated PHA content were slightly lower than the control using 2% glucose (Table 2).

Table 5. Biosynthesis of PHA in *Pseudomonas* sp. 61-3 (*phbC::tet*) harboring pJKSc54-*phab* from hydrolyzed cornstarch.

Hydrolyzed Cornstarch (g/L)	Cultivation Time (h)	Dry Cell Weight (g/L)	PHA Content (wt%)	PHA Conc. (g/L)	3HB (C4)	3HHx (C6)	3HO (C8)	3HD (C10)	3HDD (C12)	3H5DD (C12′)
10 [1]	48	0.2	2	0.004	34	0	0	25	17	24
	72	0.3	2	0.006	33	0	trace	25	15	27
100 [2]	48	2.9	57	1.7	87	trace	1	6	3	3
	72	4.1	71	2.9	86	trace	2	6	3	3
14 [3]	48	2.5	57	1.4	84	trace	2	7	4	3
	72	2.4	63	1.5	84	trace	2	7	4	3
20 [3]	48	2.5	54	1.4	83	trace	2	8	4	3
	72	2.5	55	1.4	83	trace	2	8	4	3

Cells were cultivated at 28 °C for 48 or 72 h in MS medium containing 0.05% NH$_4$Cl and hydrolyzed cornstarch (10, 14, 20 or 100 g/L) as the sole nitrogen and carbon source, respectively. [1] Cornstarch was hydrolyzed by 0.6 N H$_2$SO$_4$ at 80 °C for 5 h. [2] Cornstarch was hydrolyzed by 0.6 N H$_2$SO$_4$ at 80 °C for 8 h. [3] Cornstarch was hydrolyzed by 5 N H$_2$SO$_4$ at 90 °C for 1 h. 3HB, 3-hydroxybutyrate; 3HV, 3-hydroxyvalerate; 3HHx, 3-hydroxyhexanoate; 3HO, 3-hydroxyoctanoate; 3HD, 3-hydroxydecanoate; 3HDD, 3-hydroxydodecanoate; 3H5DD, 3-hydroxy-*cis*-5-dodecenoate.

Finally, PHA production using a mixture of SWP/HSWP and hydrolyzed cornstarch was attempted (Tables 6 and 7). The recombinant strain accumulated 64 wt% P(3HB-*co*-3HA) containing 82 mol% 3HB (0.3 g/L of PHA concentration) when 1 g/L of SWP and 13 g/L of hydrolyzed cornstarch were added to the medium as the nitrogen and the carbon sources, respectively (Table 6). From 10 g/L of SWP and 13 g/L of hydrolyzed cornstarch, the dry cell weight increased to 1.2 g/L and 0.5 g/L of PHA was obtained. However, the molar fraction of 3HB unit in the copolymer was relatively low (65 mol% 3HB). In the case where 26 g/L of hydrolyzed cornstarch was added, the dry cell weight and the PHA content decreased, resulting in a low PHA concentration (less than 0.2 g/L). When both HSWP (SWP hydrolyzed using 0.6 N H$_2$SO$_4$) and hydrolyzed cornstarch (hydrolyzed using 5 N H$_2$SO$_4$) were added to the medium, cell growth was inhibited, except when 50 g/L of HSWP only was used as the sole nitrogen and carbon source (Table 7).

Table 6. Biosynthesis of PHA in *Pseudomonas* sp. 61-3 (*phbC::tet*) harboring pJKSc54-*phab* from SWP and hydrolyzed cornstarch.

Hydrolyed Cornstarch (g/L)	SWP (g/L)	Cultivation Time (h)	Dry Cell Weight (g/L)	PHA Content (wt%)	PHA Conc. (g/L)	PHA Composition (mol%)					
						3HB (C4)	3HHx (C6)	3HO (C8)	3HD (C10)	3HDD (C12)	3H5DD (C12')
13	1	48	0.4	55	0.2	78	1	3	10	4	4
		72	0.5	64	0.3	82	1	2	8	3	4
	10	48	1.1	35	0.4	74	1	2	12	6	5
		72	1.2	44	0.5	65	1	4	16	7	7
26	1	48	0.3	46	0.1	79	1	3	9	5	3
		72	0.3	43	0.1	72	1	3	12	8	4
	10	48	0.6	15	0.1	81	trace	2	9	5	3
		72	0.7	27	0.2	73	1	3	11	7	5

Cells were cultivated at 28 °C for 48 or 72 h in nitrogen-free MS medium containing SWP and hydrolyzed cornstarch as the nitrogen and the carbon source, respectively. Cornstarch was hydrolyzed by 5 N H_2SO_4 at 90 °C for 1 h. 3HB, 3-hydroxybutyrate; 3HV, 3-hydroxyvalerate; 3HHx, 3-hydroxyhexanoate; 3HO, 3-hydroxyoctanoate; 3HD, 3-hydroxydecanoate; 3HDD, 3-hydroxydodecanoate; 3H5DD, 3-hydroxy-*cis*-5-dodecenoate.

Table 7. Biosynthesis of PHA in *Pseudomonas* sp. 61-3 (*phbC::tet*) harboring pJKSc54-*phab* from HSWP and hydrolyzed cornstarch.

HSWP (g/L)	Hydrolyzed Cornstarch (g/L)	Cultivation Time (h)	Dry Cell Weight (g/L)	PHA Content (wt%)	PHA Conc. (g/L)	PHA Composition (mol%)						
						3HB (C4)	3HV (C5)	3HHx (C6)	3HO (C8)	3HD (C10)	3HDD (C12)	3H5DD (C12')
5	20	48	0.5	47	0.2	80	1	trace	2	9	4	4
		72	0.6	56	0.3	75	1	1	3	11	5	4
	40	48	0.2	20	0.04	73	5	1	4	9	6	2
		72	0.2	22	0.04	73	3	1	4	9	7	3
10	10	48	1.0	51	0.5	79	1	1	2	9	4	4
	20	48	0.7	42	0.3	78	2	trace	2	9	5	4
50	0	48	2.2	34	0.7	85	1	trace	1	6	4	3
		72	2.3	39	0.9	84	1	trace	2	7	3	3
	15	48	0.04	0	0	0	0	0	0	trace	trace	0
		72	1.2	15	0.2	86	trace	trace	1	6	5	2
	30	48	0.02	1	trace	0	0	trace	0	100	0	0
		72	0.02	0	0	0	0	0	0	0	0	0

Cells were cultivated at 28 °C for 48 or 72 h in nitrogen-free MS medium containing the HSWP (5, 10 or 50 g/L) and cornstarch (0, 15, 20, 30 or 40 g/L) as the carbon and the nitrogen sources. Cornstarch was hydrolyzed by 5 N H_2SO_4. SWP was hydrolyzed by 0.6 N H_2SO_4. HSWP, Hydrolyzed soybean waste powder; 3HB, 3-hydroxybutyrate; 3HV, 3-hydroxyvalerate; 3HHx, 3-hydroxyhexanoate; 3HO, 3-hydroxyoctanoate; 3HD, 3-hydroxydecanoate; 3HDD, 3-hydroxydodecanoate; 3H5DD, 3-hydroxy-*cis*-5-dodecenoate.

4. Discussion

In order to effectively produce PHAs, they should be produced from inexpensive carbon sources such as waste substrates. For this reason, we elected to focus on the use of steamed soybean wastewater generated as a by-product of the soybean processing industry. In this study, P(3HB-*co*-3HA) was synthesized from SWP, HSWP and/or hydrolyzed cornstarch as the nitrogen and/or the carbon sources using *Pseudomonas* sp. 61-3 (*phbC::tet*) harboring the *phaC1* gene from *Pseudomonas* sp. 61-3 and *phbAB* genes from C. necator. *Pseudomonas* sp. 61-3 (*phbC::tet*) harboring pJKSc54-*phab* accumulated the P(3HB-*co*-3HA) copolymer from SWP and glucose as the sole nitrogen and carbon sources, respectively, but the 3HB fraction in the copolymer decreased at the amount of added SWP increased. This would be responsible for the expression level of PHA synthase gene (*phaC1*) of *Pseudomonas* sp. 61-3. The additional copies of the *phaC1* gene have been previously reported to result in an increase in the 3HB fraction in the copolymer synthesized when glucose was used as the sole carbon source [5]. This is due to the low substrate specificity of PhaC1 for (*R*)-3HB-CoA. PhaC1 synthase has been reported to have the highest activity toward (*R*)-3-hydroxydecanoate (3HD)-CoA among the C_4-C_{12} substrates [25]. Therefore, a decrease in the expression level of the *phaC1* gene leads to an increase in the 3HA fraction, especially with the 3HD unit being the main component, in the copolymer synthesized through *de novo* fatty acid synthesis pathway when unrelated carbon sources such as glucose were used. To obtain a higher 3HB composition in the copolymer, additional copies of *phaC1* are required, together with the *phbAB* genes, since PHA synthase activity has been reported to affect monomer composition in the copolymer as well as monomer supply by PhbA and PhbB when sugars were used as the sole carbon source [5]. In contrast, when fatty acids were used as carbon source, the 3HO (3-hydroxyoctanoate) fraction increased in the copolymer synthesized by a recombinant *Pseudomonas* sp. 61-3 (*phbC::tet*) strain carrying an additional *phaC1* gene compared with the strain containing only the vector [5]. The *phaC1* gene would be expected to be expressed under nitrogen-limited conditions. A sequence resembling the consensus sequence of the *Escherichia coli* σ^{54}-dependent promoter involved in expression under nitrogen-limited conditions has been found upstream of the *phaC1* gene [16] and P(3HB-*co*-3HA) was synthesized by *Pseudomonas* sp. 61-3 only under nitrogen-limited conditions [14]. When the initial molar ratio of nitrogen and carbon sources (C/N) was low, the copolymer content was also low [5]. The content (45 wt%) and the concentration (1.13 g/L) of PHA accumulated by this recombinant strain with a C/N molar ratio of 71 were the highest obtained, as reported in a previous study [5]. Thus, limitation of nitrogen source is necessary for the biosynthesis of P(3HB-*co*-3HA) by recombinant strains of *Pseudomonas* sp. 61-3, although nitrogen is essential for bacterial growth. In order to synthesize P(3HB-*co*-3HA) with a high 3HB fraction, similarly, a high expression level of *phaC1* is required. In fact, introduction of only the *phaC1* gene into *Pseudomonas* sp. 61-3 (*phbC::tet*) increased the 3HB fraction in P(3HB-*co*-3HA) from 27 mol% to 55 mol% [5]. As the amount of SWP added to the medium increased, that is, as the C/N molar ratio decreased, the 3HB fraction in the copolymer decreased (Table 2). This indicates that SWP can be utilized as a nitrogen source for PHA production by the recombinant *Pseudomonas* sp. 61-3 strain.

In this study, we also attempted to use HSWP for PHA production. Acid hydrolysis of SWP was attempted using two methods; either 0.6 N or 5 N H_2SO_4. As a result, 20–75 g/L of HSWP prepared using 0.6 N H_2SO_4, resulted in the synthesis of 32–35 wt% P(3HB-*co*-3HA) at the levels of 0.4–1.3 g/L of PHA (Table 3). However, the 3HB fraction in the copolymer decreased with increasing HSWP concentration, presumably due to the low C/N molar ratio as described previously [5]. With regard to PHA concentration and the monomer composition of the P(3HB-*co*-3HA) copolymer synthesized by the recombinant strain, 50 g/L of HSWP appeared to be the optical concentration for PHA production (1.0 g/L of PHA). In addition, the 3HB fraction in this copolymer was relatively high (80 mol%), which suggests that it would be expected to have good mechanical properties. The PHA production yield under the culture condition used here was estimated to be 20 mg-PHA/g-(H)SWP, which equates to approximately 1.0 g-PHA per liter of soybean wastewater. On the other hand, when HSWP, prepared using 5 N H_2SO_4, was used as both the nitrogen and carbon sources, the recombinant strain accumulated 26 wt% P(3HB-*co*-3HA) from 10–20 g/L of HSWP and the PHA concentration was 0.2–0.3 g/L (Table 4). Furthermore, the addition of more than 40 g/L of HSWP to the medium inhibited cell growth. This is probably due to the high concentration of salts formed by neutralization following hydrolysis. Thus, we found that both HSWP prepared using either 0.6 N or 5 N H_2SO_4 could be utilized as a nitrogen and carbon source, and HSWP prepared by hydrolysis with 0.6 N H_2SO_4 was suitable for cell growth and PHA production.

We also investigated PHA production using hydrolyzed cornstarch as a carbon source. As a result, 2.9 g/L of P(3HB-*co*-3HA) containing 86 mol% 3HB unit could be produced by the recombinant strain from 100 g/L of hydrolyzed cornstarch (Table 5).

Based on these data, we concluded that SWP and HSWP did not contain sufficient carbon to produce PHA, since the PHA contents (<35 wt%) obtained using HSWP were lower than the PHA content under control conditions (glucose and NH_4Cl) (Tables 2 and 3). Therefore, we attempted to add a mixture of SWP/HSWP and hydrolyzed cornstarch to the medium for PHA production (Tables 6 and 7). After 72 h of cultivation, the recombinant strain accumulated 64 wt% P(3HB-*co*-3HA) containing 82 mol% 3HB unit from 10 g/L of SWP and 13 g/L of hydrolyzed cornstarch as nitrogen and carbon sources, respectively (Table 6). However, the dry cell weights (less than 0.7 g/L) and PHA concentration (less than 0.2 g/L) decreased when 26 g/L of hydrolyzed cornstarch was used. In the case when both HSWP, prepared using 0.6 N H_2SO_4, hydrolyzed cornstarch were added to the medium, cell growth was inhibited with increasing concentrations of hydrolyzed cornstarch

(Table 7). The inhibition of cell growth is likely caused by the high concentration of salts formed by neutralization after hydrolysis. Therefore, the HSWP and hydrolyzed cornstarch should be desalted before use in PHA production. One alternative solution to this salt problem maybe through the use of enzymatic hydrolysis of SWP and cornstrach instead of acid hydrolysis treatment. Interestingly, the 3-hydroxyvalerate (3HV) unit was detected in the copolymer by GC analysis when HSWP was added to the medium (Tables 3, 4 and 7). Hydrolysis of SWP may produce a substrate (e.g., fatty acids with odd numbers of carbon atoms) leading to the supply of the 3HV monomer. The reason for this remains unclear.

In conclusion, P(3HB-*co*-3HA) with various monomer compositions could be synthesized from SWP, HSWP, and/or hydrolyzed cornstarch as nitrogen and/or carbon sources in this study. However, the production efficiency was found to be unsatisfactory. Although SWP could be used as a nitrogen source, the PHA concentration was less than 1.0 g/L (less than 35 wt% PHA) when only HSWP was added to the medium as both a carbon and nitrogen source (Table 7). This suggests that SWP/HSWP contains sufficient nitrogen for the recombinant *Pseudomonas* sp. 61-3 strain to produce PHA, but it is deficient as a carbon source. Further improvements will therefore be necessary to achieve effective PHA production from SWP/HSWP. Since treatment of steamed soybean wastewater by a treatment facility, for example an activated sludge, is expensive, effective utilization of this wastewater is required. The production of PHA reported here is one proposed use of this wastewater. In the future, the molecular weight and mechanical properties of the copolymer synthesized should be further investigated since the molecular weight of the polymer affects its mechanical properties in addition to the monomer composition. The utilization of waste substrates, such as steamed soybean wastewater as a nitrogen and the carbon source, could contribute significantly to reducing the costs of PHA production as well as reducing the cost of waste treatment while at the same time promoting environmental conservation.

Acknowledgments: We would like to thank Editage (www.editage.jp) for English language editing.

Author Contributions: Ayaka Hokamura analyzed data and wrote about 70% of the articles. Yuko Yunoue designed and performed the experiments, and analyzed data. Saki Goto analyzed data and wrote about 10% of the articles. Hiromi Matsusaki supervised the work and wrote about 20% of the articles.

Conflicts of Interest: The authors declare no conflict of interest.

References

1. Anderson, A.J.; Dawes, E.A. Occurrence, metabolism, metabolic role, and industrial uses of bacterial polyhydroxyalkanoates. *Microbiol. Rev.* **1990**, *54*, 450–472. [PubMed]
2. Müller, H.M.; Seebach, D. Poly(hydroxyalkanoates): A fifth class of physiologically important organic biopolymers? *Angew. Chem. Int. Ed. Engl.* **1993**, *32*, 477–502. [CrossRef]
3. Madison, L.L.; Huisman, G.W. Metabolic engineering of poly(3-hydroxyalkanoates): from DNA to plastic. *Microbiol Mol. Biol. Rev.* **1999**, *63*, 21–53. [PubMed]
4. Nomura, C.T.; Tanaka, T.; Eguen, T.E.; Appah, A.S.; Matsumoto, K.; Taguchi, S.; Ortiz, C.L.; Doi, Y. FabG mediates polyhydroxyalkanoate production from both related and nonrelated carbon sources in recombinant *Escherichia coli* LS5218. *Biotechnol. Prog.* **2008**, *24*, 342–351. [CrossRef] [PubMed]
5. Matsusaki, H.; Abe, H.; Doi, Y. Biosynthesis and properties of poly(3-hydroxybutyrate-*co*-3-hydroxyalkanoate) by recombinant strains of *Pseudomonas* sp. 61-3. *Biomacromolecules* **2000**, *1*, 17–22. [CrossRef] [PubMed]
6. Fukui, T.; Doi, Y. Efficient production of polyhydroxyalkanoates from plant oils by *Alcaligenes eutrophus* and its recombinant strain. *Appl. Microbiol. Biotechnol.* **1998**, *49*, 333–336. [CrossRef] [PubMed]
7. Wong, H.H.; Lee, S.Y. Poly-(3-hydroxybutyrate) production from whey by high-density cultivation of recombinant *Escherichia coli*. *Appl. Microbiol. Biotechnol.* **1998**, *50*, 30–33. [CrossRef] [PubMed]
8. Obruca, S.; Benesova, P.; Kucera, D.; Petrik, S.; Marova, I. Biotechnological conversion of spent coffee grounds into polyhydroxyalkanoates and carotenoids. *New Biotechnol.* **2015**, *32*, 569–574. [CrossRef] [PubMed]

9. Alsafadi, D.; Al-Mashaqbeh, O. A one-stage cultivation process for the production of poly-3-(hydroxybutyrate-co-hydroxyvalerate) from olive mill wastewater by *Haloferax mediterranei*. *New Biotechnol.* **2017**, *34*, 47–53. [CrossRef] [PubMed]
10. Poomipuk, N.; Reungsang, A.; Plangklang, P. Poly-β-hydroxyalkanoates production from cassava starch hydrolysate by *Cupriavidus* sp. KKU38. *Int. J. Biol. Macromol.* **2014**, *65*, 51–64. [CrossRef] [PubMed]
11. Valentino, F.; Morgan-Sagastume, F.; Campanari, S.; Villano, M.; Werker, A.; Majone, M. Carbon recovery from wastewater through bioconversion into biodegradable polymers. *New Biotechnol.* **2017**, *37*, 9–23. [CrossRef] [PubMed]
12. Koller, M.; Maršálek, L.; de Sousa Dias, M.M.; Braunegg, G. Producing microbial polyhydroxyalkanoate (PHA) biopolyesters in a sustainable manner. *New Biotechnol.* **2017**, *37*, 24–38. [CrossRef] [PubMed]
13. Zhu, C.; Chiu, S.; Nakas, J.P.; Nomura, C.T. Bioplastics from waste glycerol derived from biodiesel industry. *J. Appl. Polym. Sci.* **2013**, *130*, 1–13. [CrossRef]
14. Kato, M.; Bao, H.J.; Kang, C.K.; Fukui, T.; Doi, Y. Production of a novel copolyester of 3-hydroxybutyric acid and medium-chain-length 3-hydroxyalkanoic acids by *Pseudomonas* sp. 61-3. *Appl. Microbiol. Biotechnol.* **1996**, *45*, 363–370. [CrossRef]
15. Kato, M.; Fukui, T.; Doi, Y. Biosynthesis of polyester blends by *Pseudomonas* sp. 61-3 from alkanoic acids. *Bull. Chem. Soc. Jpn.* **1996**, *69*, 515–520. [CrossRef]
16. Matsusaki, H.; Manji, S.; Taguchi, K.; Kato, M.; Fukui, T.; Doi, Y. Cloning and molecular analysis of the poly(3-hydroxybutyrate) and poly(3-hydroxybutyrate-co-3-hydroxyalkanoate) biosynthesis genes in *Pseudomonas* sp. strain 61-3. *J. Bacteriol.* **1998**, *180*, 6459–6467. [PubMed]
17. Matsumoto, K.; Matsusaki, H.; Taguchi, K.; Seki, M.; Doi, Y. Isolation and characterization of polyhydroxyalkanoates inclusions and their associated proteins in *Pseudomonas* sp. strain 61-3. *Biomacromolecules* **2002**, *3*, 787–792. [CrossRef] [PubMed]
18. Kimura, I.; Matsubara, Y.; Shimasaki, H. Utilization of waste water from soy-bean cooking (in Japanese). *J. Brew. Jpn.* **1997**, *7*, 478–485. [CrossRef]
19. Matsubara, Y.; Iwasaki, K.; Nakajima, M.; Nabetani, H.; Nakao, S. Recovery of oligosaccharides from steamed soybean waste water in *tofu* processing by reverse osmosis and nanofiltration membranes. *Biosci. Biotech. Biochem.* **1996**, *60*, 421–428. [CrossRef] [PubMed]
20. Liu, W.; Zhang, H.X.; Wu, Z.L.; Wang, Y.J.; Wang, L.J. Recovery of isoflavone aglycones from soy whey wastewater using foam fractionation and acidic hydrolysis. *J. Agric. Food Chem.* **2013**, *61*, 7366–7372. [CrossRef] [PubMed]
21. Hongyang, S.; Yalei, Z.; Chunmin, Z.; Xuefei, Z.; Jinpeng, L. Cultivation of *Chlorella pyrenoidosa* in soybean processing wastewater. *Bioresour. Technol.* **2011**, *102*, 9884–9890. [CrossRef] [PubMed]
22. Zhu, G.-F.; Li, J.-Z.; Liu, C.-X. Fermentative hydrogen production from soybean protein processing wastewater in an anaerobic baffled reactor (ABR) using anaerobic mixed consortia. *Appl. Biochem. Biotechnol.* **2012**, *168*, 91–105. [CrossRef] [PubMed]
23. Zhu, J.; Zheng, Y.; Xu, F.; Li, Y. Solid-state anaerobic co-digestion of hay and soybean processing waste for biogas production. *Bioresour. Technol.* **2014**, *154*, 240–247. [CrossRef] [PubMed]
24. Liu, S.; Zhang, G.; Zhang, J.; Li, X.; Li, J. Performance, 5-aminolevulic acid (ALA) yield and microbial population dynamics in a photobioreactor system treating soybean wastewater: Effect of hydraulic retention time (HRT) and organic loading rate (OLR). *Bioresour. Technol.* **2016**, *210*, 146–152. [CrossRef] [PubMed]
25. Takase, K.; Matsumoto, K.; Taguchi, S.; Doi, Y. Alteration of substrate chain-length specificity of type II synthase for polyhydroxyalkanoate biosynthesis by in vitro evolution: in vivo and in vitro enzyme assays. *Biomacromolecules* **2004**, *5*, 480–485. [CrossRef] [PubMed]

bioengineering

MDPI

Article

The Molecular Level Characterization of Biodegradable Polymers Originated from Polyethylene Using Non-Oxygenated Polyethylene Wax as a Carbon Source for Polyhydroxyalkanoate Production

Brian Johnston [1,*], Guozhan Jiang [1], David Hill [1], Grazyna Adamus [2], Iwona Kwiecień [2], Magdalena Zięba [2], Wanda Sikorska [2], Matthew Green [3], Marek Kowalczuk [1,2] and Iza Radecka [1,*]

[1] Wolverhampton School of Biology, Chemistry and Forensic Science, Faculty of Science and Engineering, University of Wolverhampton, Wolverhampton WV1 1LY, UK; Guozhan.jiang@wlv.ac.uk (G.J.); D.Hill@wlv.ac.uk (D.H.); M.Kowalczuk@wlv.ac.uk (M.K.)
[2] Centre of Polymer and Carbon Materials, Polish Academy of Sciences, 41-800 Zabrze, Poland; Grazyna.Adamus@cmpw-pan.edu.pl (G.A.); ikwiecien@cmpw-pan.edu.pl (I.K.); mzieba@cmpw-pan.edu.pl (M.Z.); wsikorska@cmpw-pan.edu.pl (W.S.)
[3] Recycling Technologies Ltd., South Marston Industrial Park, Swindon SN3 4WA, UK; Matt.Green@recyclingtechnologies.co.uk
* Correspondence: B.Johnston@wlv.ac.uk (B.J.); I.Radecka@wlv.ac.uk (I.R.); Tel.: +44-1902-322-366 (B.J. & I.R.)

Academic Editor: Martin Koller
Received: 2 August 2017; Accepted: 24 August 2017; Published: 28 August 2017

Abstract: There is an increasing demand for bio-based polymers that are developed from recycled materials. The production of biodegradable polymers can include bio-technological (utilizing microorganisms or enzymes) or chemical synthesis procedures. This report demonstrates the corroboration of the molecular structure of polyhydroxyalkanoates (PHAs) obtained by the conversion of waste polyethylene (PE) via non-oxygenated PE wax (N-PEW) as an additional carbon source for a bacterial species. The N-PEW, obtained from a PE pyrolysis reaction, has been found to be a beneficial carbon source for PHA production with *Cupriavidus necator* H16. The production of the N-PEW is an alternative to oxidized polyethylene wax (O-PEW) (that has been used as a carbon source previously) as it is less time consuming to manufacture and offers fewer industrial applications. A range of molecular structural analytical techniques were performed on the PHAs obtained; which included nuclear magnetic resonance (NMR) and electrospray ionisation tandem mass spectrometry (ESI-MS/MS). Our study showed that the PHA formed from N-PEW contained 3-hydroxybutyrate (HB) with 11 mol% of 3-hydroxyvalerate (HV) units.

Keywords: polyhydroxyalkanoates; PHAs; non-oxidized PE wax; N-PEW; *Cupriavidus necator* H16

1. Introduction

Polyhydroxyalkanoates (PHAs) are a family of polyhydroxyesters with 3-, 4-, 5-, and 6-hydroxyalkanoic acids that are biodegradable, non-toxic, biocompatible organic polyesters synthesized by some species of bacteria [1]. Currently there is a great deal of interest in PHAs due to their far-reaching applications, which include being used as biological implants, pharmacological delivery systems, packaging materials, and many more, as they can provide unique properties that do not exist in some synthetic polymers [1]. Traditional plastics, however, originate from petrochemical sources and have been a vital business commodity since their first industrial production. They are lightweight, cheap to manufacture, strong, flexible, and adaptable, but petro-based polymers form a growing proportion of the municipal solid waste (MSW) produced world-wide annually [2] and their waste materials are extremely harmful on a microscopic scale. These persistent plastics infiltrate the food chain as they are mechanically broken down (but not completely) in our oceans. Microplastics, defined as fragments that are less than five millimetres in diameter, can range from the size of an insect down to the size of a virus [3]. These particles, and larger plastics, also have a deadly impact on sea-life through entanglement or ingestion. It has been stated that plastics now account for somewhere between 60 and 80% of all ocean and shoreline debris [3–6]. The challenge, for both economic and environmental reasons, is to develop novel strategies that can create alternative materials to match or surpass the physical and commercial appeal of current petro-based plastic, with no adverse environmental consequences.

Waste polyethylene (PE) is a potential carbon source that could be utilized to make value-added biopolymers, particularly as it is the most commonly-produced plastic, making up over 29% of worldwide plastic manufacture, while only 10% of it is recycled [7–10]. Unfortunately, the high molecular weight, hydrophobicity, and chemically-stable hydrocarbon covalent bonds of PE make it difficult for bacteria to metabolize effectively [11–15]. However, in the melt-form after oxidative degradation with free radical initiators and then sonication with an additional fermentation medium, PE waxes can be metabolized by bacteria and PHAs can be produced [16,17]. These polar PE waxes already have commercial uses; they can be used to produce PVC, as components of aqueous emulsions, lubricants, additives for varnishes, road marking paint, and adhesives [16,18–20]. The alternative non-polar, non-oxidized PE waxes (N-PEWs) have fewer industrial uses [7]. They are also cheaper and relatively easier to produce, which suggests that N-PEWs provide an alternative carbon source to O-PEWs.

We propose that N-PEWs could be a pragmatic application of waste PE as a carbon source for PHA synthesis using *Cupriavidus necator* H16. The pyrolysis of PE samples in anaerobic conditions yields a complex mixture of relatively low molecular weight hydrocarbons with a molecular mass of 200–1000 [7,16]. Due to the hydrophobic nature of the non-oxidized waxes, ultrasound sonication can be performed to breakdown and distribute the wax as an emulsion so that it will be readily metabolized by bacteria [16,21]. For alkane degradation (with oxygen), the most widely-used pathway is the oxidation of the terminal methyl group into a carboxylic acid, via an alcohol intermediate, followed by complete mineralization through β-oxidation [15,22]. The intercellularly-synthesized PHAs are then held within bacterial species like *Cupriavidus necator* (formerly *Ralstonia eutropha* or *Alcaligenes eutropha*) as granules. *C. necator* is a Gram-negative, hydrogen-oxidizing bacterium ("knallgas") that is able to grow at the interface of anaerobic and aerobic environments [12]. These microbes can synthesize PHA in the presence of excess amounts of carbon and limited nutrients [11] and they have been shown to metabolize a variety of carbon sources including fatty acids, hemicellulose, crude glycerol, methane, and even liquefied wood, for PHA production [11,23–25]. They are also able to accumulate up to 85% PHA per dry cell weight, often with 8–12 biopolymer granules per bacterium [16]. To extract the PHAs held within *C. necator*, their biomass is lyophilized and then hot solvent extraction is used, followed by precipitation in ethanol or hexane [16,23].

Here, we report on the production and molecular-level structural analysis of PHAs formed using *C. necator*, with non-oxidized wax as a carbon source in a nitrogen-rich tryptone soya broth (TSB) growth media. This is a novel use of N-PEWs with the aid of *C. necator*, which was selected due to its yield potential, limited resistance to copper metals (a possible contaminant from some wax production methods), growth rate at relatively low temperature, and its well-documented genetic profile and gene stability (for future modification and enhancement purposes) [16,25–27].

2. Materials and Methods

2.1. Microorganism

The bacterial species used for PHA production with N-PEWs and TSB or basal salts medium (BSM), was *C. necator* H16 (NCIMB 10442, ATCC 17699). This organism was obtained from the

University of Wolverhampton stock culture (freeze-dried and kept at $-20\,°C$). Previous to the study, cultures were revived and grown overnight at $30\,°C$ (optimum) in TSB at 150 rpm. The microorganism was then sub-cultured on tryptone soya agar (TSA) plates and incubated at $30\,°C$ for 24 h.

2.2. Carbon Source and Chemicals

The non-oxidized wax was kindly provided by Recycling Technologies Ltd. (Swindon, UK) from waste plastics that were scanned and separated from contaminants (such as glass or stone) and then shredded and dried before chemical treatment [28]. These remnants were then passed to a thermal cracker, where the long hydrocarbon chains in the plastics were cracked into shorter chains. The hot hydrocarbon vapour leaving the reactor was then filtered and treated for impurities. At this stage the refined gas was condensed into a wax [28]. The wax was used as received without any further purification (more N-PEW structural details are found in Section 3.1).

2.3. Media

TSB and TSA were purchased from Lab M Ltd. (Lancashire, UK). Both of these growth media were prepared under aseptic conditions, using the instructions of the manufacturer. The basal salts medium (BSM) used contained distilled water, 1 g/L K_2HPO_4, 1 g/L KH_2PO_4, 1 g/L KNO_3, 1 g/L $(NH_4)_2SO_4$, 0.1 g/L $MgSO_4 \cdot 7H_2O$, 0.1 g/L NaCl, 10 mL/L trace elements solution. The trace element solution contained: 2 mg/L $CaCl_2$, 2 mg/L $CuSO_4 \cdot 5H_2O$, 2 mg/L $MnSO_4 \cdot 5H_2O$, 2 mg/L $ZnSO_4 \cdot 5H_2O$, 2 mg/L $FeSO_4$, 2 mg/L $(NH_4)_6Mo_7O_{24} \cdot 4H_2O$. BSM salts were purchased from BDH Chemicals Ltd. (Poole, UK). Ringer's solution was also purchased from Lab M Ltd. (Lancashire, UK). A 1/4 strength tablet was used in 500 mL of deionized water; then it was completely dissolved. All media were sterilized by being autoclaved at standard conditions ($121\,°C$ for 15 min).

2.4. Fermentation Procedure

Starter cultures were prepared using 20 mL TSB (in a 50 mL flask) inoculated with a single *C. necator* colony from a TSA spread plate. That culture was then incubated (aerobically) for 24 h at $30\,°C$ and 150 rpm in a rotary incubator (Incu-Shake MIDI, Shropshire, UK). After 24 h these cultures were checked for contamination by Gram staining and microscope observation.

Shake flask fermentation analysis was performed in triplicate using 500 mL wide neck Erlenmeyer flasks. Each flask used 1 g of N-PEW that was first put into a 50 mL beaker, covered with foil, and melted at $70\,°C$. Then 20 mL of sterile TSB or BSM was added to the melted wax, which caused the wax to become solid again. The temperature was then increased to re-melt the wax. The waxes were then sonicated (in TSB) for 8 min at 0.5 active and passive intervals with a power of 70% using a Bandelin Electronic sonicator, (Berlin, Germany) to form a TSB/wax emulsion or BSM/wax emulsion. This emulsion was tested for sterility by spread plating. The sterile emulsion was then added to 210 mL sterile TSB or BSM in a 500 mL flask, followed by 20 mL of the starter culture, giving a total volume of 250 mL. The experimental control was 230 mL TSB or BSM, inoculated with 20 mL of starter culture with no PE wax included. The flasks were incubated in a rotary incubator under the same conditions mentioned for 48 h.

Viable cell counts were done using the methodology of Miles and Misra [29]. Briefly, 5 mL samples were aseptically collected from the flask cultures at 0, 3, 24, and 48 h. Serial dilutions of each sample were performed to 10^{-8} and then 20 µL of each dilution was pipetted onto a standard TSA plate in triplicate. These plates were incubated at $30\,°C$ for 48 h and the colonies were counted and expressed in $\log_{10}$ CFU mL^{-1}.

2.5. PHA Extraction Procedure

PHA extraction was performed after the 48 h fermentation time period had elapsed. The flasks were removed from their incubators and the flask contents were filtered using a sieve (to remove any conjugates of wax) and separated into centrifuge tubes, usually 35–40 mL per tube. The media was then centrifuged in a Sigma 6-16KS centrifuge for 10 min at 4500 rpm. At this point, there was only the pellet containing the biomass and the emulsion layer visible. The supernatant in each tube was discarded and the biomass was obtained and frozen overnight a −20 °C. This was followed by lyophilization using an Edwards freeze-drier (Modulyo, Crawley, UK) for 48 h at a temperature of −40 °C and at a pressure of 5 MBAR. The dry biomass was then weighed and recorded as cell dry weight (CDW) before being placed into an extraction thimble and, using Soxhlet extraction with HPLC grade chloroform for 48 h, the PHA was collected as a chloroform/biopolymer mixture. Then rotation evaporation was used (at 50 °C) to remove the chloroform. Polymer precipitation using n-hexane was performed in a round-bottom flask and then this was further separated by filtration (Watman No. 1 paper). If required, hexane was used to rinse the product further to remove any residue low molecular weight wax. The sample was then left in a fume cupboard to dry for up to five hours and the yield was recorded using:

$$\text{Percentage yield of PHA} = (\text{weight of extracted polymer})/(\text{cell dry weight}) \times 100$$

2.6. Characterization

2.6.1. FTIR

Fourier transform infrared spectroscopy (FTIR) was performed using a DuraScope (Genesis II) spectrophotometer (Smiths Detection Inc., Danbury, CT, USA). After a background scan, samples of N-PEW were placed under the lens. FTIR spectra were recorded in the transmittance mode with a resolution of 1 cm^{-1} in the range of 4000–400 cm^{-1}.

2.6.2. GPC

Gel permeation chromatography (GPC) was used to find the molecular weight and molecular weight distribution of the wax and polymers produced during this study. The analysis was conducted using a TOSOH EcoSec HLC/GPC 8320 (Tosoh Bioscience, Yamaguchi, Japan) system equipped with a RI and a UV detector (Tosoh Bioscience, Yamaguchi, Japan) operating at a temperature of 40 °C. The column used was TSKgel HZM-N (Tosoh Bioscience, Yamaguchi, Japan), calibrated against polystyrene standards. The UV detector was set at a wavelength of 254 nm. Chloroform was used as the eluent, at a flow rate of 0.25 mL/min. A sample size of 2 μL was injected into the system using an autosampler.

The wax GPC analysis included: (i) wax pre-sonication; (ii) post-sonication; (iii) post-fermentation (wax and TSB) and the control; (iv) post-shake flask N-PEW in TSB. The number-average molar mass (Mn) and the molecular mass distribution index (Mw/Mn) of the N-PEWs were determined in a CHCl$_3$ solution at 35–60 °C heating to melt the waxes, then 10 min of cooling, to ascertain whether the waxes were fully dissolved in the selected volume of HPLC-grade chloroform. This N-PEW/chloroform solution was carefully pipetted into a syringe and passed through a filter into a GPC ampoule.

2.6.3. NMR Analysis

Proton nuclear magnetic resonance (^{1}H-NMR) spectra were recorded with a Bruker Avance II (Bruker, Rheinstetten, Germany) operating at 600 MHz, with 64 scans, 2.65 s acquisition time, and an 11 μs pulse width. ^{13}C-NMR spectra were recorded with a Bruker Avance II operating at 150.9 MHz, with 20,480 scans, 0.9088 s acquisition times, and 9.40 μs pulse width. The ^{1}H-NMR and ^{13}C-NMR spectra were run in CDCl$_3$ at room temperature with tetramethylsilane (TMS) as an internal standard.

2.6.4. ESI-MS/MS Analysis and Identification of PHA at the Molecular Level

Electrospray mass spectrometry analysis was performed using a Finnigan LCQ ion trap mass spectrometer (Thermo Finnigan LCQ Fleet, San Jose, CA, USA). The oligomer samples, prepared as described by Kawalec et al. [30], were dissolved in a chloroform/methanol system (1:1 v/v), and the solutions were introduced into the ESI source by continuous infusion using the instrument syringe pump at a rate of 5 µL/min. The LCQ ESI source was operated at 4.5 kV, and the capillary heater was set to 200 °C; the nebulizing gas applied was nitrogen. For ESI-MS/MS experiments, the ions of interest were isolated monoisotopically in the ion trap and were activated by collisions. The helium damping gas that was present in the mass analyser acted as a collision gas. The analysis was performed in the positive-ion mode.

PHA with low molar mass was obtained via thermal degradation of bacterial PHA in the presence of potassium hydrogen carbonate ($KHCO_3$). The respective amount of polyester and salt, at a ratio PHA/$KHCO_3$ equal to 5, was introduced in a vial with ethanol. The mixture was then stirred for 30 min to provide a homogeneous suspension. The experiment was carried out in an oven at 180 °C. The length of time for the process was dependent on the molar mass plain PHA. For the PHAs with a mass bellow 100,000 it was 20 min, and for polymers with a mass between 200,000 and 400,000, it was 30 min. This time regime allowed for the obtaining of oligomers with a molar mass of around 2000. After the experiment, oligomers were protonated with the Dowex® 50WX4 in hydrogen form. The oligomer and Dowex® (1:1 wt%) was dissolved in chloroform and the mixture was stirred for four hours. Next, the Dowex® was removed by filtration. The oligomers obtained were characterized by GPC and ^{1}H-NMR spectrometry, as well as their structure at a molecular level using ESI-MS/MS. The ^{1}H-NMR analysis revealed the presence of characteristic signals corresponding to the protons of 3-hydroxybutyrate (HB) (and 3-hydroxyvalerate (HV)) repeating units, and also the signals attributed to the crotonate end groups.

3. Results

3.1. N-PEW Initial FTIR and NMR Analysis

The non-oxidized wax obtained for this study was tested to confirm its structure and chemical properties. The waxes were produced from waste plastic underwent thermal cracking to produce a refined gas that was then condensed into a wax [28]. FTIR was performed on O-PEW and N-PEW for comparison and to determine the presence or absence of functional groups associated with oxidation (Figure 1).

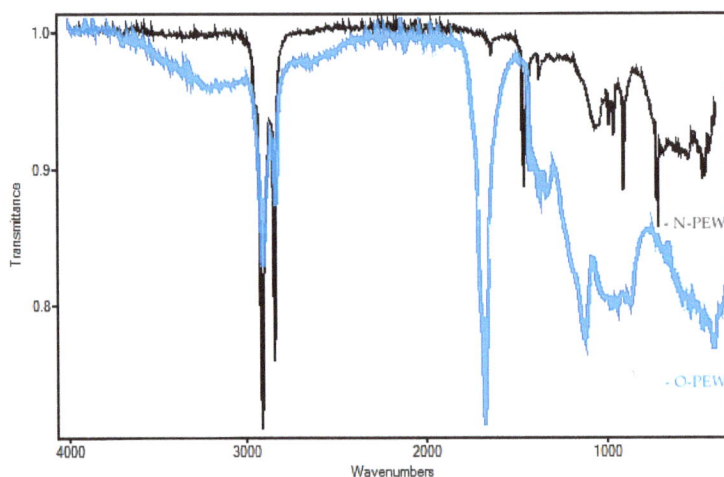

Figure 1. FTIR of O-PEW (blue trace) and N-PEW (black trace) used in this study.

The N-PEW (black) spectrum showed a strong absorption at 2800 cm^{-1}, an indication of a C-H stretch, which confirms saturated chains. There was no absorbance at 3500 cm^{-1} (C=O stretch overtone) and weak absorbance at 1790–1700 cm^{-1} (C=O stretch), as compared with the spectrum of the oxidized wax, further confirmation of hydrocarbon chains with a very small degree of oxidation. When these results are compared to the O-PEW (blue) reading (a wax with an acid number of 197), there was less absorbance at 2800 cm^{-1} than what is displayed in the N-PEW reading. There was also a greater absorbance at 1700 cm^{-1} and broad absorbencies at 1160–1060 cm^{-1} which are likely due to R-O-R aliphatic bonds.

The structure of the non-oxidized PE wax was further studied using ^{1}H-NMR and ^{13}C-NMR, as shown in Figure 2a,b. In the ^{1}H-NMR, there are strong signals at 0.88 and 1.25 ppm, which are attributed to CH$_3$ and CH$_2$. The signals at 1.6 and 2.0 ppm may be due to the allylic hydrogen or hydrogen adjacent to the C=O group. There are signals at 4.9 ppm, 5.4 ppm, and 5.8 ppm in the ^{1}H spectrum, and there are signals at 114 ppm and 139 ppm in the ^{13}C spectrum.

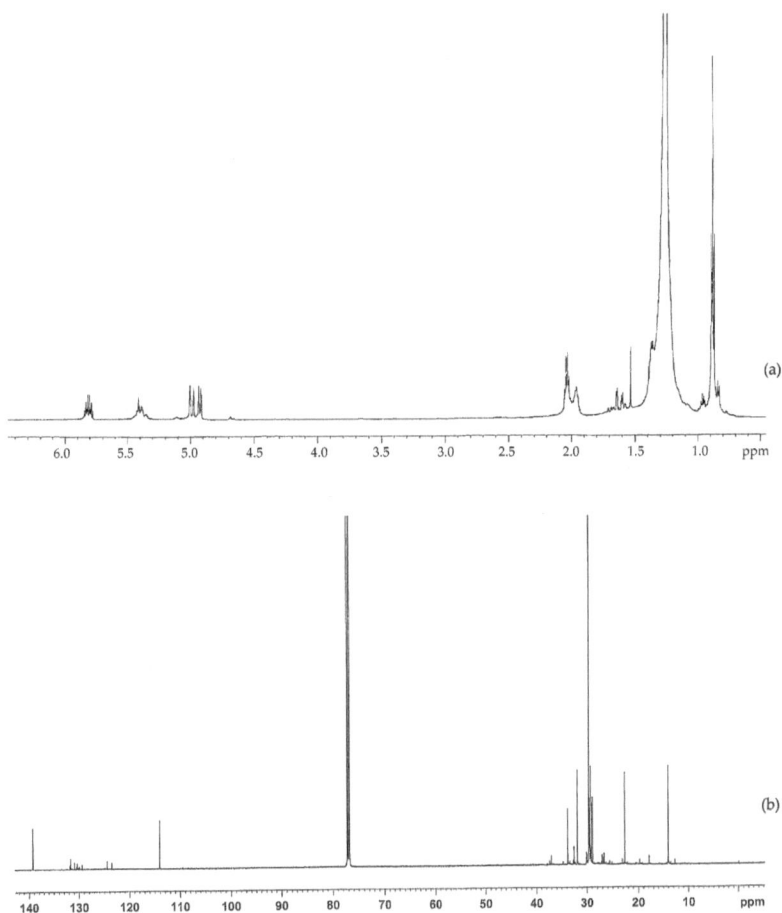

Figure 2. (a) ^{1}H-NMR and (b) ^{13}C-NMR spectra for the non-oxidized polyethylene wax used in this study.

These signals indicate that there are very small amounts of oxygenated functional groups in the non-oxidized wax. However, the amount of oxygenated functional groups are much less than those in the oxidized wax.

3.2. Growth of Bacteria and PHA Yield

The growth of *C. necator* H16 in TSB only (control), BSM only, and TSB/BSM supplemented with N-PEW is shown in Figure 3.

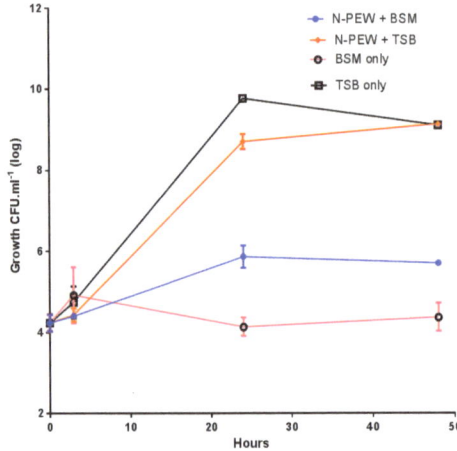

Figure 3. Small-scale fermentations of *C. necator* H16 with 4 g/L of N-PEW in TSB or BSM. Data points are arithmetic means of triplicates, while error bars denote the SE of the mean.

The bacteria were inoculated at approximately 4.1 $\log_{10}$ CFU mL^{-1} and grew to 9.1 $\log_{10}$ CFU mL^{-1} with TSB only and TSB with N-PEW as an additional carbon source. There was no significant difference ($p < 0.5$) between these growth curves, however, when the growth curves with and without N-PEW and BSM are examined, it is evident that there is a difference in final counts. There was little growth when BSM only was used, cell numbers were merely maintained at 4.1–4.3 $\log_{10}$ CFU mL^{-1}) over 48 h, but when N-PEW was added to BSM, there was more growth, an increase of 1.8 $\log_{10}$ CFU mL^{-1}, and the counts ended at 5.8 $\log_{10}$ CFU mL^{-1} rather than 4.3 $\log_{10}$ CFU mL^{-1} in BSM only. In addition, there was no product yield with BSM only and with BSM with N-PEW, suggesting that the wax, alone, is not sufficient for PHA production, but it can sustain cells (Table 1). When TSB only and TSB with N-PEW were used the product was visually different; the PHA produced from TSB only was paper-like in texture and white, while the PHA produced from the N-PEW with TSB was more brittle.

Table 1. The amount of PHA synthesized by *C. necator* using 4 g/L N-PEW. This experiment was conducted over a 48 h incubation period in TSB, TSB with N-PEW, and BSM with N-PEW at 30 °C (150 rpm).

Media	Average CDW (g/L)	Average PHA (g/L)	PHA (%w/w)
TSB only	0.98 ± 0.05	0.20 ± 0.05	20%
TSB with N-PEW	1.42 ± 0.20	0.46 ± 0.20	32%
BSM only	0.16 ± 0.06	ND	ND
BSM with N-PEW	0.20 ± 0.10	ND	ND

ND = non-detected; CDW = cellular dry weight.

3.3. GPC of PHA and N-PEW

GPC analysis of the PHAs obtained using N-PEW and TSB as growth medium showed that the number-average molar mass (Mn) was in the range of 64,000 g/mol with a molecular distribution of

9.7 (Mw/Mn). The Mn results for polyesters synthesized using TSB only were lower, in the range of 52,000 g/mol with a distribution of 5.7 (Mw/Mn).

GPC analysis of the waxes under different conditions during this investigation is reported in Table 2. The N-PEW results showed the number-average molar mass of normal N-PEWs used was (Mn) was 1600 g/mol with a weight average (Mw) of 5400 g/mol.

Table 2. GPC results of unprocessed, pre-sonicated, post-sonicated, post-fermentation, and N-PEWs after shake flask exposure for 48 h without bacterial inoculation.

N-PEW Conditions	Number-Average Molar Mass (Mn) in g/mol	Molecular Mass Distribution Index (Mw/Mn)
Normal N-PEW (pre-sonication)	1600	3.5
Post-sonication N-PEWs	1200	1.2
Post-fermentation	1100	1.3
Post-shake flask (no bacteria)	1100	1.3

The GPC results showed that before sonication there were two distinct types of N-PEW in the sample: high molecular weight (the first peak) and lower molecular weight chains (second peak). After the sonication procedure (Section 2.4) the higher molecular weight peak was removed as the high molecular weight chains were broken down to a Mn of 1200 g/mol with a molar mass index of 1.2. However, any further reduction in the Mn value cannot be attributed to bacterial activity as the flasks without bacteria produced similar Mn values of 1100 g/mol after 48 h.

3.4. PHA Identification and Characterization

The ^{1}H-NMR results of the PHAs formed from shake-flask fermentation experiments with TSB and N-PEW using *C. necator* is shown in Figure 4.

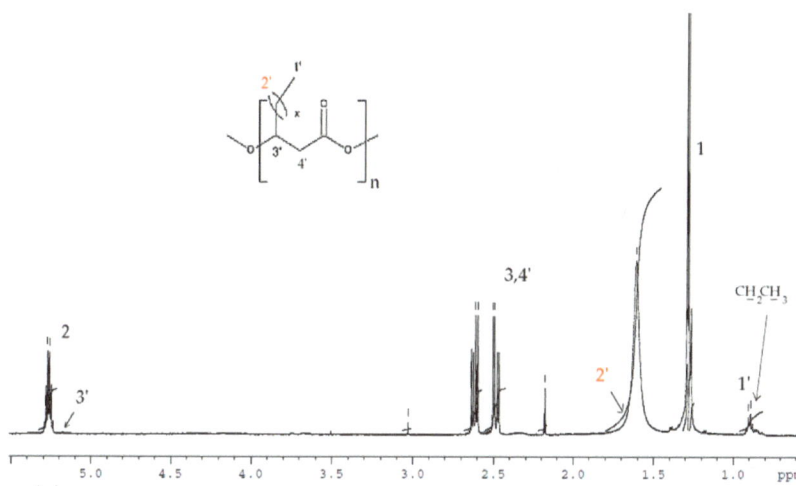

Figure 4. ^{1}H-NMR spectra of PHAs produced from small-scale fermentations with *C. necator* H16 using TSB media supplemented with N-PEW as a carbon source.

In the spectrum the signals corresponded to the protons of the HB repeating units (labelled 1, 2, and 3) and additional signals at $\delta = 0.9$ (-CH$_2$CH$_3$) and $\delta = 5.17$ (-CH) correspond to HV-repeated units that were observed. After integrating these signals, the content of monomer units other than

3-hydroxybutyrate (3-HB) has been estimated at 11 mol% 3-hydroxyvalerate (HV). Due to some of overlapping of signals, the protons of 2' are not visible.

3.5. ESI-MS/MS

The controlled thermal degradation of PHAs synthesized from N-PEW, induced by $KHCO_3$, was performed according to the procedure referred to in Section 2.6.4 [30]. This kind of E1cB degradation leads to PHA oligomers with unsaturated and carboxylic end groups. The results of PHAs formed from TSB with N-PEW are displayed in Figures 5–7. Scheme 1 shows the general structure of the ions presented in the ESI-MS spectra.

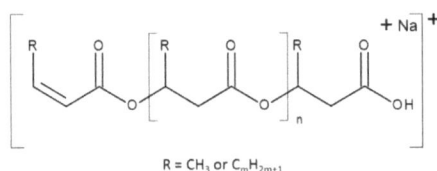

$R = CH_3$ or C_mH_{2m+1}

Scheme 1. The general formula of ions observed in the ESI-MS spectra.

There are two series of ions visible in the mass spectrum. The main series of the ions at *m/z* 1571, 1485, 1399, 1313, 1277, 1141, 1055, 969, 883, and 797 corresponds to the sodium adduct of oligomers with crotonate and carboxyl end groups, calculated according to the formula m/z 86 + (86 × n) + 23, as shown in Scheme 1. There are also a small series of ions besides the main series. The m/z difference is 14 between the main series and the small series, which corresponds to a CH_2 group.

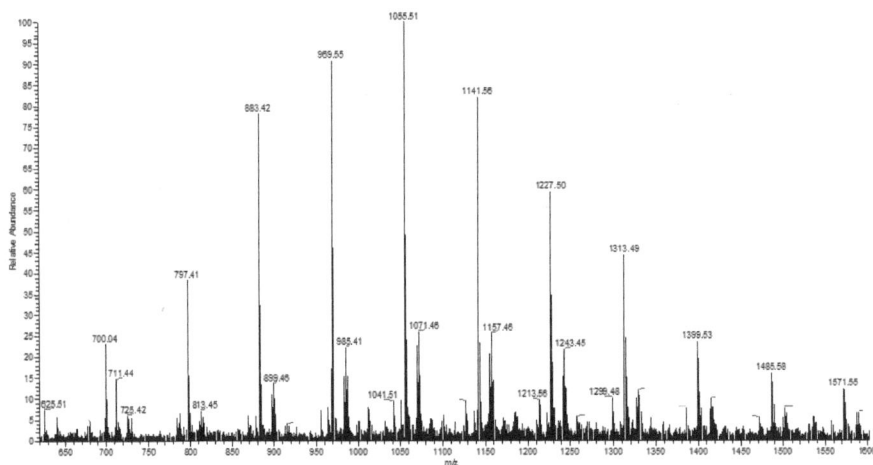

Figure 5. The ESI-MS spectrum (positive-ion mode) of the PHA oligomers, obtained via partial thermal degradation of the biopolyester produced by *C. necator* H16 utilizing TSB and N-PEW as an additional carbon source.

The ESI-MS^2 spectrum of the sodium adduct of the biopolyester oligomers at *m/z* 1055 is presented in Figure 6. The product ions at *m/z* 969, 883, 797, 711, 625, 539, 453, and 367 correspond to the oligo (3-hydroxybutyrate) terminated with carboxyl end groups.

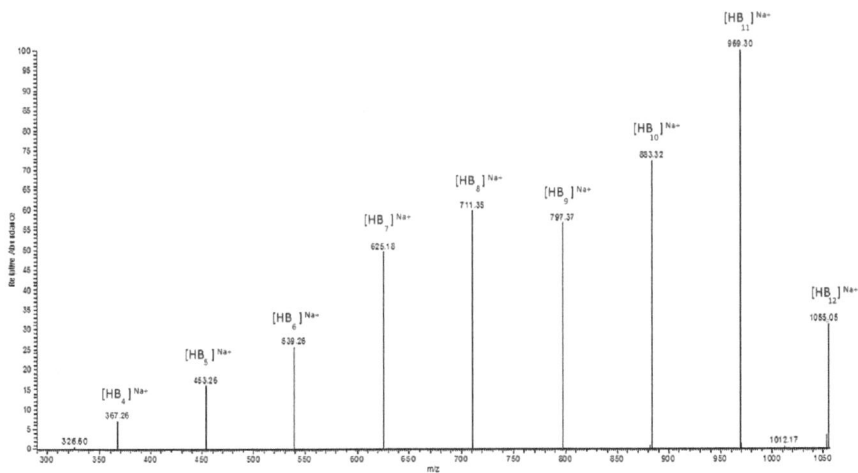

Figure 6. ESI-MS/MS spectrum for the selected sodium adduct ion of oligomers of [HB$_{12}$ + Na]$^+$ at 1055 m/z.

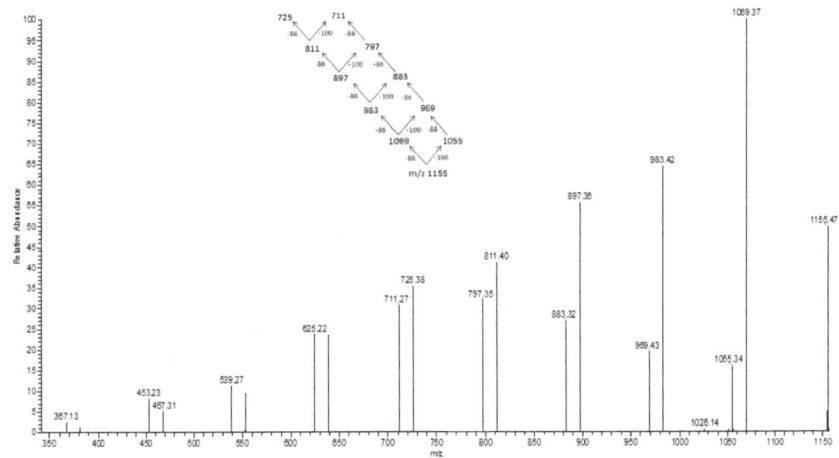

Figure 7. ESI-MS/MS spectrum obtained for the selected sodium adduct of oligomers of [HB$_{13}$HV + Na]$^+$ at m/z at 1155 m/z.

In Figure 7 the fragmentation of the sodium adduct ion of oligomer of [HB$_{13}$HV + Na]$^+$ at m/z at 1155 m/z is presented. This parent ion produces fragment ions of subsequent oligomers according to the theoretical fragmentation pathway diagram. The loss of 86 Da equates to the loss of crotonic acid (e.g., 1155–1069 m/z), and the loss of 100 Da equates to an HV unit. The fragmentation spectrum of these ions confirms that the most intensive ions in the clusters correspond to sodium adducts of 3-hydroxybutyrate oligomers. Therefore, the fragmentation spectrum for the precursor ion at m/z 1155 confirms the presence of 3-hydroxybutyrate and randomly-distributed 3-hydroxyvalerate co-monomer units in polyester chains produced from N-PEW and TSB.

4. Discussion

The FTIR spectra showed that the N-PEWs received were non-polar, saturated hydrocarbons (Figure 1). However, the NMR spectra revealed that there were traces of oxygenated functional groups

Bioengineering **2017**, *4*, 73

(Figure 2). The GPC analysis of the N-PEWs showed that they have a lower molecular mass than O-PEWs, which could make it a more accessible carbon source [16]. Although the hydrocarbon chains are hydrophobic, their shorter chains (after sonication [21]) would be able to be distributed within the TSB or BSM media, and from the growth curves in Figure 3, the addition of the wax made a difference ($+1.8$ log10 CFU mL^{-1}) when it was added to BSM fermentations. The cell growth analysis also showed that N-PEWs do not have an adverse effect upon the growth rate of *C. necator* H16 in either TSB (not significant) or BSM. This was also reported by Radecka et al. [16] with O-PEW. In an experiment conducted by Zhang et al. [31] it was stated that PE has very low antimicrobial properties, limiting growth in Gram-positive and Gram-negative bacterial cells. However, in this study there was no evidence of antimicrobial activity. The low antimicrobial activity reported by Zhang et al. could have been due to the presence of antimicrobial moieties, such as chlorine or fluorine held within the polymer chains, rather than the PE itself. These contaminants could have been freed during the oxidation process and then dispersed into the medium to inhibit microbial growth. It is, therefore, likely that, in a similar way to O-PEW, N-PEWs are being successfully metabolized as fatty acids via β-oxidation. However, the GPC data was not able to fully support this theory as any reduction in Mn or Mw occurred in the controls without bacteria. Those smaller hydrocarbon chains that were not successfully recovered after the 48 h fermentation could have either been metabolized (in the inoculated flask) or they were fully dissolved in the medium. Another possibility is that some of the wax attached to the sides of the flasks as a thin film. For these reasons it was not possible to give a definitive value of the wax metabolized by the bacteria; a possible answer to this would be use a labelled variant of the wax. The presence of accessible carbon sources, such as fatty and carboxylic acids from O-PEW means that more carbon is present in the wax-supplemented cultures in a form that promotes the synthesis of PHAs [16]. It is possible that some of the larger alkane chains from N-PEWs that are more inaccessible for bacterial metabolism could be acting as oxygen inhibitors, limiting distribution within the system. This could then contribute to the stress conditions that push the growth culture towards the production of PHA [11,16].

The PHA yield of TSB cultures supplemented with N-PEW was 32% with a cell dry weight 1.42 g/L compared to shake flask experiments using TSB exclusively, which produced a lower yield of 20% and a dry weight of 0.98 g/L. There was growth of *C. necator* with N-PEW in BSM (shown by cell counts), however, no PHAs were recovered after 48 h. This is an indication that the addition of N-PEW into TSB has an effect upon the PHA production and the structure, with a Mn of 64,000 g/mol with a molecular distribution of 9.7 (Mw/Mn). The Mn results for PHAs produced with TSB only were lower, in the range of 52,000 g/mol with a distribution of 5.7 (Mw/Mn). These dispersity values indicate that N-PEW with TSB produces a more varied range of polyesters, and it is further evidence that N-PEWs are being metabolized.

^{1}H-NMR was used to determine the structure of the copolymers produced. It was found that PHB was produced in the TSB only fermentation (data not shown) [16]. In the spectrum (Figure 4) the signals corresponded to the protons of the HB repeating units (labelled 1, 2, and 3) and additional signals at $\delta = 0.9$ (-CH$_2$CH$_3$) and $\delta = 5.17$ (-CH) correspond to HV repeated units that were observed. These signals are typical for PHAs with longer aliphatic groups in the β-position [32,33]. After integrating these signals, the content of monomer units other than 3-hydroxybutyrate (3-HB) has been estimated at 11 mol% 3-hydroxyvalerate (HV). However, due to some overlapping of certain signals, the protons of 2' were not visible, although they have been reported in that location in PHBV previously [33].

In the N-PEW supplemented cultures, electrospray mass spectrometry (ESI-MS) determined copolymers were present. This method is known to provide more detail than GC and NMR when used for PHBV characterization. This is because it can corroborate whether a copolymer is a co-block polymer or randomly distributed. Therefore, it is feasible to find the precise sequence distribution of the oligomers generated and construct their sequences [34]. The oligomers that were obtained by partial depolymerisation with the controlled thermal degradation of the PHAs produced were induced by KHCO$_3$ (Figures 5–7). This procedure was conducted following the methodology of Kawalec et al. [30]. The E1cB degradation method displayed end groups with both unsaturated and carboxylic acid properties. Based on the mass assignment of single charged ions in the mass spectrum, the structure of the end groups and repeating units can be presumed to be sodium adducts of individual co-oligoester chains of 3-hydroxybutyrate (HB) and 3-hydroxyvalerate (HV). The fragmentation of these ions resulted from the random breakage of ester bonds along both sides of the oligomer chains and the results were consistent with previous studies [16,32,34]. The fragmentation data also suggests that the presence of N-PEW as a carbon source produces different PHAs to those synthesized from media supplemented with O-PEW, which has been reported to produce hydroxyhexanoate (HH) comonomer units, in addition to PHB and PHBV [16,32–36]. This is significant because an increase in the ratio of 3HV to 3HB can decrease the melting point and improve the mechanical strength of the polymers produced [37].

5. Conclusions

N-PEW, a substance produced from waste PE materials is a potential carbon source for PHA production with *C. necator*. The N-PEWs did not mix with growth media, such as TSB, due to their hydrophobic nature; however, with ultrasound sonication, these waxes formed an emulsion with reduced-length hydrocarbon chains that did provide an increased cell count, clearly acting as a viable carbon source. The structural and thermal characterization of the biopolymers produced showed that by adjusting the carbon source, the types of polymer synthesized can be changed. This variation affects the molecular masses and properties of the PHB or PHBV biopolmyers. N-PEW does not limit the growth rate of bacterial cells; in fact the cultures with N-PEW as an additional carbon source produced a higher yield of PHAs. BSM supplemented with N-PEW produced no PHAs, while TSB and N-PEW can produce yields of up to 40% (*w/w*), TSB only experiments gave a yield of 20% (*w/w*).

The ^{1}H-NMR and ESI-MS/MS analysis indicated that in the presence of TSB/N-PEW the chains of the resulting biopolyester contained 3-HB repeating units with 11 mol% of 3-HV units randomly distributed along the copolymer chain. This is different to the PHAs reported from O-PEW/TSB fermentations, that produced PHBV and 3HH repeating units. The N-PEW/TSB media also yielded polymers with greater polymer diversity. Ultimately this analytical technique has enabled the characterization of PHAs synthesized from N-PEW/TSB media using *C. necator* H16, at the molecular level even when the contents of 3HB is high and the levels of other PHA oligomers (such as HV) are limited. This study demonstrates that plastic manufacturing companies could recycle more and utilize a wider range of materials for the production of value-added bioplastics and possibly direct the type of polyesters formed by selecting the level of wax oxidation. By using N-PEWs, food-based materials previously used for biopolymer synthesis, such as corn seed or sugar cane, could be reserved. These N-PEWs are a viable carbon source that can yield truly biodegradable, bio-compatible, non-toxic bioplastics for diverse applications.

Acknowledgments: This research was funded by the Research Investment Fund, University of Wolverhampton, Faculty of Science and Engineering, UK. This work was also partially supported under the EU 7FP BIOCLEAN Project, Contract No. 312100, "New biotechnological approaches for biodegrading and promoting the environmental biotransformation of synthetic polymeric materials".

Author Contributions: Matthew Green was responsible for N-PEW production. Brian Johnston, Iza Radecka, and David Hill were responsible for the bacterial production of PHA and PHA initial characterization. Brian Johnston and Guozhan Jiang were responsible for further GPC and NMR analysis. Grazyna Adamus, Iwona Kwiecień, Magdalena Zięba, Wanda Sikorska, and Marek Kowalczuk were responsible for PHA characterization using ESI-MS/MS. Iza Radecka, Marek Kowalczuk, and Brian Johnston were the main persons involved in the planning of the experiments and interpretation of the data of PHA characterization.

Conflicts of Interest: The authors declare no conflict of interest.

References

1. Chen, G.Q. A microbial polyhydroxyalkanoates (PHA) based bio- and materials industry. *Chem. Soc. Rev.* **2009**, *38*, 2434–2446. [CrossRef] [PubMed]
2. Al-Salem, S.M.; Lettieri, P.; Baeyens, J. Recycling and recovery routes of plastic solid waste (PSW): A review. *Waste Manag.* **2009**, *29*, 2625–2643. [CrossRef]
3. Bondareff, J.M.; Carey, M.; Lyden-Kluss, C. Plastics in the Ocean: The Environmental Plague of Our Time. *Roger Williams UL Rev.* **2017**, *22*, 360–506.
4. Derraik, J.G.B. The pollution of the marine environment by plastic debris: A review. *Mar. Pollut. Bull.* **2002**, *44*, 842–852. [CrossRef]
5. Barnes, D.K.A.; Galgani, F.; Thompson, R.C.; Barlaz, M. Accumulation and fragmentation of plastic debris in global environments. *Philos. Trans. R. Soc. Lond. B* **2009**, *364*, 1985–1998. [CrossRef] [PubMed]
6. Hopewell, J.; Dvorak, R.; Kosior, E. Plastics recycling: Challenges and opportunities. *Philos. Trans. R. Soc. Lond. B* **2009**, *364*, 2115–2126. [CrossRef]
7. Guzik, M.; Kenny, S.; Duane, G.; Casey, E.; Woods, T.; Babu, R.; Nikodinovic-Runic, J.; Murray, M.; O'Connor, K. Conversion of post consumer polyethylene to the biodegradable polymer polyhydroxyalkanoate. *Appl. Microbiol. Biotechnol.* **2014**, *98*, 4223–4232. [CrossRef] [PubMed]
8. Miskolczi, N.; Bartha, L.; Deak, G.; Jover, B. Thermal degradation of municipal plastic waste for production of fuel-like hydrocarbons. *Polym. Degrad. Stabil.* **2004**, *86*, 357–366. [CrossRef]
9. Global Demand for Polyethylene to Reach 99.6 Million Tons in 2018. Available online: https://pgjonline.com/2014/12/10/global-demand-for-polyethylene-to-reach-99-6-million-tons-in-2018/ (accessed on 12 July 2017).
10. Wei, R.; Zimmermann, W. Biocatalysis as a Green Route for Recycling the Recalcitrant Plastic Polyethylene Terephthalate. Available online: http://onlinelibrary.wiley.com/doi/10.1111/1751-7915.12714/full (accessed on 12 July 2017).
11. Verlinden, R.A.J.; Hill, D.J.; Kenward, M.A.; Williams, C.D.; Radecka, I. Bacterial synthesis of biodegradable polyhydroxyalkanoates. *J. Appl. Microbiol.* **2007**, *102*, 1437–1449. [CrossRef]
12. Shah, A.A.; Hasan, F.; Hameed, A.; Ahme, S. Biological degradation of plastics: A comprehensive review. *Biotechnol. Adv.* **2008**, *26*, 246–265. [CrossRef] [PubMed]
13. Philip, S.; Keshavartz, T.; Roy, I. Polyhydroxyalkanoates: Biodegradable polymers with a range of applications. *J. Chem. Technol. Biotechnol.* **2007**, *82*, 233–247. [CrossRef]
14. Verlinden, R.A.J.; Hill, D.J.; Kenward, M.A.; Williams, C.D.; Radecka, I. Production of polyhydroxyalkanoates from waste frying oil by *Cupriavidus necator*. *AMB Express* **2011**, *1*, 1–11. [CrossRef] [PubMed]
15. Leja, K.; Lewandowicz, G. Polymer biodegradation and biodegradable polymers—A review. *Pol. J. Environ. Stud.* **2010**, *19*, 255–266.
16. Radecka, I.; Irorere, V.; Jiang, G.; Hill, D.; Williams, C.; Adamus, G.; Kwiecień, M.; Marek, A.A.; Zawadiak, J.; Johnston, B.; et al. Oxidized Polyethylene Wax as a Potential Carbon Source for PHA Production. *Materials* **2016**, *9*, 367. [CrossRef] [PubMed]
17. Zawadiak, J.; Orlinska, B.; Marek, A.A. Catalytic oxidation of polyethylene with oxygen in aqueous dispersion. *J. Appl. Polym. Sci.* **2013**, *127*, 976–981. [CrossRef]
18. Benefield, R.E.; Boozer, C.E. Thermoplastic Polyolefin and Ethylene Copolymers with Oxidized Polyolefin. Patent US 4,990,568 A, 5 February 1991.
19. Seven, M.K. Process for Oxidizing Linear Low Molecular Weight Polyethylene. Patent WO 2005066219 A1, 21 July 2005.
20. Basstech International. Available online: http://basstechintl.com/products/polyethylene-wax/ (accessed on 12 July 2017).
21. Malykh, N.V.; Petrov, V.M.; Mal'tzev, L.I. Ultrasonic and hydrodynamic cavitation and liquid hydrocarbon cracking. *XX Sess. Russ. Acoust. Soc.* **2008**, *10*, 345–348.
22. Gautam, R.; Bassi, A.S.; Yanful, E.K.; Cullen, E. Biodegradation of automotive waste polyester polyurethane foam using Pseudomonas chlororaphis ATCC55729. *Int. Biodeterior. Biodegrad.* **2007**, *60*, 245–249. [CrossRef]
23. Jiang, G.; Hill, D.J.; Kowalczuk, M.; Johnston, B.; Adamus, G.; Irorere, V.; Radecka, I. Carbon Sources for Polyhydroxyalkanoates and an Integrated Biorefinery. *Int. J. Mol. Sci.* **2016**, *17*, 1157. [CrossRef] [PubMed]

24. Chaijamrus, S.; Udpuay, N. Production and Characterization of Polyhydroxybutyrate from Molasses and Corn Steep Liquor produced by Bacillus megaterium ATCC 6748. *Agric. Eng. Int. CIGR J.* **2008**, *X*, Manuscript FP 07 030.

25. Koller, M.; Sousa Dias, M.M.; Rodríguez-Contreras, A.; Kunaver, K.; Žagar, E.; Kržan, A.; Braunegg, G. Liquefied Wood as Inexpensive Precursor-Feedstock for Bio-Mediated Incorporation of (R)-3-Hydroxyvalerate into Polyhydroxyalkanoates. *Materials* **2015**, *8*, 6543–6557. [CrossRef] [PubMed]

26. Casida, L.E., Jr. Response in soil of *Cupriavidus necator* and other copper-resistant bacterial predators of bacteria to addition of water, soluble nutrients, various bacterial species, or Bacillus thuringiensis spores and crystals. *Appl. Environ. Microbiol.* **1988**, *54*, 2161–2166. [PubMed]

27. Schwartz, E.; Henne, A.; Cramm, R.; Eitinger, T.; Friedrich, B.; Gottschalk, G. Complete nucleotide sequence of pHG1: A Ralstonia eutropha H16 megaplasmid encoding key enzymes of H_2-based lithoautotrophy and anaerobiosis. *J. Mol. Biol.* **2003**, *332*, 369–383. [CrossRef]

28. Recycling Technologies. Available online: http://recyclingtechnologies.co.uk/technology/the-rt7000/#thermal-cracking (accessed on 27 July 2017).

29. Miles, A.A.; Misra, S.S.; Irwin, J.O. The estimation of the bactericidal power of the blood. *Epidemiol. Infect.* **1938**, *38*, 732–749. [CrossRef]

30. Kawalec, M.; Sobota, M.; Scandola, M.; Kowalczuk, M.; Kurcok, P. A convenient route to PHB macromonomers via anionically controlled moderate-temperature degradation of PHB. *J. Polym. Sci. Polym. Chem.* **2010**, *48*, 5490–5497. [CrossRef]

31. Zhang, W.; Luo, Y.; Wang, H.; Jiang, J.; Pu, S.; Chu, P.K. Ag and Ag/N2 plasma modification of polyethylene for the enhancement of antibacterial properties and cell growth/proliferation. *Acta Biomater.* **2008**, *4*, 2028–2036. [CrossRef] [PubMed]

32. Zagar, E.; Krzan, A.; Adamus, G.; Kowalczuk, M. Sequence distribution in microbial poly(3-hydroxybutyrate-co-3-hydroxyvalerate) co-polyesters determined by NMR and MS. *Biomacromolecules* **2006**, *7*, 2210–2216. [CrossRef] [PubMed]

33. Adamus, G.; Sikorska, W.; Janeczek, H.; Kwiecien, M.; Sobota, M.; Kowalczuk, M. Novel block copolymers of atactic PHB with natural PHA for cardiovascular engineering: Synthesis and characterization. *Eur. Polym. J.* **2012**, *48*, 621–631. [CrossRef]

34. Wei, L.; Guho, N.M.; Coats, E.R.; McDonald, A.G. Characterization of Poly(3-hydroxybutyrate-co-3-hydroxyvalerate) Biosynthesized by Mixed Microbial Consortia Fed Fermented Dairy Manure. *J. Appl. Polym. Sci.* **2014**. [CrossRef]

35. Adamus, G.; Sikorska, W.; Kowalczuk, M.; Noda, I.; Satkowski, M.M. Electrospray ion-trap multistage mass spectrometry for characterisation of co-monomer compositional distribution of bacterial poly(3-hydroxybutyrate-co-3-hydroxyhexanoate) at the molecular level. *Rapid Commun. Mass Spectrom.* **2003**, *17*, 2260–2266. [CrossRef] [PubMed]

36. Adamus, G. Aliphatic polyesters for advanced technologies structural characterization of biopolyesters with the aid of mass spectrometry. *Macromol. Symp.* **2006**, *239*, 77–83. [CrossRef]

37. Rudnik, E. Biodegradable Polymers from Renewable Sources. In *Composite Polymer Materials*, 1st ed.; Elsevier: Amsterdam, The Netherlands, 2008; p. 21.

bioengineering

MDPI

Article

The Bistable Behaviour of *Pseudomonas putida* KT2440 during PHA Depolymerization under Carbon Limitation

Stephanie Karmann [1,2], Sven Panke [2] and Manfred Zinn [1,*]

[1] Institute of Life Technologies, University of Applied Sciences and Arts Western Switzerland (HES-SO Valais), Route du Rawyl 47, 1950 Sion, Switzerland; stephanie.karmann@bsse.ethz.ch
[2] Department of Biosystems Science and Engineering, ETH Zurich (ETHZ), Mattenstrasse 26, 4058 Basel, Switzerland; sven.panke@bsse.ethz.ch
*   Correspondence: manfred.zinn@hevs.ch; Tel.: +41-27-606-86-66

Academic Editor: Martin Koller
Received: 14 May 2017; Accepted: 13 June 2017; Published: 19 June 2017

Abstract: Poly(hydroxyalkanoates) (PHAs) are bacterial polyesters offering a biodegradable alternative to petrochemical plastics. The intracellular formation and degradation of PHAs is a dynamic process that strongly depends on the availability of carbon and other nutrients. Carbon excess and nitrogen limitation are considered to favor PHA accumulation, whereas carbon limitation triggers PHA depolymerization when all other essential nutrients are present in excess. We studied the population dynamics of *Pseudomonas putida* KT2440 at the single cell level during different physiological conditions, favoring first PHA polymerization during growth on octanoic acid, and then PHA depolymerization during carbon limitation. PHAs accumulate intracellularly in granules, and were proposed to separate preferentially together with nucleic acids, leading to two daughter cells containing approximately equal amounts of PHA. However, we could show that such *P. putida* KT2440 cells show bistable behavior when exposed to carbon limitation, and separate into two subpopulations: one with high and one with low PHA. This suggests an asymmetric PHA distribution during cell division under carbon limitation, which has a significant influence on our understanding of PHA mobilization.

Keywords: *Pseudomonas putida*; flow cytometry; polyhydroxyalkanoate; PHA mobilization; PHA depolymerization; starvation; carbon limitation; BODIPY 493/503

1. Introduction

In nature, bacteria rarely live under constant nutrient conditions. It is a clear advantage for single cell organisms to accumulate and store carbon if other nutrients are growth-limiting. This stored carbon can later be mobilized as an energy source for growth when all other essential nutrients are present again. Many bacteria can store carbon intracellularly in the form of poly(hydroxyalkanoate) (PHA) granules [1,2]. This polyester receives much attention as a biodegradable alternative to petrochemical plastics [3,4]. PHAs are applied as biodegradable packaging materials in the food industry [5], or in the biomedical field as biocompatible medical implants, biodegradable sutures, skin substitutes, and much more [6].

Pseudomonas putida contains two independent enzymes that are responsible for PHA polymerization, PhaC1 and PhaC2, and one responsible for depolymerization, the depolymerase PhaZ. The polymerizing and depolymerizing enzymes are directly associated with the PHA granule and form the so-called carbonosomes [7]. The PHA polymerization and depolymerization processes are closely linked to enable the immediate adaptation to changes in the chemical composition of the cells' environment [8,9].

The intracellular polymer content of a bacterial culture can be quantified with different techniques, such as gas chromatography (GC) [10,11], fluorescence microscopy [12–14], or flow cytometry (FCM) [15,16]. A comprehensive review about qualitative and quantitative methods for PHA analytics has been published recently [17]. Most quantification methods allow for determining an average amount of PHA per cell dry weight or similar. FCM is the only method that enables a quantification of intracellular PHA at the level of subpopulations and single cells. Single cell analysis is of particular interest for the characterization of biological samples such as PHA producing bacteria, since even genetically identical cells show considerable diversity in their actual composition (e.g., RNA and protein level), and effectively different physiological states can co-exist (bistability) [18]. Such bistability has been reported for *Bacillus subtilis* and its expression of the sporulation gene *spoIIA*. The FCM measurements of green fluorescent protein expressed under a *spoIIA* promotor revealed two distinct populations with a high and a low level of SpoIIA when entering the stationary phase. Effectively, only the high level SpoII population entered sporulation [19].

There are several publications about intracellular PHA quantification by FCM [15,16,20,21]. Interestingly, none of them compared the distribution pattern of PHA granules among the cells under different growth conditions. Typically, the PHA granule-associated phasin proteins are involved in the structural aspects of the PHA granule [22–24]. In particular, it has been suggested for *P. putida* that the PhaF phasin plays a crucial role for the distribution of PHA granules into the two new daughter cells upon cell division. PhaF has a DNA and PHA binding domain, and was suggested to ensure that the PHA is separated together with the chromosomes in a homogeneous way [25]. A similar behavior was reported for poly(hydroxybutyrate) (PHB) producing *Ralstonia eutropha* wild type cells, where the granules were located in the vicinity of the nucleoid region and separated in a homogeneous way to the daughter cells [26].

Here, we show by a FCM-based single cell analysis of *P. putida* KT2440 cultures that the onset of carbon limitation leads in fact to the appearance of two subpopulations of bacteria: one which seems to retain the PHA granules, and one that seems to receive none or at least substantially less than the first one. This suggests that, at least at this stage, PHA distribution is in fact highly asymmetrical, which casts doubt on the generality of the previously postulated equal distribution mechanism.

2. Materials and Methods

2.1. Bacterial Strains, Growth Conditions, and Media

All experiments were performed with *P. putida* KT2440 (ATCC 47054) stored in 1.5 mL 16% glycerol stocks at $-80\ ^{\circ}$C. The strain was grown on LB plates [27] to inoculate precultures as described previously [15]. The modified mineral medium E2 with a pH of 7 [28] and a reduced ammonium content [15] was used for liquid cultivation. Per L of dH_2O were added: 7.5 g K_2HPO_4, 3.7 g KH_2PO_4, and 1.49 g $NaNH_4HPO_4 \cdot 4H_2O$. Trace elements and $MgSO_4 \cdot 7H_2O$ were added as published [28]. The carbon sources were supplied depending on the application of the medium: the preculture medium contained 1.89 $g \cdot L^{-1}$ of citrate and 0.144 $g \cdot L^{-1}$ of octanoic acid (C8), whereas the initial medium for fed-batch fermentations contained 0.189 $g \cdot L^{-1}$ of citrate.

2.2. Bioreactor Settings and Growth Conditions for Fed-Batch Fermentations

The fed-batch fermentation was carried out in a 3.6 L bioreactor (Labfors 5, Infors AG, Bottmingen, Switzerland) with an initial working volume of 2.5 L. Online probes were continuously recording pH and pO_2 (Hamilton Bonaduz AG, Bonaduz, Switzerland). The pH was maintained at 7.0 ± 0.05 by the automated addition of 3 M KOH or 3 M H_3PO_4. The culture was agitated with two six-blade Rushton turbines (diameter of 54 mm) and aerated with normal air at 1 $L \cdot L^{-1} \cdot min^{-1}$. To maintain a pO_2 above 45%, the agitation was programmed in a cascade mode allowing a stirrer speed between 600 and 1000 rpm. An aliquot of 1 mL PPG2000 antifoam was added to the initial medium to avoid foam formation. The temperature was kept constant at 30 $^{\circ}$C.

To inoculate the bioreactor, 150 mL of a preculture with an OD_{600} of 2.0 were harvested, centrifuged (4400 g, 4 °C, 30 min), and resuspended in 20 mL 0.1 M K_2HPO_4 buffer of pH 7. The initial OD_{600} in the bioreactor was 0.1.

The fed-batch fermentation consisted of three different phases: first, an exponential feeding, followed by a linear feeding, and finally no feeding at all in order to reach carbon limitation. The exponential feeding of an aqueous solution containing 0.75 M C8 and equimolar KOH was controlled by a liquid mass flow controller (mini Cori Flow M12, Bronkhorst Cori-Tech B.V., Ruurlo, The Netherlands). The exponential feed rate was programmed with the system software Iris (Infors AG), leading to a specific growth rate of 0.3 h^{-1}, which corresponded to 50% of the maximum specific growth rate (μ_{max}) of *P. putida* KT2440 on C8, as determined in previous experiments with *P. putida* KT2440 [15]. After 12.75 h, nitrogen limitation was reached and the feed rate was kept constant for 3.75 h at 1.48 $g \cdot h^{-1}$ of C8. The feeding resulted in a carbon to nitrogen (C/N) ratio of 26 $g \cdot g^{-1}$. The third phase, without carbon feeding, started after 16.3 h of fermentation. At the same time, an ammonium pulse was applied in the form of $Na(NH_4)HPO_4 \cdot 4H_2O$ (20.9 g in 100 mL dH_2O) that was aseptically added through a sterile 0.22 μm filter (Sarstedt, Nümbrecht, Germany). This pulse reduced the C/N ratio to 4 $g \cdot g^{-1}$.

Frequent samples were taken throughout the bioprocess. The optical density and FCM measurements were done immediately with fresh culture. The supernatant samples to determine the ammonium nitrogen (NH_4^+-N) or C8 content were stored at −20 °C. The cell pellets for cellular PHA quantification by GC were stored at −80 °C.

The fed-batch was run twice with the same feeding strategy, resulting in two identical data sets. For clarity, only one is shown.

2.3. Analytics

2.3.1. Biomass, NH_4^+-N Quantification

To follow the growth of the culture, the cellular dry weight (CDW) was measured by weighing a triplicate of 2 mL dried culture broth as described previously [29], and the optical density was measured at 600 nm (OD_{600}) [15]. The NH_4^+-N in the culture supernatant was determined with a test kit (Ammonium-Test Spectroquant, Merck, Darmstadt, Germany) [15].

2.3.2. Quantification of Octanoic Acid (C8)

The C8 content in the supernatant was determined by high pressure liquid chromatography (HPLC). A volume of 10 μL of undiluted supernatant (centrifuged 30,670 g, at 4 °C for 5 min) was injected into a chromatograph (1200 Series, Agilent, Santa Clara, CA, USA) equipped with a Zorbax Eclipse XDB-C18, 150 × 4.6 mm^2, 5 μm (Agilent Nr 993967-902) and a 12.5 × 4.6 mm^2, 5 μm guard column (Agilent Nr 820950-925). A gradient elution with 0.1% v/v formic acid in acetonitrile (eluent A) and 0.1% v/v formic acid in milliQ water (eluent B) was applied with a linear gradient from 30 to 100% of eluent A within the first 10 min followed by 15 min of elution with 100% of eluent A. The temperature of the column was kept at 25 °C. Finally, the detection was performed with a 210/4 nm UV detector.

2.3.3. Quantification of Volumetric Total Cell Count (vTCC) and PHA by FCM

The *P. putida* KT2440 culture was sampled in duplicates, diluted, and then double-stained with BODIPY 493/503 and SYTO 62 according to a previously published protocol [15]. FCM was carried out with a BD Accuri C6 flow cytometer (BD Bioscience, Erembodegem, Belgium). The results are shown as mean values from the two samples, which had a maximal difference of 10%.

2.3.4. PHA Quantification by GC

The samples of the culture broth with a volume between 10 and 300 mL were centrifuged (3900 *g*, 4 °C, 10 min), and washed once with 0.9% aqueous NaCl. The cell pellet was stored at −80 °C for at least 24 h, then freeze-dried and later methanolyzed as described earlier [11,15].

3. Results and Discussion

3.1. Fed-Batch with Feeding Phase and Subsequent Carbon Limitation Inducing PHA Polymerization and Depolymerization

We performed a fed-batch process with *P. putida* KT2440 that consisted of an initial PHA accumulation phase (carbon excess and nitrogen limitation, 0–16.3 h), followed by a depolymerization phase (nitrogen excess and later carbon limitation, ~20–40 h). During this fermentation, the growth and the PHA-based population structure were analyzed by FCM (Figure 1).

In general, after 12.3 h of exponential C8 feeding, an OD_{600} of 5.7 was reached. Subsequently, a linear carbon feed was applied over 4 h to increase PHA accumulation under nitrogen limitation. During this time, a total amount of 2.67 $g \cdot L^{-1}$ carbon was supplied to the culture, leading to a reduction of the nitrogen concentration in the culture supernatant below the detection limit after 13 h. A pulse of ammonium was added after 16.3 h, resulting in a concentration in the culture broth of 0.3 g NH_4^+-N L^{-1}. With this nitrogen excess, the remaining carbon source was completely consumed after 24 h. The cultivation continued for 17 h under the carbon-limited condition.

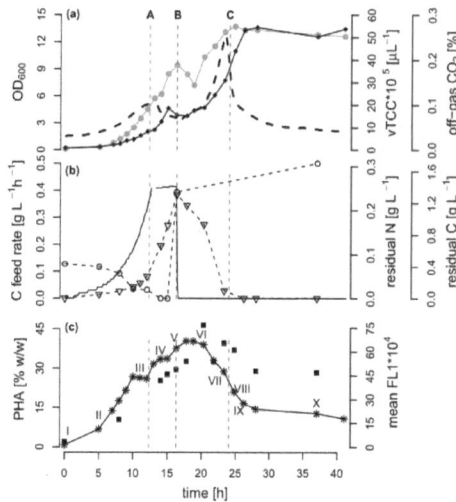

Figure 1. Bioprocess data of the fed-batch fermentation. A fed-batch cultivation of *Pseudomonas putida* KT2440 was carried out with an exponential feed of octanoic acid (C8), followed by a poly(hydroxyalkanoate) (PHA) depolymerization phase under carbon limitation. The dashed lines indicate A: the time point 12.3 h when the feed changed from exponential to linear mode, B: the time point 16.3 h when the carbon feeding was stopped and an ammonium pulse was added, and C: the time point 24 h when the carbon source was depleted. (**a**) Optical density (●, OD_{600}), volumetric total cell count (♦ vTCC), CO_2 content in the off-gas (–); (**b**) Carbon (C) feed rate (-), carbon and nitrogen content in the culture supernatant (∇ residual C, ○ residual N); (**c**) PHA quantity as determined by gas chromatography (GC) analysis (■) and by measuring the mean green fluorescence (✱ mean FL1) of BODIPY 493/503 stained cells with flow cytometry (FCM). The Roman numbers (I–X) in proximity of the mean FL1 values mark the FCM samples depicted in Figure 3 in Section 3.3.

3.2. Comparison of PHA Quantification by FCM and GC during PHA Polymerization and Depolymerization

The cellular PHA content was followed by FCM and GC throughout all of the fed-batch phases. In FCM-based single cell analytics, the dye BODIPY 493/503 can be used to stain the PHA granules so that the green fluorescence (mean FL1) value per cell can be taken as a measure of the accumulated PHA [15]. As expected, the mean green fluorescence increased as long as C8 but no nitrogen was present in the medium, suggesting that the PHA amount per cell increased. When nitrogen was available again, and later when the carbon source was depleted, the PHA-derived fluorescence decreased, and reached a plateau at around 30% of the maximum value after 27 h of cultivation. The GC analysis of selected biomass samples confirmed that the PHA content was maximal between 17.5 and 20 h of cultivation (46% w/w). The accumulated PHA was a co-polymer of 3-hydroxyoctanoate and 10–11 mol % 3-hydroxyhexanoate.

Towards the end of the process, the GC analysis suggested that the cellular PHA content remained at a high level (almost 30% w/w), whereas a content of approximately 10% w/w would be expected based on the fluorescence data alone (Figure 1). Figure 2 shows that the correlation between the FCM and GC measurements was consistent with the linear correlation equation identified earlier [15], confirming that the FCM method is a robust method with hardly any inter-experimental variations. However, the samples taken in the experiments presented here during the depolymerization phase separated clearly from this line (Figure 2). The results indicate that either the FCM method underestimates the actual PHA content during the depolymerization phase, or the GC-based method overestimates it.

An explanation for the mismatch between the FCM and GC-based PHA amounts might be presented by the different mechanisms for PHA determination. A GC analysis will also include PHA degradation products, such as 3-hydroxyoctanoic acid monomers accumulated intracellularly or attached to the cell membrane, as well as PHA granules released from lysed cells. In order to avoid that GC measurements could be biased in such a way, the washing of the cell pellet prior to freeze-drying was an important step of the sample's preparation protocol. However, the efficiency of the removal of water insoluble particles was not quantified, and it cannot be excluded as a source of error. In contrast, the double-staining-based FCM method applied in this study only considers PHA in intact cells. Released PHA granules and cell debris are ignored in the FCM measurement, as they do not contain any DNA, and consequently are not stained in the double staining procedure with the red fluorescent SYTO 62 [15].

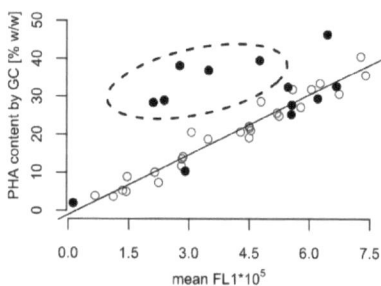

Figure 2. Correlation between poly(hydroxyalkanaote) (PHA) contents determined by gas chromatography (GC) and flow cytometry (FCM). The plot contains PHA content data measured by GC and FCM from the fed-batch shown in Figure 1 (•), and from previously published fed-batches with *P. putida* KT2440 grown on octanoic acid without a carbon-limited phase (o) [15]. The FCM data are given as mean green fluorescence derived from BODIPY 493/503 stained intracellular PHA (mean FL1). The dashed ellipse comprises all samples taken during the carbon-limited phase of the fed-batch fermentation (Figure 1), which significantly differ from the samples taken during growth with carbon in excess.

3.3. Population Dynamics of P. putida KT2440 during PHA Polymerization and Degradation Determined by FCM

The analysis of FCM fluorescence histograms from selected time points acquired during different phases of the cultivation suggested that the PHA content changed quite drastically over time (Figure 3). During growth under carbon excess (data points I to VI), the peak of PHA-derived fluorescence in the histogram moved, as expected, from low FL1 intensities (gate G1) towards higher fluorescence intensities (gate G2). During this phase, the coefficient of variation (CV) of FL1, a value that gives information about the width of the peak and therefore about the spread of cellular PHA contents, resulted to be relatively small (<100%), suggesting a narrow distribution of intracellular PHA content across the population of cells (see supplementary materials Figure S1). In contrast, after the carbon source C8 was depleted (data points VII to X), a separation into two distinct populations took place. One cell cluster remained in the high FL1 intensity gate (G2), indicating a high PHA content, and a second cluster formed in the lower FL1 intensity gate (G1), suggesting the formation of a distinct population of cells with low or no intracellular PHA (Figure 3). Fluorescence scatter plots and several values for cell counts, fluorescence intensities, as well as the CV values for the overall measurements and for the gates G1 and G2 are shown in supplementary materials Figure S1.

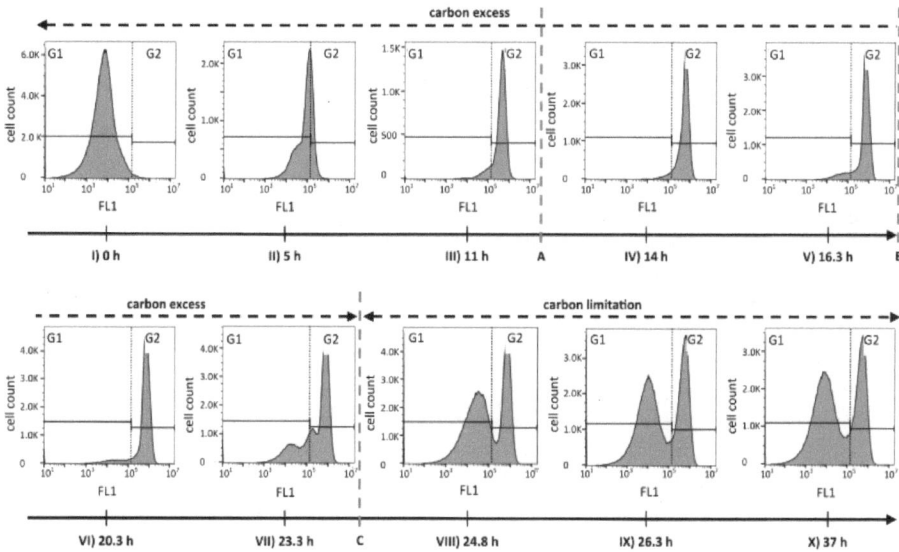

Figure 3. Histograms of fluorescence data (FL1) representing the PHA distribution among *P. putida* KT2440 cells during a fed-batch with a phase of carbon excess and a phase of carbon limitation. The green fluorescence of BODIPY 493/503 stained cells was recorded by flow cytometry during the bioprocess that is depicted in Figure 1. The dashed lines A, B, and C correspond to the time points (**A**) 12.3 h feed changed from exponential to linear mode; (**B**) 16.3 h the carbon feeding was stopped and an ammonium pulse was added; and (**C**) 24 h the carbon source was depleted. The gates G1 and G2 (black step lines and dotted vertical line) separate low and high FL1 values, respectively, and were defined based on the histogram from PHA-free cells at the time of inoculation, to visualize the evolution of the FL1 signal.

In order to further analyze the mechanism behind the formation of subpopulations, we re-plotted the data of Figure 1 in terms of the volumetric total cell count (vTCC), FCM-determined average PHA content per cell (mean FL1, which would be the signal most similar to the PHA content determined by GC), and total volumetric PHA content (vFL1, the product of vTCC and mean FL1) with a focus on the depolymerization phase (Figure 4). Next, we repeated the data analysis, but used the cell numbers and the fluorescence values of the two subpopulations in the G1 and G2 gates (see Figure 3, supplementary materials Figure S1). This should allow for an understanding of the development of the total cell number and mean PHA in the total population in terms of the development of the two contributing subpopulations. Interestingly, the number of cells in G2 (high PHA content) showed only a small increase between 16.3 h (end of carbon feed) and 24 h (carbon source consumed), and then remained essentially constant. In contrast, the number of cells in G1 (low PHA content) increased more than 10-fold, until the total number of cells in G1 was about twice the number of cells in G2, about 2.3 h after the carbon source in the supernatant had been consumed ($t = 26.3$ h). This clearly suggests that the overall increase of the vTCC after 16.3 h (Figures 1a and 4a) is mostly due to the formation of a subpopulation containing cells with a low PHA content. Furthermore, the PHA content (mean FL1) of the cells belonging to the high-PHA gate G2 remained constant between 16.3 h and 21 h, and then decreased by approximately 30% until the end of the experiment (Figure 4b). As a result, the volumetric PHA content increased proportionally with the number of cells in the high-PHA gate G2 until 24 h, and then decreased with the decreasing PHA content per cell, again in the G2 gate. However, it is striking that even though the PHA seems to have been mobilized to some extent in the G2 population, it was only the G1 population that increased in terms of vTCC, whereas the vTCC of the high-PHA content cells remained constant (Figure 4). These results strongly suggest that the apparent decrease in average PHA content between 20 h and 28 h, as shown in Figure 1c, is nearly entirely due to the formation of a low-PHA or even PHA-free subpopulation under conditions of carbon limitation.

We see three possible explanations for the observed behavior that are also illustrated in Figure 5. First, low-PHA cells that were present at very small concentrations already during the polymerization phase might have preferentially continued to divide at a higher growth rate when the lack of nitrogen did no longer inhibit cell division. The carbon source needed for the growth of this G1 population would have come from the small amount of the intracellular PHA of these cells, and from recycled 3-hydroxyalkanoic acid monomers secreted by cells of the G2 population as soon as the remaining carbon source in the supernatant was consumed ($t = 24$ h) (Figure 5a). Second, the ability to degrade intracellular PHA rapidly might have been distributed in an uneven manner among the PHA-containing population. This would mean that some cells would have degraded PHA in order to divide, whereas others would not have done so (Figure 5b). Indeed, in the critical phase when the carbon source starts to be limiting (Figure 3, sample VII at 23.3 h), we clearly see a transition peak of cells with an intermediate PHA content that is supposedly the basis of the G1 population that shows an increasing vTCC during this phase of the process. However, this transition population was only present during a very short period of time. Finally, PHA-containing cells might distribute the granules in which PHA is stored asymmetrically among the daughter cells when dividing (Figure 5c). This would lead in the extreme to one daughter cell with PHA and the other one without any or with only little PHA that would then appear in the G1 gate of the fluorescence histogram or scatterplot (Figure 3, supplementary materials Figure S1).

In this context, it is interesting to note that the heterogeneous distribution of PHA during cell division has been observed before, specifically in exponentially growing *phaF* deletion mutants of *P. putida* KT2442 [25,30], and in *phaM* deletion mutant *R. eutropha* cells [31,32]. PhaF and PhaM are proteins that are involved in the distribution of PHA granules into daughter cells during a cell division. However, in the case presented here, it is very unlikely that an asymmetric cell division occurred due to a mutation in PhaF, since cells were grown for only approximately 10 generations, and because of the reproducibility of the phenotype. It seems, therefore, much more plausible that the different growth condition, namely carbon limitation, favors a bistable growth and PHA distribution pattern according to one or several of the hypotheses mentioned above. Several studies mentioned above assessed the distribution of PHA granules under PHA accumulating conditions under carbon excess [25,26,30–32].

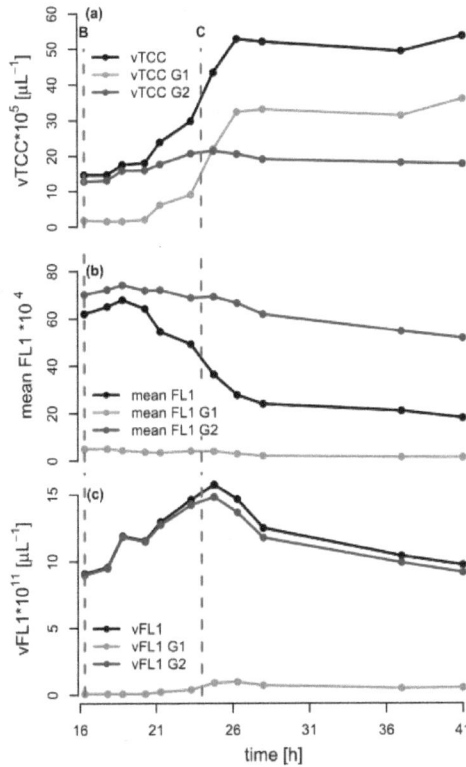

Figure 4. Analysis of the flow cytometry (FCM) data during the depolymerization phase of the fed-batch experiment with *P. putida* KT2440 on octanoic acid (C8). The data from the FCM analysis during the second phase of fermentation (t = 16.3–41 h) are shown for the total population, as well as for the populations of gates G1 and G2 (low/no PHA and high PHA content, respectively). The dashed lines B and C indicate the time points when the C8 feeding was stopped and the ammonium pulse added (16.3 h), and when C8 was completely consumed (t = 24 h), respectively. (**a**) Volumetric total cell count (vTCC); (**b**) Mean FL1 fluorescence (PHA content); (**c**) Volumetric fluorescence (vFL1, computed by multiplying vTCC by FL1).

In general, PHA depolymerization is undesirable in industrial PHA production, since it causes a fast loss of valuable product. Hence, it is crucial to harvest the cells as close as possible to the time point when the PHA content is maximal. This time point can be determined by a sudden increase of the pO$_2$ signal in aerobic processes, or in general with the here applied FCM-based analysis method. According to the data presented above, a first indication of PHA depolymerization is not necessarily detectable as a decrease in mean FL1, as would be expected, but rather by the formation of a low PHA subpopulation. In fact, it is clear that the vTCC in the G1 subpopulation starts to increase after 23.3 h, whereas only a minor change in the mean FL1 can be observed at this time (Figure 4). Interestingly, 23.3 h corresponds to the time point a few minutes prior to complete depletion of the carbon substrate. Therefore, a process operator can decide based on a fast FCM-based analysis (total analysis time in the order of 5 min) if the production process should be terminated, or, during the process, if the feeding rate needs to be adjusted. In comparison, the determination of the substrate concentration in the culture supernatant by HPLC takes at least 30 min (see Section 2.3.2), during which valuable PHA could already be lost by depolymerization.

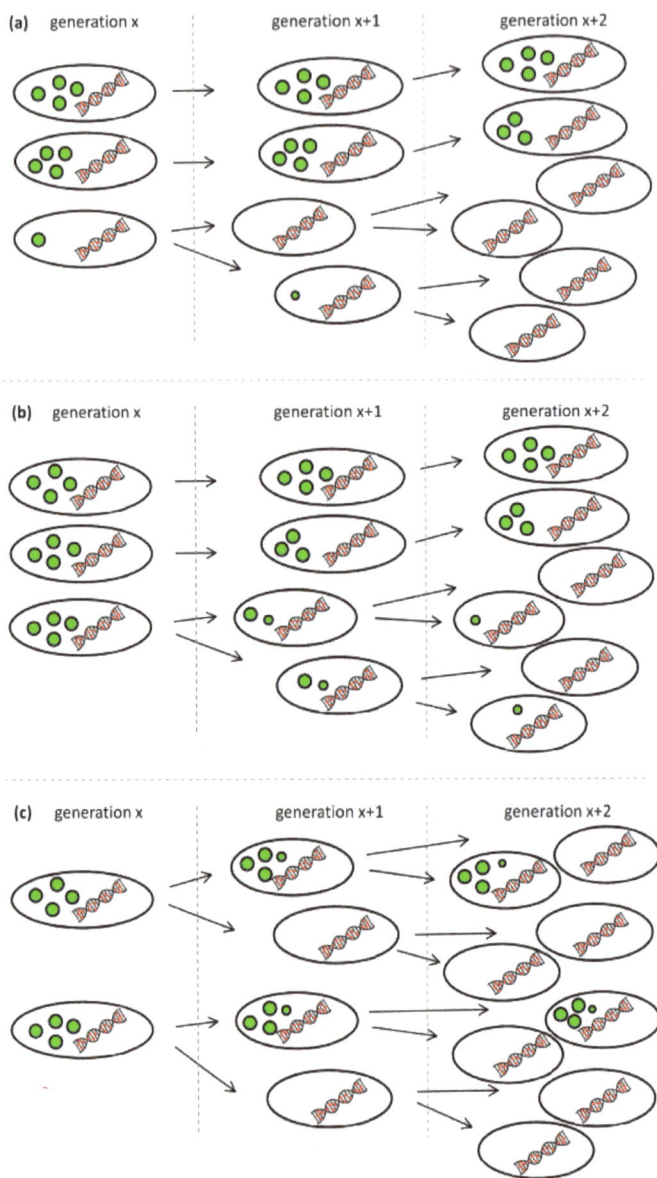

Figure 5. Possible scenarios leading to the occurrence of a low-PHA subpopulation at the onset of carbon limitation. (**a**) Low-PHA or PHA-free cells that are present in small concentrations divide faster than the cells that accumulated a lot of PHA and consume fatty acid monomers secreted from the PHA containing subpopulation; (**b**) Not all cells have the same propensity to recycle their PHA for further growth; (**c**) PHA granules are distributed asymmetrically when cells are dividing under conditions of carbon limitation. The colors were chosen according to the fluorescent staining for the analysis by flow cytometery: green represents the BODIPY 493/503-stained PHA granules, and red represents SYTO 62-stained nucleic acids. The amount and the size of the green granules symbolize the PHA content per cell.

Bioengineering **2017**, *4*, 58

4. Conclusions

Single cell analysis by flow cytometry of *P. putida* K2440 cells revealed that the nutritional state significantly influenced the distribution of PHA granules among the population. Expectedly, when cells were grown under carbon excess, PHA was accumulated and fluorescence measurements of stained, intracellular PHA showed a very homogeneous population in terms of PHA contents. However, bistability became apparent when the culture was subsequently exposed to carbon limitation and the population clearly separated into two main clusters. One cluster remained with a high amount of PHA, whereas a second cluster of cells with no or very little PHA was rising in number. This suggests either that PHA-free cells were able to grow faster by consuming fatty acid monomers released from PHA-degrading cells or that the cells were dividing in an asymmetric way into one daughter cell without PHA and the other one keeping all PHA of the mother cell.

Supplementary Materials: The following are available online at http://www.mdpi.com/2306-5354/4/2/58/s1. Figure S1: Fluorescence scatter plots and histograms from flow cytometry measurements during the PHA polymerization-depolymerization experiment.

Acknowledgments: The research leading to these results has received funding from the European Union's Seventh Framework Programme for research, technological development, and demonstration under grant agreement no. 311815. We thank Antoine Fornage for support with GC analytics.

Author Contributions: Stephanie Karmann designed and performed the experiment and analyzed the data. Stephanie Karmann, Sven Panke and Manfred Zinn wrote the paper.

Conflicts of Interest: The authors declare no conflict of interest. The founding sponsors had no role in the design of the study; in the collection, analyses, or interpretation of data; in the writing of the manuscript, and in the decision to publish the results.

References

1. Lemoigne, M. Produit de deshydration et de polymerisation de l'acide beta-oxybutyrique. *Bull. Soc. Chim. Biol.* **1926**, *8*, 770–782.

2. Kadouri, D.; Jurkevitch, E.; Okon, Y.; Castro-Sowinski, S. Ecological and agricultural significance of bacterial polyhydroxyalkanoates. *Crit. Rev. Microbiol.* **2005**, *31*, 55–67. [CrossRef] [PubMed]

3. Keshavarz, T.; Roy, I. Polyhydroxyalkanoates: Bioplastics with a green agenda. *Curr. Opin. Microbiol.* **2010**, *13*, 321–326. [CrossRef] [PubMed]

4. Sudesh, K.; Abe, H.; Doi, Y. Synthesis, structure and properties of polyhydroxyalkanoates: Biological polyesters. *Prog. Polym. Sci.* **2000**, *25*, 1503–1555. [CrossRef]

5. Byun, Y.; Kim, Y.T. Utilization of bioplastics for food packaging industry. In *Innovations in Food Packaging*, 2nd ed.; Elsevier Ltd.: San Diego, CA, USA, 2013; pp. 369–390.

6. Jiang, L.; Zhang, J. Biodegradable and biobased polymers. In *Applied Plastics Engineering Handbook*; Elsevier Inc.: Waltham, MA, USA, 2017; pp. 127–143.

7. Jendrossek, D. Polyhydroxyalkanoate granules are complex subcellular organelles (carbonosomes). *J. Bacteriol.* **2009**, *191*, 3195–3202. [CrossRef] [PubMed]

8. Arias, S.; Bassas-Galia, M.; Molinari, G.; Timmis, K.N. Tight coupling of polymerization and depolymerization of polyhydroxyalkanoates ensures efficient management of carbon resources in *Pseudomonas putida*. *Microb. Biotechnol.* **2013**, *6*, 551–563. [CrossRef] [PubMed]

9. Ren, Q.; de Roo, G.; Ruth, K.; Witholt, B.; Zinn, M.; Thöny-Meyer, L. Simultaneous accumulation and degradation of polyhydroxyalkanoates: Futile cycle or clever regulation? *Biomacromolecules* **2009**, *10*, 916–922. [CrossRef] [PubMed]

10. Riis, V.; Mai, W. Gas chromatographic determination of poly-β-hydroxybutyric acid in microbial biomass after hydrochloric acid propanolysis. *J. Chromatogr.* **1988**, *445*, 285–289. [CrossRef]

11. Furrer, P.; Hany, R.; Rentsch, D.; Grubelnik, A.; Ruth, K.; Panke, S.; Zinn, M. Quantitative analysis of bacterial medium-chain-length poly([R]-3-hydroxyalkanoates) by gas chromatography. *J. Chromatogr. A* **2007**, *1143*, 199–206. [CrossRef] [PubMed]

12. Berlanga, M.; Montero, M.T.; Fernández-Borrell, J.; Guerrero, R. Rapid spectrofluorometric screening of polyhydroxyalkanoate-producing bacteria from microbial mats. *Int. Microbiol.* **2006**, *9*, 95–102. [PubMed]

13. Degelau, A.; Scheper, T.; Bailey, J.E.; Guske, C. Fluorometric measurement of poly-β hydroxybutyrate in *Alcaligenes eutrophus* by flow cytometry and spectrofluorometry. *Appl. Microbiol. Biotechnol.* **1995**, *42*, 653–657. [CrossRef]

14. Gorenflo, V.; Steinbüchel, A.; Marose, S.; Rieseberg, M.; Scheper, T. Quantification of bacterial polyhydroxyalkanoic acids by Nile Red staining. *Appl. Microbiol. Biotechnol.* **1999**, *51*, 765–772. [CrossRef] [PubMed]

15. Karmann, S.; Follonier, S.; Bassas-Galia, M.; Panke, S.; Zinn, M. Robust at-line quantification of poly(3-hydroxyalkanoate) biosynthesis by flow cytometry using a BODIPY 493/503-SYTO 62 double-staining. *J. Microbiol. Methods* **2016**, *131*, 166–171. [CrossRef] [PubMed]

16. Kacmar, J.; Carlson, R.; Balogh, S.J.; Srienc, F. Staining and quantification of poly(3-hydroxybutyrate) in *Saccharomyces cerevisiae* and *Cupriavidus necator* cell populations using automated flow cytometry. *Cytometry A* **2006**, *69*, 27–35. [CrossRef] [PubMed]

17. Koller, M.; Rodríguez-Contreras, A. Techniques for tracing PHA-producing organisms and for qualitative and quantitative analysis of intra- and extracellular PHA. *Eng. Life Sci.* **2015**, *15*, 558–581. [CrossRef]

18. Veening, J.-W.; Smits, W.K.; Kuipers, O.P. Bistability, epigenetics, and bet-hedging in bacteria. *Annu. Rev. Microbiol.* **2008**, *62*, 193–210. [CrossRef] [PubMed]

19. Veening, J.W.; Hamoen, L.W.; Kuipers, O.P. Phosphatases modulate the bistable sporulation gene expression pattern in *Bacillus subtilis*. *Mol. Microbiol.* **2005**, *56*, 1481–1494. [CrossRef] [PubMed]

20. Saranya, V.; Poornimakkani; Krishnakumari, M.S.; Suguna, P.; Binuramesh, C.; Abirami, P.; Rajeswari, V.; Ramachandran, K.B.; Shenbagarathai, R. Quantification of intracellular polyhydroxyalkanoates by virtue of personalized flow cytometry protocol. *Curr. Microbiol.* **2012**, *65*, 589–594. [CrossRef] [PubMed]

21. Vidal-Mas, J.; Resina-Pelfort; Haba, E.; Comas, J.; Manresa, A.; Vives-Rego, J. Rapid flow cytometry-Nile Red assessment of PHA cellular content and heterogeneity in cultures of *Pseudomonas aeruginosa* 47T2 (NCIB 40044) grown in waste frying oil. *Antonie Van Leeuwenhoek* **2001**, *80*, 57–63. [CrossRef] [PubMed]

22. Steinbüchel, A.; Aerts, K.; Babel, W.; Follner, C.; Liebergesell, M.; Madkour, M.H.; Mayer, F.; Pieper-Fürst, U.; Pries, A.; Valentin, H.E. Considerations on the structure and biochemistry of bacterial polyhydroxyalkanoic acid inclusions. *Can. J. Microbiol.* **1995**, *41* (Suppl. S1), 94–105. [CrossRef] [PubMed]

23. Pötter, M.; Steinbüchel, A. Poly(3-hydroxybutyrate) granule-associated proteins: Impacts on poly(3-hydroxybutyrate) synthesis and degradation. *Biomacromolecules* **2005**, *6*, 552–560. [CrossRef] [PubMed]

24. Pötter, M.; Müller, H.; Reinecke, F.; Wieczorek, R.; Fricke, F.; Bowien, B.; Friedrich, B.; Steinbüchel, A. The complex structure of polyhydroxybutyrate (PHB) granules: Four orthologous and paralogous phasins occur in *Ralstonia eutropha*. *Microbiology* **2004**, *150*, 2301–2311. [CrossRef] [PubMed]

25. Galán, B.; Dinjaski, N.; Maestro, B.; Eugenio, L.I.; de Escapa, I.F.; Sanz, J.M.; García, J.L.; Prieto, M.A. Nucleoid-associated PhaF phasin drives intracellular location and segregation of polyhydroxyalkanoate granules in *Pseudomonas putida* KT2442. *Mol. Microbiol.* **2011**, *79*, 402–418. [CrossRef] [PubMed]

26. Wahl, A.; Schuth, N.; Pfeiffer, D.; Nussberger, S.; Jendrossek, D. PHB granules are attached to the nucleoid via PhaM in *Ralstonia eutropha*. *BMC Microbiol.* **2012**, *12*, 262. [CrossRef] [PubMed]

27. Miller, J. *Experiments in Molecular Genetics*; Cold Spring Harbour Laboratory Press: Cold Spring Harbour, NY, USA, 1972.

28. Durner, R.; Zinn, M.; Witholt, B.; Egli, T. Accumulation of poly[(R)-3-hydroxyalkanoates] in *Pseudomonas oleovorans* during growth in batch and chemostat culture with different carbon sources. *Biotechnol. Bioeng.* **2001**, *72*, 278–288. [CrossRef]

29. Karmann, S.; Follonier, S.; Egger, D.; Hebel, D.; Panke, S.; Zinn, M. Tailor-made PAT platform for safe syngas fermentations in batch, fed-batch and chemostat mode with *Rhodospirillum rubrum*. *Microb. Biotechnol.* **2017**. [CrossRef] [PubMed]

Bioengineering **2017**, *4*, 58

30. Prieto, A.; Escapa, I.F.; Martínez, V.; Dinjaski, N.; Herencias, C.; de la Peña, F.; Tarazona, N.; Revelles, O. A holistic view of polyhydroxyalkanoate metabolism in *Pseudomonas putida*. *Environ. Microbiol.* **2016**, *18*, 341–357. [CrossRef] [PubMed]

31. Jendrossek, D.; Pfeiffer, D. New insights in the formation of polyhydroxyalkanoate granules (carbonosomes) and novel functions of poly(3-hydroxybutyrate). *Environ. Microbiol.* **2014**, *16*, 2357–2573. [CrossRef] [PubMed]

32. Pfeiffer, D.; Wahl, A.; Jendrossek, D. Identification of a multifunctional protein, PhaM, that determines number, surface to volume ratio, subcellular localization and distribution to daughter cells of poly(3-hydroxybutyrate), PHB, granules in *Ralstonia eutropha* H16. *Mol. Microbiol.* **2011**, *82*, 936–951. [CrossRef] [PubMed]

bioengineering

MDPI

Article

The Evolution of Polymer Composition during PHA Accumulation: The Significance of Reducing Equivalents

Liliana Montano-Herrera [1], Bronwyn Laycock [1], Alan Werker [2] and Steven Pratt [1,*

[1] School of Chemical Engineering, University of Queensland, St Lucia QLD 4072, Australia;
liliana.montano@usys.ethz.ch (L.M.-H.); b.laycock@uq.edu.au (B.L.)

[2] Veolia Water Technologies AB—AnoxKaldnes, Klosterängsvägen 11A SE-226 47 Lund, Sweden;
alan@werker.se

* Correspondence: s.pratt@uq.edu.au; Tel.: +61-7-3365-4943

Academic Editor: Martin Koller
Received: 6 January 2017; Accepted: 23 February 2017; Published: 7 March 2017

Abstract: This paper presents a systematic investigation into monomer development during mixed culture Polyhydroxyalkanoates (PHA) accumulation involving concurrent active biomass growth and polymer storage. A series of mixed culture PHA accumulation experiments, using several different substrate-feeding strategies, was carried out. The feedstock comprised volatile fatty acids, which were applied as single carbon sources, as mixtures, or in series, using a fed-batch feed-on-demand controlled bioprocess. A dynamic trend in active biomass growth as well as polymer composition was observed. The observations were consistent over replicate accumulations. Metabolic flux analysis (MFA) was used to investigate metabolic activity through time. It was concluded that carbon flux, and consequently copolymer composition, could be linked with how reducing equivalents are generated.

Keywords: PHA; monomer evolution; mixed culture; modeling; polymer composition; biopolymer

---

1. Introduction

Polyhydroxyalkanoates (PHAs) are biobased and biodegradable polyesters. PHA copolymers, such as poly(3-hydroxybutyrate-*co*-3-hydroxyvalerate) (PHBV), are of particular interest as they are the basis for biomaterials with desirable mechanical properties. These copolymers can be produced in mixed microbial cultures [1]. However, predicting and controlling the copolymer composition can be challenging.

PHAs are most typically synthesized in mixed microbial cultures from volatile fatty acids (VFAs), through well-described metabolic pathways [2,3]. In the specific case of PHBV, short chain acids such as acetic and propionic acids are transported though the cell membrane and converted into acetyl-CoA and propionyl-CoA respectively. PHA synthesis then takes place in three steps. Firstly, two acyl-CoA molecules are condensed in a reaction catalyzed by a thiolase to produce various intermediates. For example, two acetyl-CoA monomers form acetoacetyl-CoA (a 3-hydroxybutyrate (3HB) precursor), while one acetyl-CoA and one propionyl-CoA combine to form ketovaleryl-CoA (a 3-hydroxyvalerate (3HV) precursor) [4], Escapa et al. 2012). In addition, it has been observed that a portion of the propionyl-CoA produced is converted into acetyl-CoA though different pathways [5]. Secondly, a reduction, catalyzed by a reductase, produces 3-hydroxyalkanoate (3HA) monomers, with the reducing power to support PHA production being generated during anabolic pathways for cell growth, as well as in reactions related to the tricarboxylic acid (TCA) cycle [6]. Finally, a polymerase adds 3HA monomers to the PHA polymer. As such, the flux of carbon through the acyl-CoA intermediates influences the resulting polymer composition.

The fraction of 3HV monomer units in the final PHBV copolymer can be manipulated by adjusting the proportion of even-chain (i.e., acetic acid) to odd-chain (i.e., propionic acid) fatty acids in the feed composition [7–9], since odd-chain fatty acids are generally required for the formation of propionyl-CoA, which is the precursor of 3HV monomer. Diverse monomer compositions and sequence distributions of PHBV copolymers produced by mixed microbial cultures have been achieved using different feeding strategies with acetic and propionic acid mixtures as model substrates [10,11].

Most mixed microbial culture accumulation studies have been applied under conditions of some form of nutrient starvation to inhibit cell growth and favor PHA synthesis [12–14]. In contrast, a recent study has shown that PHA storage can occur concurrently with active biomass growth. Valentino et al. [15] achieved a consistent improvement of PHA productivity when N and P were supplied in an optimal C:N:P ratio. It is important to consider that shifts in the active biomass growth rates may influence carbon flux through the acyl-CoA intermediates and the availability of reducing equivalents for PHA synthesis, and therefore affect polymer production and composition.

Literature on mixed microbial culture PHA production coupled with high rate cell growth is scant. Simulations of existing models have successfully fitted data on PHA productivity and even monomer composition evolution in some cases [4,16,17]; however, these models apply only for scenarios of negligible growth. In addition, in these experiments the feedstock composition was kept constant during the accumulation process resulting in a polymer with a constant ratio of 3HB:3HV. These existing model frameworks are in contrast to some published experimental data that do show shifts in copolymer composition during fed-batch mixed culture PHA accumulation, even under non-growing conditions [11]. Such data indicate that the 3HB:3HV ratio during accumulation is not simply dependent on the feedstock but is also affected by the history of the accumulation and the resulting metabolic activity in the biomass. The potential for biomass growth and other processes to directly influence the composition of intracellular acyl-CoA reservoirs and hence copolymer composition has not been examined.

The aim of this paper is to examine 3HB and 3HV monomer evolution through PHA accumulation, giving consideration to the effect of biomass growth and alternating feedstocks on this process. To this end, monomer development through four sets of PHA accumulation experiments (based on the feeding regime) is investigated: Set 1: acetic acid (HAc) feed; Set 2: propionic acid (HPr) feed; Set 3: mixed HAc and HPr feed; and Set 4: alternating HAc/HPr feed. Concurrent biomass growth and carbon storage is encouraged in each set. Metabolic Flux Analysis (MFA) is used to quantify metabolic pathway activity through the accumulations.

2. Materials and Methods

2.1. Experimental Set-Up

PHA was produced at pilot scale at AnoxKaldnes AB (Lund, Sweden) using a three stage process that had been in continuous operation from 2008 to 2013 [11]. The first stage (acidogenic fermentation) was performed in a 200 L continuous stirred tank reactor under anaerobic conditions and fed with cheese whey permeate, producing a mixture comprising 35% ± 4% acetic, 4% ± 1% propionic, 49% ± 4% butyric, 4% ± 1% valeric and 8% ± 3% caproic acids. The second stage was carried out in a sequential batch reactor (SBR) operated under Aerobic Dynamic Feeding (ADF) conditions with nutrient addition (COD:N of 200:5). The excess biomass with a high PHA storage capacity (as enriched in stage two) was used to produce PHA-rich biomass in the third stage, in a reactor operated in fed batch mode. The details of this process and analytical methods can be found in Janarthanan et al. [18].

2.2. Fed-Batch PHA Production

PHA was accumulated in batches of 100 L harvested SBR mixed liquor by means of a 150 L (working volume) aerated reactor. Aeration provided mixing as well as oxygen supply. Acetic and propionic acids (HAc and HPr, respectively) were fed using different HAc:HPr ratios and feeding strategies. The microbial community was dominated by the genera *Flavisolibacter* and *Zoogloea* [18].

The carbon source concentration for pulse-wise substrate addition was ~100 gCOD/L (see Table 1), with pH adjusted to 4 and additions of nitrogen and phosphorus for nutrient limitation to give COD:N:P of 200:2:1 [15]. N and P additions were 3.82 g/L NH_4Cl and 0.22 g/L KH_2PO_4, respectively. For the fed-batch accumulations, a pulse-wise feedstock addition was applied for feed-on-demand [19] controlled by the biomass respiration response as measured by dissolved oxygen (DO) trends [7,11]. Semi-continuous (pulse-wise) additions of feedstock aliquots were made targeting peak COD concentrations of between 100 and 200 mg-COD/L. Feedstock additions were triggered by a measured relative decrease in biomass respiration rate [7]. pH was monitored but not controlled. The fed-batch accumulations were run over 20–25 h with samples taken at selected times for analyses, including VSS (volatile suspended solids), TSS (total suspended solids), PHA content and composition, soluble COD, volatile fatty acids (VFAs), and nutrients (nitrogen and phosphorus).

Table 1. Experimental conditions in the PHA (Polyhydroxyalkanoates) fed-batch accumulations.

Experiment set	Substrate Composition and Feeding Strategy (gCOD Basis)	Experiment Label	Process Time (h)	Initial VSS (g·L^{-1})	Total substrate added (gCOD)		Feed Concentration (gCOD·L^{-1})	Total Number of Pulses	
					HAc	HPr		HAc	HPr
1	100% Acetic acid	Exp 1	23.2	1.2	1374	-	96	138	-
		Exp 1'	21.9	1.4	1684	-	98	147	-
2	100% Propionic acid	Exp 2	23.3	1.2	-	939	106	-	77
		Exp 2'	24.6	1.7	-	1439	102	-	137
3	50% Acetic/50% propionic acid	Exp 3	24.2	1.5	600	600	98	95	
		Exp 3'	20.4	1.9	817	817	98	140	
4	100% Acetic acid—100% propionic acid (alternating)	Exp 4	22.3	1.4	1429	982	101/103	106	105
		Exp 4'	22.5	1.7	570	552	94/97	57	56

2.3. Analytical Methods

Total concentrations were analyzed from well-mixed grab samples and soluble concentrations were analyzed after filtering the aqueous samples with 1.6 μm pore size (Ahlstrom Munktell, Falun, Sweden) filters. Volatile fatty acid concentrations were quantified by gas chromatography [20]. Solids analyses (total and volatile suspended solids, or TSS/VSS) were performed according to Standard Methods [21].

Hach Lange™ kits were used for the determination of soluble COD (sCOD) (LCK 114), NH_4–N (LCK 303), NO_3–N (LCK 339), soluble total phosphorus (LCK 349) and soluble total nitrogen (LCK 138). PHA content and monomeric composition (3HB and 3HV) of samples was determined using the gas chromatography method described in [11] using a Perkin-Elmer gas chromatograph (GC) (Perkin Elmer, Inc., Waltham, MA, USA). Quantitative ^{13}C high resolution NMR spectra were acquired on a Bruker Avance 500 spectrometer (Bruker, Billerica, MA, USA) as described by Arcos-Hernandez et al. [11] to determine polymer microstructure (details of polymer structure can be found in Supplementary Materials).

2.4. Experimental Design for PHA Accumulations

The full set of PHA accumulation (third stage) experiments are summarized in Table 1. For this work, four experiments (replicated, with replicates denoted using the symbol ') were considered. Experiment set 1 used a single acetic acid (HAc) substrate; Experiment set 2 used a single propionic acid (HPr) substrate; Experiment set 3 used a mixed HAc and HPr substrate fed simultaneously in equal COD ratios; while in Experiment set 4 the acids were supplied in alternating pulses.

2.5. Rate and Yield Calculations

For PHA concentration at a given time (in g PHA/L), only the PHA produced during the accumulation process was considered. Therefore, the measured PHA concentration (*PHA*) was corrected by subtracting the initial measured PHA content (*PHA₀*). Typically the initial biomass PHA content (*%PHA₀* in wt %) was between 0% and 4%. Active biomass ($CH_{1.4}O_{0.4}N_{0.2}$) at a given time (*X*, recorded in g/L) was determined from the total concentration of biomass, measured as volatile suspended solids (*VSS* in g VSS/L), subtracting the produced PHA concentration (*PHA*):

$$X = VSS \ (g \ VSS/L) - PHA \ (g \ PHA/L)$$

PHA intracellular content (*%PHA*) was calculated as the PHA concentration divided by the volatile suspended solids concentrations on a mass basis.

$$\%PHA \ (gPHA/g \ VSS) = \frac{PHA \ (g \ PHA/L)}{VSS \ (g \ VSS/L)}$$

The PHA fraction (f_{PHA}) was measured as PHA concentration divided by active biomass concentration on a COD basis.

$$f_{PHA} = \frac{PHA \ (in \ gCOD \ PHA/L)}{X \ (in \ mCOD \ X/L)}$$

Experimental data for the total amount of VFA consumed, PHA polymer (*PHA*) produced and active biomass (*X*) produced were fitted using global nonlinear regression in GraphPad Prism (v.6.0.5). This analysis was performed using an exponential growth model (one phase association) [22]. The batch process mass balance accounted for input feed dosing volumes as well as sampling withdrawal volumes. Kinetic rates and yields were calculated from fitted data as follows:

Acetic (q_{HAc}) and propionic acid (q_{HPr}) specific consumption rates and specific monomer 3HB and 3HV production rates: q_{HB} and q_{HV}, respectively, for the *i*th uptake of each acid or production of each monomer were calculated with reference to active biomass (*X*) concentration:

$$q_{HAc} = \frac{(HAc_i - HAc_{i-1})}{(t_i - t_{i-1}) \cdot X_i} \quad q_{HPr} = \frac{(HPr_i - HPr_{i-1})}{(t_i - t_{i-1}) \cdot X_i}$$

$$q_S \left(Cmol \ VFA \cdot Cmol \ X^{-1} \cdot h^{-1} \right) = q_{HAc} + q_{HPr}$$

$$q_{HB} = \frac{(3HB_i - 3HB_{i-1})}{(t_i - t_{i-1}) \cdot X_i} \quad q_{HV} = \frac{(3HV_i - 3HV_{i-1})}{(t_i - t_{i-1}) \cdot X_i}$$

$$q_{PHA} \left(Cmol \ PHA \cdot Cmol \ X^{-1} \cdot h^{-1} \right) = q_{HB} + q_{HV}$$

where *t* is time; *HAc* and *HPr* are the moles of acetic and propionic acids in solution; *3HB* and *3HV* are the moles of 3HB and 3HV, respectively; and q_S is the specific consumption rate of substrate (S). The instantaneous relative rate change in 3HV monomers ($\%3HV^{inst}$) was calculated relative to the total PHA specific production rate on a mole basis.

$$\%3HV^{inst} = \frac{q_{HV} \ (mol \ 3HV \cdot h^{-1} \cdot X^{-1})}{q_{PHA} \ (mol \ PHA \cdot h^{-1} \cdot X^{-1})}$$

As previously reported by Janarthanan et al. [18], a linear correlation was obtained between gCOD PHA produced versus total substrate consumed (also in gCOD) and the yield ($Y_{PHA/S}$) in gCOD PHA/gCOD S at 20 h was determined ($0.968 < r^2 < 0.998$). This time point was selected for consistent comparison between runs as all accumulations had reached at least 98% of plateau PHA content by this time. Likewise, plots of active biomass (in gCOD X) versus time were represented by linear regression to a linear quadratic equation, and the yield ($Y_{X/S}$) in gCOD X/gCOD S at 20 h was determined. The 95% confidence intervals associated with all the determined stoichiometric and kinetic parameters were estimated using error propagation formulae. The values were also converted to Cmol basis.

Maximum specific growth rate (μ_{max}) was calculated according to the re-parameterization of the empirical expression applied to growth curves developed by Gompertz [23]. Analysis was performed in SigmaPlot (Systat Software, v.12) plotting $\ln(X/X_0)$ versus time ($0.938 < r^2 < 0.989$).

The maximum specific VFA consumption rate (q_S, Cmol VFA/(Cmol X·h)) and maximum specific PHA storage rate (q_{PHA}, Cmol VFA/(Cmol X·h)), were determined from the trends in the experimental data during the exponential growth phase. The ratio of PHA concentration and total VFA consumed divided by the active biomass concentration at that time were plotted over time, calculating the first derivative.

2.6. Metabolic Flux Analysis (MFA)

MFA was performed in order to investigate the effect of VFA composition and the feeding strategy on active biomass growth and PHA (3HV and 3HB) monomer formation kinetics assuming a pseudo-steady state. The metabolic network used in this work is based on previously published models [4,17] and summarized in Figure 1. The reactions R_9 and R_{10} (Figure 1) describe the conversion of acetyl-CoA and propionyl-CoA into PHA precursors, where acetyl-CoA* and propionyl-CoA* are representations of molecules which have undergone the first two steps of PHA synthesis (condensation and reduction) [2,4]. Subsequently, PHA precursors are polymerized to form the biopolymer (PHB and PHV), with two units of acetyl-CoA* forming one 3HB molecule, and one unit of acetyl-CoA* and one of propionyl-CoA* forming one molecule of 3HV. The cells obtain energy from adenosine triphosphate (ATP), which is generated by the oxidation of NADH, and the efficiency of ATP production is represented by the phosphorylation efficiency (P/O) ratio (δ). The maximum theoretical P/O ratio is 3 mol-ATP/mol-NADH$_2$ in bacteria growing under aerobic conditions [24].

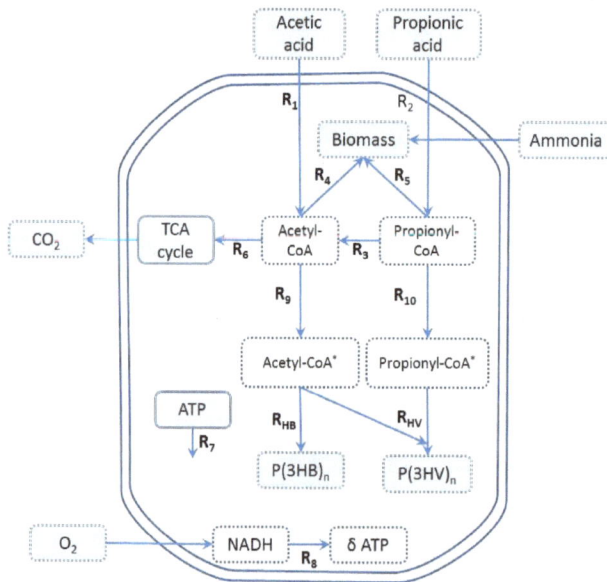

Figure 1. Metabolic network for PHBV synthesis and biomass production, adapted from [17], with permission from © 2013 Elsevier. Light blue dotted squares represent external metabolites; white dotted squares represent internal metabolites.

The metabolic model consists of 12 reactions, 6 intracellular metabolites (acetyl-CoA, propionyl-CoA, acetyl-CoA*, propionyl-CoA*, ATP, and NADH), 4 substrates (HAc, HPr, O_2, and NH_4), and 4 end products (3HB, 3HV, X, and CO_2). The system of equations has six degrees of freedom [25], and a total of seven rates were measured (VFA consumption rates, PHA monomers storage rate, oxygen uptake rate, active biomass synthesis rate, and ammonium consumption). Therefore, the system is overdetermined, with one degree of redundancy, which made it possible to estimate the experimental errors in measurements.

The following constraints and assumptions were set for MFA:

- Active biomass can be formed either from acetyl-CoA or propionyl-CoA. Previous models were performed under ammonia limiting conditions with negligible cellular growth [4,13]. In the present work, we assumed that fluxes of acetyl-CoA and propionyl-CoA used for active biomass (v_4, v_5) synthesis are proportional to the consumption rate of acetic and propionic acid (v_{HAc}, v_{HPr}) [16].

$$f_{Pr} = \frac{v_{HPr}}{v_{HAc} + v_{HPr}}$$

$$\frac{v_4}{v_5} = \frac{1 - f_{Pr}}{f_{Pr}}$$

- Reactions R_9 and R_{10} are reversible, the rest are irreversible reactions.
- The maintenance requirement (v_7) was an estimated flux while the P/O ratio was fixed ($\delta = 3$).
- PHA depolymerization was not considered.

MFA was performed using the CellNetAnalyzer (v. 2014.1, Max Planck Institute, Magdeburg, Germany) toolbox for Matlab [26]. To evaluate the consistency of experimental data with the assumed biochemistry and the pseudo-steady state assumption a *chi-squares-test* was carried out. The flux distributions calculated were found to be reliable given that the consistency index (h) values were below a reference chi-squared test function ($\chi^2 = 3.84$ for a 95% confidence level and 1 degree of redundancy) [27]. The stoichiometry of the metabolic reactions is provided in the Supplementary Materials.

3. Results and Discussion

3.1. Biomass Growth and PHA Content

The experiments were designed to follow the time evolution of PHA storage and active biomass growth during the third stage of the PHA-production system and representative Experiment sets 3 and 4 are shown in Figure 2 (sets 1 and 2 can be seen in the Supplementary Materials). The extent of production of active biomass was variable between the experiments, but higher maximum specific growth rates were achieved for accumulations where acetic acid was fed (Experiment set 1 with 100% HAc, Experiment set 3 with 50%/50% HAc/HPr, and Experiment set 4 with alternating substrates) (Table 2). However, active biomass growth rates attenuated sooner for those accumulations where acetic acid was present at all times (Experiment set 1 and Experiment set 3), while the highest biomass production (X/X_0) was achieved in Exp 4 (Figure 2, Table 2). With regard to PHA fraction evolution, a similar PHA content at plateau was achieved for all experiments. However, PHA content and yield tended to be higher in those accumulations with alternating substrates (Experiment set 4), with one experiment (Exp 4) maintaining an increasing PHA fraction even after 22 h of accumulation. This observation fits with the interpretations from other works that it is possible to stimulate PHA storage with concurrent cellular growth by supplying an optimal nutrient ratio [6,15,18].

Table 2. PHA accumulation yields and kinetic parameters.

Experiment Label	f_{Ac} Consumed (mol HAc/mol VFA)	f_{Pr} Consumed (mol HPr/mol VFA)	%PHA Plateau (gPHA/gVSS)	%3HV (mol 3HV/mol PHA) at 20 h	$Y_{PHA/S}$ (gCOD PHA/gCOD VFA)	$Y_{X/S}$ (gCOD X/gCOD VFA)	μ_{max} (h^{-1})	Final X/X_0 (gCOD/gCOD)	$-q_{Smax}$ (Cmol VFA/Cmol X·h)	q_{PHAmax} (Cmol PHA/Cmol X·h)
Exp 1	1.0	0	0.56 ± 0.04	0	0.48 ± 0.02	0.17 ± 0.03	0.27 ± 0.03	2.20	0.75	0.33
Exp 1'	1.0	0	0.48 ± 0.06	0	0.38 ± 0.04	0.15 ± 0.01	0.21 ± 0.02	2.28	0.80	0.36
Exp 2	0	1.0	0.40 ± 0.04	74	0.31 ± 0.03	0.16 ± 0.03	0.18 ± 0.02	2.13	0.33	0.11
Exp 2'	0	1.0	0.48 ± 0.03	80	0.40 ± 0.07	0.18 ± 0.03	0.13 ± 0.02	2.09	0.33	0.18
Exp 3	0.64	0.36	0.48 ± 0.06	40	0.39 ± 0.03	0.17 ± 0.05	0.35 ± 0.05	1.88	0.63	0.19
Exp 3'	0.64	0.36	0.52 ± 0.03	42	0.45 ± 0.03	0.12 ± 0.05	0.27 ± 0.05	1.75	0.40	0.19
Exp 4	0.72	0.28	0.59 ± 0.03	34	0.52 ± 0.03	0.18 ± 0.02	0.22 ± 0.02	3.08	0.37	0.23
Exp 4'	0.64	0.36	0.52 ± 0.06	36	0.49 ± 0.03	0.20 ± 0.02	0.19 ± 0.02	1.98	0.45	0.20

All data in table recorded as ± 95% confidence interval where possible.

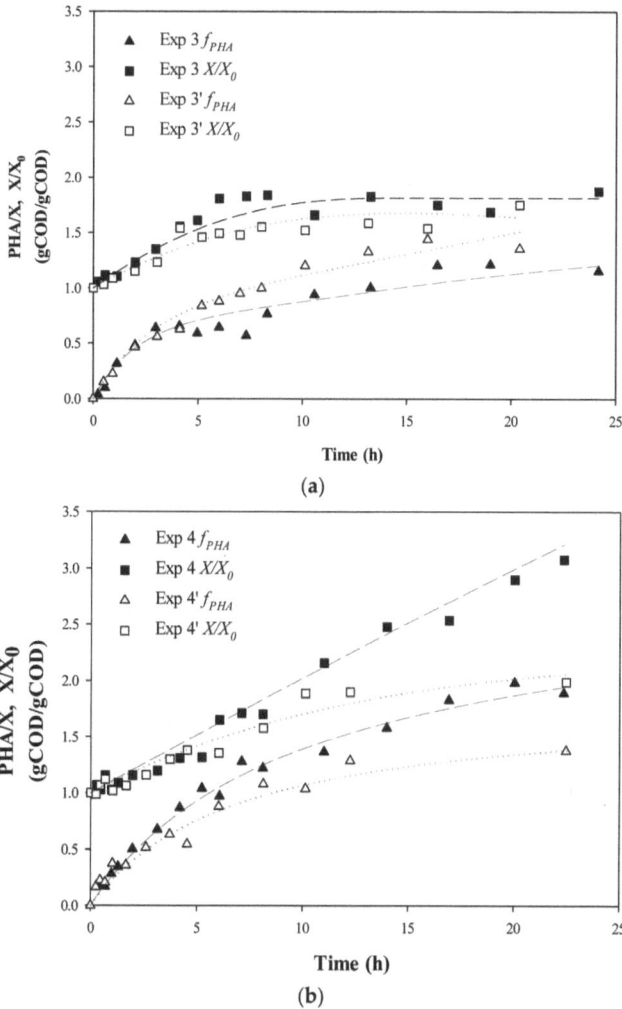

Figure 2. Experimental data for PHA fraction and relative active biomass production: (**a**) Experiment set 3: 50% acetic acid and 50% propionic acid fed simultaneously; and (**b**) Experiment set 4: 100% acetic acid alternating with 100% propionic acid. ▲, △, PHA fraction respect to active biomass concentration f_{PHA} (PHA/X); ■, □, Active biomass concentration respect to the initial biomass concentration (X/X_0). Dotted lines represent fitted data. Coefficients are given on gCOD basis.

3.2. Monomer Development

The polymer composition over time during the accumulations is shown in Figure 3a, while the flow of carbon to 3HV relative to PHA overall at each time point (the instantaneous 3HV fraction) is shown in Figure 3b. The trends of replicate runs all matched well with the originals in terms of monomer development, although the final 3HV content differed slightly from run to run. The highest values of 3HV content were achieved in accumulations with propionic acid present at all times; Experiment set 2 (100% HPr) and Experiment set 3 (50%/50% HAc/HPr) reached a maximum %3HV content of 0.90 and 0.72 (on a mole % basis) at 6 and 2 h, respectively (Figure 3a). Although the formation rates of 3HV units relative to the formation of PHA dominated in the early stages of the accumulation for Sets 2 and 3, a sharp decrease in the instantaneous 3HV fraction was identified (Figure 3b). In contrast, the instantaneous 3HV fraction ($\%3HV^{inst}$) in Experiment set 4, which followed an alternating pulse feeding strategy of acetic and propionic acids, did not show any remarkable change with time. However, it should be noted that $\%3HV^{inst}$ steadily decreased during Exp 4' but gradually increased during Exp 4 (Figure 3b). The trend in $\%3HV^{inst}$ in Exp 4 coincided with high production of active biomass in that system (Figure 3a). Overall, concurrent PHA storage and active biomass growth resulted in a dynamic trend in polymer composition, except for the alternating feeding strategy.

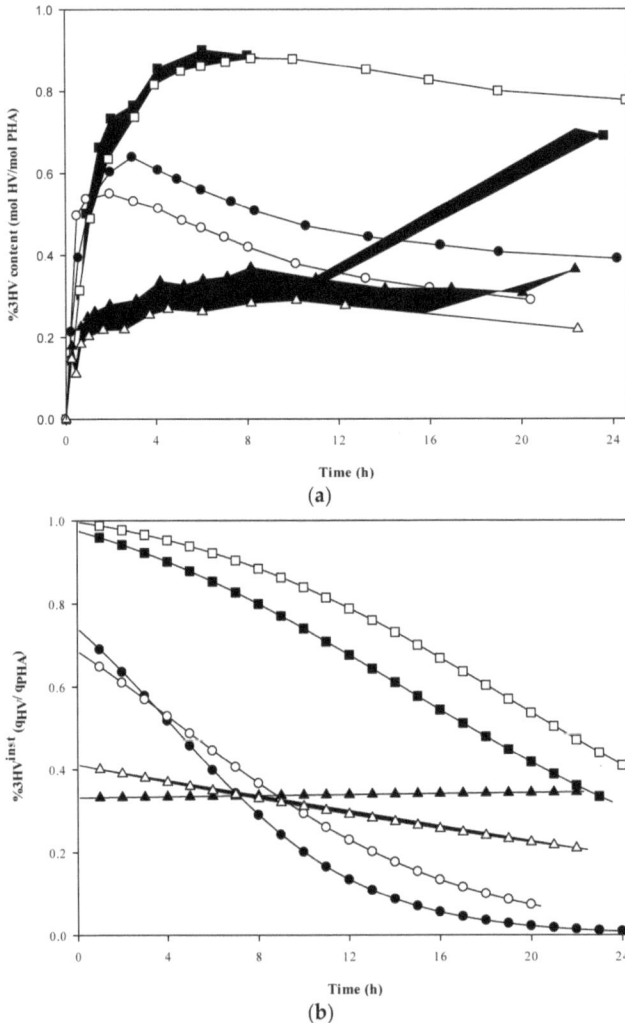

(a)

(b)

Figure 3. Evolution of 3HV fraction during PHA accumulation using different feeding strategies: (a) accumulated 3HV fraction; and (b) instantaneous 3HV fraction calculated from data regression. Experiment set 2: 100% propionic acid (■, Exp 2; and □, Exp 2'). Experiment set 3: 50% acetic acid and 50% propionic acid fed simultaneously (●, Exp 3; and ○, Exp 3'). Experiment set 4: 100% acetic acid alternating with 100% propionic acid (▲, Exp 4; and △, Exp 4').

Current models for mixed culture PHA production using acetic and propionic acids as substrates consider that the proportion of 3HV monomer units in the copolymer obtained changes in proportion to the relative composition of the carbon feeds. This assumption has often adequately predicted 3HV:3HB molar composition in cultures with negligible cellular growth. However, when a shifting substrate strategy was applied, and moreover when cellular growth was maintained, even while maintaining a constant feed composition, then the published models cannot predict the observations of the present investigation. Metabolic analysis of the carbon flux distribution through time reveals why this would be so.

3.3. Carbon Flux to PHA, Biomass and CO_2

Figure 4 shows the calculated carbon flux distribution to PHA monomers, active biomass and carbon dioxide for the four experiment sets at various time points. Of note is an increase in the proportion of carbon flux directed to carbon dioxide (CO_2) production over time, mostly due to the carbon flux through the TCA cycle (R_6) in all experiments, while VFAs were being consumed despite a reduction of active biomass synthesis rate and PHA production rate. According to Escapa et al. [28], the uptake of carbon source in cultures of *Pseudomonas putida* with no PHA synthase activity remains active and the excess is directed to the TCA cycle which produces CO_2, as a way to dissipate the carbon surplus.

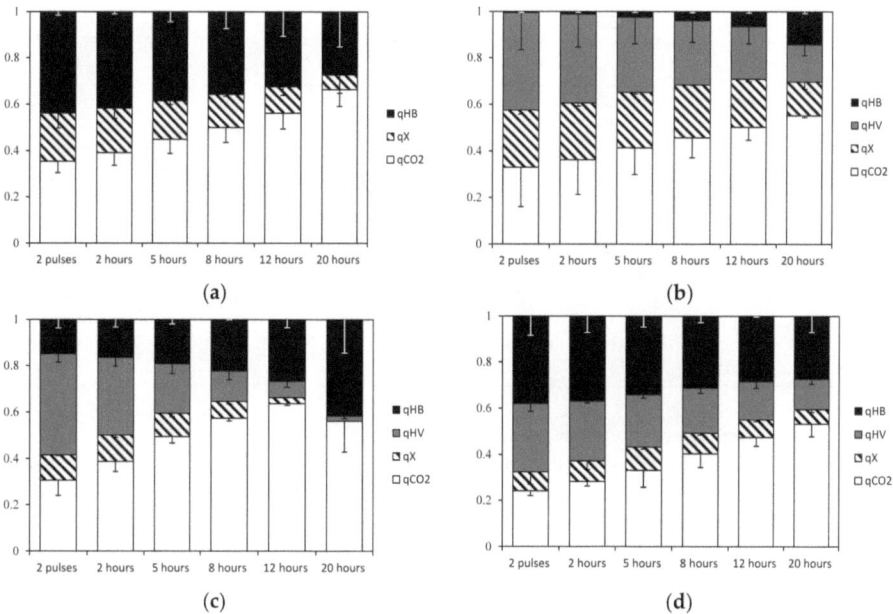

Figure 4. Carbon flux distribution (q_{PHB}, q_{PHV}, q_X, q_{CO2}) normalised respect to substrate uptake rate: (**a**) Experiment set 1: 100% acetic acid; (**b**) Experiment set 2: 100% propionic acid; (**c**) Experiment set 3: 50% acetic acid and 50% propionic acid fed simultaneously; and (**d**) Experiment set 4: 100% acetic acid alternating with 100% propionic acid.

In the model used in the present study, it was considered that the extra carbon consumed was passed to the TCA cycle for production of high energy molecules (ATP and NADH), and dissipated as ATP (Table 3). When the VFA consumption rate exceeded the respiratory capacity, a slightly decreasing trend in the respiratory quotient was observed. It should also be noted that no pathways for polymer consumption were included, although it is known that polymerization and PHA consumption can occur simultaneously [29]. It has been demonstrated that PHA operon proteins, including PHA depolymerase, are expressed from the start of the growth phase in *Pseudomonas putida* [30]. The present model was found to be feasible, if this extra carbon either goes to a depolymerization pathway or if it is spilled to produce ATP in R_7 which accounts for non-growth associated ATP maintenance. In both scenarios the same flux of CO_2 is predicted. Further model improvements that measure probable depolymerization subproducts and CO_2 production rates would be necessary to confirm the balances.

Table 3. Propionyl-CoA decarboxylation fraction, respiratory quotient and energy dissipated estimated by MFA (with standard deviation in brackets).

Experiment Set	Elapsed Duration	%Conversion PrCoA to AcCoA (mmol/mmol)		RQ (Cmmol CO_2/mol O_2)		ATP Dissipated (molATP/Cmmol VFA Consumed)	
1	2 pulses	-	-	1.19	(0.03)	0.57	(0.45)
	2 h	-	-	1.16	(0.03)	0.87	(0.46)
	5 h	-	-	1.13	(0.03)	1.32	(0.46)
	8 h	-	-	1.10	(0.04)	1.71	(0.45)
	12 h	-	-	1.07	(0.03)	2.34	(0.24)
	20 h	-	-	1.05	(0.04)	3.11	(0.44)
2	2 pulses	0.49	(0.11)	0.79	(0.03)	1.83	(1.36)
	2 h	0.51	(0.10)	0.80	(0.02)	2.09	(1.19)
	5 h	0.55	(0.07)	0.81	(0.01)	2.50	(0.90)
	8 h	0.59	(0.05)	0.81	(0.01)	2.84	(0.67)
	12 h	0.65	(0.03)	0.82	(0.01)	3.21	(0.43)
	20 h	0.75	(0.01)	0.83	(0.01)	3.68	(0.08)
3	2 pulses	0.27	(0.04)	1.08	(0.06)	0.62	(0.38)
	2 h	0.39	(0.02)	1.01	(0.03)	1.17	(0.26)
	5 h	0.55	(0.02)	0.99	(0.02)	1.99	(0.18)
	8 h	0.69	(0.02)	0.95	(0.01)	2.66	(0.08)
	12 h	0.85	(0.03)	0.95	(0.00)	3.37	(0.04)
	20 h	0.97	(0.01)	1.01	(0.08)	0.46	(0.28)
4	2 pulses	0.44	(0.05)	1.14	(0.11)	0.27	(0.04)
	2 h	0.46	(0.01)	1.10	(0.10)	0.46	(0.19)
	5 h	0.48	(0.07)	1.05	(0.08)	0.76	(0.59)
	8 h	0.54	(0.07)	1.01	(0.05)	1.30	(0.49)
	12 h	0.60	(0.07)	0.99	(0.03)	1.83	(0.35)
	20 h	0.67	(0.05)	0.97	(0.01)	2.22	(0.68)

Comparing experiments with single substrates, more CO_2 is generated when acetic acid is the only substrate compared with when propionic acid is used. The TCA cycle was calculated to be more active in Experiment set 1 (100% acetic acid), because more energy is needed to metabolize acetic acid (given 1 mol ATP is necessary to produce 1 Cmol acetyl-CoA, while activating 1 Cmol of propionic acid consumes only 0.67 mol ATP, see stoichiometry in the Supplementary Materials). On the other hand, propionic acid was found to have a higher oxygen demand when acetyl-CoA is formed from propionyl-CoA decarboxylation [31], leading to a decreased respiratory quotient (RQ). The MFA results were in agreement with this expectation. Experiment set 2, fed with propionic acid as a single substrate, had a lower RQ compared with Experiment set 1 (Table 3). However, for feeding strategies with mixed substrates, (Experiment set 3 and Experiment set 4), the RQ was very similar and remained at similar values throughout the accumulation experiments. The latter observations agreed with the composition data: similar molar fractions of propionic acid in the feed for Experiment set 3 and Experiment set 4 resulted in a constant molar fraction of propionic acid uptake relative to total carbon uptake flux (f_{Pr}).

Concerning active biomass synthesis, reaction stoichiometry indicates that propionyl-CoA gives higher theoretical growth yields compared with acetyl-CoA (1.06 mol PrCoA produces 1 mol X, and 1.27 mol AcCoA generates 1 mol X, see Supplementary Materials). According to this MFA analysis, propionyl-CoA was diverted to cell growth and PHA production during the exponential phase growth

(0–8 h) in those accumulations that had propionic acid present at all times (see Experiment sets 2 and 3 in Figure 4b,d as examples), with propionyl-CoA having been shown to be the preferred substrate for active biomass growth by Lemos et al. [32] and Jiang et al.[16]. However, at the end of all experiments, when the decarboxylation rate was high, acetyl-CoA units were converted into 3HB monomers. As a consequence, 3HV monomers dominate in the early stages of accumulation, with their formation rate decreasing over time. To understand this result, there needs to be some consideration of the role that generation of reducing equivalents plays in controlling the pathways.

3.4. Pathways for Generation of Reducing Power

There are three important cofactors involved in PHA synthesis and regulation: coenzyme-A, NADH/NAD$^+$ and NADPH/NADP$^+$ [28,33]. The relative concentrations of acyl-CoA and free coenzyme-A are critical in controlling metabolic pathways, in particular PHA storage. NADH participates in catabolic reactions, while NADPH has an important role in reductive biosynthesis such as PHA biopolymers and active biomass [33]. In reduced metabolic networks, NADPH is not considered separately. The model used in the present study, developed by Pardelha et al. [17], works under the assumption that there exists free interchange between the reducing equivalents NADPH and NADH [34].

PHA production is favored when NADPH concentrations and NADPH/NADP$^+$ ratios are high [13]. In the model used in this current study, both PHA accumulation and energy production during oxidative phosphorylation require a source of reducing power. Therefore, the metabolic processes where reducing equivalents (NADH) are formed become key factors. Considering the processes outlined in Figure 1, these are reactions related to cellular growth (R$_4$ and R$_5$), the TCA cycle (R$_6$) and decarboxylation of propionyl-CoA to acetyl-CoA (R$_3$). In this sense, fluctuations in active biomass growth and PHA synthesis activities could be related to changes in the flux though the TCA cycle and/or decarboxylation of propionyl-CoA. On the other hand, VFA uptake and cell growth have an ATP requirement which is met by the TCA cycle and the electron transport of the respiratory chain [35].

One limitation for maximizing PHA yields is the regeneration of reducing equivalents (NADPH/NADH). MFA results showed that most of NADH was generated by the TCA cycle. According to the metabolic model, less reductive equivalents are necessary to produce 1 Cmol of propionyl-CoA* than 1 Cmol of acetyl-CoA* (0.167 vs. 0.25 mol NADH, respectively). Therefore, in order to produce 3HB monomer units, a greater amount of carbon must be directed to TCA cycle for NADH generation reducing equivalents (NADH).

To investigate the role of these pathways for generation of reducing equivalents in evolution of copolymer composition, metabolic flux analysis (MFA) was performed at different stages of the accumulations for all the feed regimes tested (Experiment sets 1, 2, 3, and 4) (see Figure 5).

In cultures fed with alternating substrates (Exp 4 and 4′), carbon fluxes for active biomass formation and 3HV monomer production remained constant during the accumulations (Figure 4). As shown in Figure 5b, NADH generation rate by propionyl-CoA decarboxylation (v_3) was kept low, to a level to cover energy production by TCA cycle requirements. On the other hand, in experiments with mixed acetic and propionic feeds (Exp 3 and 3′), the decarboxylation rates increased markedly when the active biomass growth rates attenuated, and thus more acetyl-CoA units became available and the 3HB production rate could increase as a result.

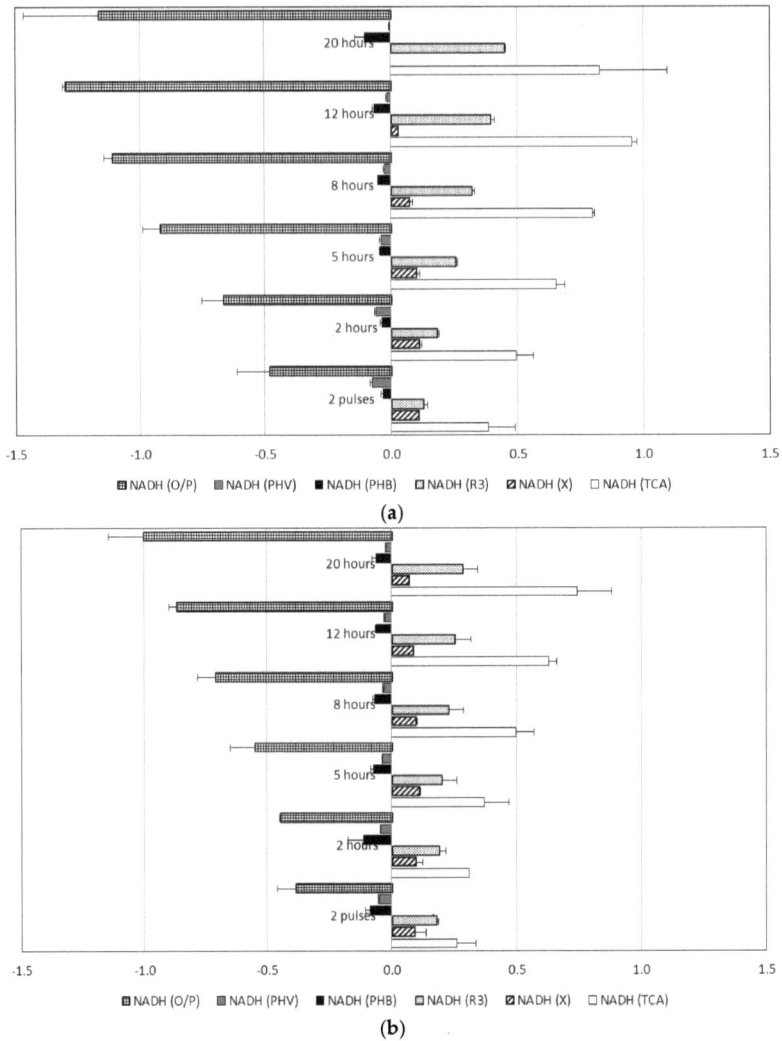

Figure 5. NADH generated and consumed at different stages of culture obtained by metabolic flux analysis (MFA): (**a**) Experiment set 3: 50% acetic acid and 50% propionic acid fed simultaneously; and (**b**) Experiment set 4: 100% acetic acid alternating with 100% propionic acid. (NADH was considered as internal metabolite).

As mentioned previously, propionic acid has been shown to be the preferred substrate for active biomass formation. Production of pure 3HV was not possible; however, it was close to 100% at the beginning of accumulations based on propionic acid alone (Experiment set 2). It has been demonstrated that keeping an optimal active biomass specific growth rate enhances PHBV copolymer synthesis in *Cupriavidus necator* [36]. However, at high specific cell growth rates, more substrate is used for active biomass formation and less is available for PHA production. The specific cell growth rate in the present study was relatively low compared to pure cell cultures, and it was found that the higher the specific growth rate, the higher the %3HV (fitted data for specific growth rate and monomer specific synthesis rates are in the Supplementary Materials). Grousseau et al. [6] found that a higher PHB yield on substrate was obtained when the Entner–Doudoroff pathway was active. This pathway produces NADPH and is linked to anabolic requirements. It supports the idea that maintaining cellular growth offers an alternative pathway to TCA cycle for NADPH generation and it favours PHA synthesis.

When cells experienced a degree of cell growth limitation, a larger proportion of 3HB monomer units compared to 3HV monomers units was produced. In those cultures fed continuously with acetic acid (Experiment sets 1 and 3), reductive equivalents can be directly generated by acetyl-CoA pathway through the TCA cycle, which favors PHA production. Similar PHA fluxes were obtained using acetic or propionic acid as single substrates (Figure 4). However, total PHA yield on substrate was higher for cultures fed exclusively with acetic acid than cultures fed with propionic acid as sole substrate (Table 2). However, compared with accumulations fed with propionic acid exclusively (Experiment set 2) or periodically (Experiment set 4), acetic acid as a feed did not stimulate concurrent active growth and storage as much as propionic acid. Previous MFA studies have suggested that when acetic and propionic acids are fed simultaneously, the catabolic activity (TCA) primarily depends on acetic acid uptake [4]. According to the metabolic model, active biomass synthesis has a higher demand of ATP and NADH when it is generated from acetic acid rather than propionic acid. In further MFA calculations, the rate of active biomass synthesis is not considered as being proportional to the consumption rate of acetic and propionic acids. This has resulted in a non-redundant MFA system, which showed that when acetic acid and propionic acids are fed simultaneously, most of the new active biomass is synthesized from propionic acid uptake. Future studies where more experimental rates are available (such as CO_2 evolution) would be required to test this hypothesis.

4. Conclusions

This study presented an analysis of PHA accumulation processes by mixed cultures when adopting different feeding strategies which favor concurrent cellular growth and carbon storage. Although higher maximum specific growth rates were achieved in cultures fed continuously with acetic acid as a sole substrate or as part of a mixture, constant cell growth was not achieved. Significant changes through time to the instantaneous 3HV content were observed under most accumulation conditions, and such changes cannot be adequately described by existing metabolic models. An alternating feeding strategy resulted in constant instantaneous 3HV content, despite the decarboxylation rate increasing with time. Overall, the 3HV monomer production rate is high. Finally, the incorporation of cellular metabolism in the evaluation of process performance for PHA production by mixed cultures offers an opportunity to help understand PHA polymer composition fluctuations and the carbon flux distribution in different cell physiological states. In this way, metabolic models can help improve the estimation of process final concentrations and yields. However, a better description of the 3HB:3HV fluctuations relative to active biomass growth needs a more detailed metabolic network to take into account the reactions in which NADPH and NADH are formed.

Supplementary Materials: The following are available online at www.mdpi.com/2306-5354/4/1/20/s1, Figure S1: Experimental data for PHA fraction and relative active biomass production; (a) Experiment set 1: 100% acetic acid; (b) Experiment set 2: 100% propionic acid. Figure S2: Fitted data for specific growth rate and monomer specific synthesis rate. Experiment set 1: 100% acetic acid; Experiment set 2: 100% propionic acid; Experiment set 3: 50% acetic acid and 50% propionic acid fed simultaneously; Experiment set 4: 100% acetic acid alternating with 100% propionic acid, Figure S3: NADH generated and consumed at different stages of culture obtained by metabolic flux analysis (MFA). (a) Experiment set 1: 100% acetic acid; (b) Experiment set 2: 100% propionic acid, Figure S4: Internal carbon flux distribution (v3, v6, v9, v10, vX) normalised respect to substrate uptake rate. (a) Experiment set 1: 100% acetic acid; (b) Experiment set 2: 100% propionic acid; (c) Experiment set 3: 50% acetic acid and 50% propionic acid fed simultaneously; (d) Experiment set 4: 100% acetic acid alternating with 100% propionic acid, Table S1: Metabolic network for PHA processes by microbial mixed cultures used in the present work, Table S2: Microstructure of PHA copolymer samples by 13C NMR analysis.

Acknowledgments: The authors would like to acknowledge Australian Research Council for funding this work through project LP0990917. The authors also thank AnoxKaldnes Sweden for funding through grant ARC LP0990917, and acknowledge gratefully the support provided by Monica Arcos-Hernandez, Lamija Karabegovic, Per Magnusson and Anton Karlsson in helping with the running of the pilot plant, accumulation studies and sample analysis. The authors confirm that there are no conflict of interests to declare.

Author Contributions: Liliana Montano-Herrera, Bronwyn Laycock, Steven Pratt and Alan Werker all conceived and designed the experiments; Liliana Montano-Herrera performed the experiments; Liliana Montano-Herrera, Bronwyn Laycock, Steven Pratt and Alan Werker all analyzed the data; Liliana Montano-Herrera wrote the paper, which was reviewed and edited by the other authors.

Conflicts of Interest: The authors declare no conflict of interest.

Abbreviations

%3HV	3HV fraction in total PHA (mol 3HV/mol PHA)
%3HV^{inst}	Instantaneous 3HV content (mol 3HV/mol PHA)
%PHA	PHA intracellular content (gPHA/gVSS)
%PHA$_0$	Initial PHA intracellular content (gPHA/gVSS)
f_{Pr}	Fraction of propionic acid uptake to total carbon uptake flux
3HB	3-Hydroxybutyrate
3HV	3-Hydroxyvalerate
AcCoA	Acetyl-CoA
ATP	Adenosine triphosphate
CO_2	Carbon dioxide
CoA	Coenzyme A
COD	Chemical oxygen demand
f_{PHA}	PHA fraction with respect to active biomass (PHA/X)
HAc	Acetic acid
HPr	Propionic acid
NAD$^+$	Nicotinamide adenine dinucleotide (oxidised)
NADH	Nicotinamide adenine dinucleotide (reduced)
NADP$^+$	Nicotinamide adenine dinucleotide phosphate (oxidised)
NADPH	Nicotinamide adenine dinucleotide phosphate (reduced)
PHA	Poly(3-hydroxyalkanoate) or PHA concentration at certain time
PHA$_0$	Initial PHA concentration
PHB	Poly(3-hydroxybutyrate)
PHV	Poly(3-hydroxyvalerate)
PrCoA	Propionyl-CoA
q_{HAc}	Acetic acid specific consumption rate (Cmol HAc/Cmol X·h)
q_{HB}	Specific 3HB monomer synthesis rate (Cmol 3HB/Cmol X·h)
q_{HPr}	Propionic acid specific consumption rate (Cmol HPr/Cmol X·h)
q_{HV}	Specific 3HV monomer synthesis rate (Cmol 3HV/Cmol X·h)
q_{PHA}	Specific PHA synthesis rate (Cmol PHA/Cmol X·h)
q_S	Specific VFA consumption rate (Cmol VFA/Cmol X·h)
RQ	Respiratory quotient
TCA	Tricarboxylic acid cycle
v	Reaction rate (Cmol/Cmol X·h)
VFA	Volatile fatty acid
VSS	Volatile suspended solids
X	Active biomass
X$_0$	Initial active biomass

References

1. Laycock, B.; Halley, P.; Pratt, S.; Werker, A.; Lant, P. The chemomechanical properties of microbial polyhydroxyalkanoates. *Prog. Polym. Sci.* **2013**, *38*, 536–583. [CrossRef]

2. Filipe, C.D.M.; Daigger, G.T.; Grady, C.P.L. A metabolic model for acetate uptake under anaerobic conditions by glycogen accumulating organisms: Stoichiometry, kinetics, and the effect of pH. *Biotechnol. Bioeng.* **2001**, *76*, 17–31. [CrossRef] [PubMed]

3. Taguchi, K.; Taguchi, S.; Sudesh, K.; Maehara, A.; Tsuge, T.; Doi, Y. Metabolic pathways and engineering of polyhydroxyalkanoate biosynthesis. *Biopolym. Online* **2005**, *3*. [CrossRef]

4. Dias, J.M.L.; Oehmen, A.; Serafim, L.S.; Lemos, P.C.; Reis, M.A.M.; Oliveira, R. Metabolic modelling of polyhydroxyalkanoate copolymers production by mixed microbial cultures. *BMC Syst. Biol.* **2008**, *2*, 59. [CrossRef] [PubMed]

5. Lemos, P.C.; Serafim, L.S.; Santos, M.M.; Reis, M.A.M.; Santos, H. Metabolic pathway for propionate utilization by phosphorus-accumulating organisms in activated sludge: C-13 labeling and in vivo nuclear magnetic resonance. *Appl. Environ. Microb.* **2003**, *69*, 241–251. [CrossRef]

6. Grousseau, E.; Blanchet, E.; Deleris, S.; Albuquerque, M.G.E.; Paul, E.; Uribelarrea, J.L. Impact of sustaining a controlled residual growth on polyhydroxybutyrate yield and production kinetics in *Cupriavidus necator*. *Bioresour. Technol.* **2013**, *148*, 30–38. [CrossRef] [PubMed]

7. Gurieff, N. Production of Biodegradable Polyhydroxyalkanoate Polymers Using Advanced Biological Wastewater Treatment Process Technology. Ph.D. Thesis, The University of Queensland, Brisbane, Australia, 2007.

8. Serafim, L.S.; Lemos, P.C.; Torres, C.; Reis, M.A.M.; Ramos, A.M. The influence of process parameters on the characteristics of polyhydroxyalkanoates produced by mixed cultures. *Macromol. Biosci.* **2008**, *8*, 355–366. [CrossRef] [PubMed]

9. Albuquerque, M.G.E.; Martino, V.; Pollet, E.; Averous, L.; Reis, M.A.M. Mixed culture polyhydroxyalkanoate (PHA) production from volatile fatty acid (VFA)-rich streams: Effect of substrate composition and feeding regime on pha productivity, composition and properties. *J. Biotechnol.* **2011**, *151*, 66–76. [CrossRef] [PubMed]

10. Ivanova, G.; Serafim, L.S.; Lemos, P.C.; Ramos, A.M.; Reis, M.A.M.; Cabrita, E.J. Influence of feeding strategies of mixed microbial cultures on the chemical composition and microstructure of copolyesters p(3HB-*co*-3HV) analyzed by NMR and statistical analysis. *Magn. Reson. Chem.* **2009**, *47*, 497–504. [CrossRef] [PubMed]

11. Arcos-Hernandez, M.V.; Laycock, B.; Donose, B.C.; Pratt, S.; Halley, P.; Al-Luaibi, S.; Werker, A.; Lant, P.A. Physicochemical and mechanical properties of mixed culture polyhydroxyalkanoate (PHBV). *Eur. Polym. J.* **2013**, *49*, 904–913. [CrossRef]

12. Johnson, K.; Van Loosdrecht, M.C.M.; Kleerebezem, R. Influence of ammonium on the accumulation of polyhydroxybutyrate (PHB) in aerobic open mixed cultures. *J. Biotechnol.* **2010**, *147*, 73–79. [CrossRef] [PubMed]

13. Pardelha, F.; Albuquerque, M.G.E.; Reis, M.A.M.; Dias, J.M.L.; Oliveira, R. Flux balance analysis of mixed microbial cultures: Application to the production of polyhydroxyalkanoates from complex mixtures of volatile fatty acids. *J. Biotechnol.* **2012**, *162*, 336–345. [CrossRef] [PubMed]

14. Serafim, L.S.; Lemos, P.C.; Oliveira, R.; Ramos, A.M.; Reis, M.A.M. High storage of PHB by mixed microbial cultures under aerobic dynamic feeding conditions. *Eur. Symp. Environ. Biotechnol.* **2004**, 479–482.

15. Valentino, F.; Karabegouic, L.; Majone, M.; Morgan-Sagastume, F.; Werker, A. Polyhydroxyalkanoate (PHA) storage within a mixed-culture biomass with simultaneous growth as a function of accumulation substrate nitrogen and phosphorus levels. *Water Res.* **2015**, *77*, 49–63. [CrossRef] [PubMed]

16. Jiang, Y.; Hebly, M.; Kleerebezem, R.; Muyzer, G.; van Loosdrecht, M.C.M. Metabolic modeling of mixed substrate uptake for polyhydroxyalkanoate (PHA) production. *Water Res.* **2011**, *45*, 1309–1321. [CrossRef] [PubMed]

17. Pardelha, F.; Albuquerque, M.G.E.; Reis, M.A.M.; Oliveira, R.; Dias, J.M.L. Dynamic metabolic modelling of volatile fatty acids conversion to polyhydroxyalkanoates by a mixed microbial culture. *New Biotechnol.* **2014**, *31*, 335–344. [CrossRef] [PubMed]

18. Murugan Janarthanan, O.; Laycock, B.; Montano-Herrera, L.; Lu, Y.; Arcos-Hernandez, M.V.; Werker, A.; Pratt, S. Fluxes in PHA-storing microbial communities during enrichment and biopolymer accumulation processes. *New Biotechnol.* **2016**, *33*, 61–72. [CrossRef] [PubMed]

19. Werker, A.G.; Bengtsson, S.O.H.; Karlsson, C.A.B. Method for Accumulation of Polyhydroxyalkanoates in Biomass with on-Line Monitoring for Feed Rate Control and Process Termination. WO 2011070544 A2, 16 June 2011.

20. Morgan-Sagastume, F.; Pratt, S.; Karlsson, A.; Cirne, D.; Lant, P.; Werker, A. Production of volatile fatty acids by fermentation of waste activated sludge pre-treated in full-scale thermal hydrolysis plants. *Bioresour. Technol.* **2011**, *102*, 3089–3097. [CrossRef] [PubMed]

21. American Public Health Association (APHA). *Standard Methods for the Examination of Water and Wastewater*; American Public Health Association: Washington, DC, USA, 1995.

22. Motulsky, H.J. *Prism5 Statistics Guide*; Graphpad Software Inc.: San Diego, CA, USA, 2007.
23. Zwietering, M.H.; Jongenburger, I.; Rombouts, F.M.; Vantriet, K. Modeling of the bacterial growth curve. *Appl. Environ. Microb.* **1990**, *56*, 1875–1881.
24. Third, K.A.; Newland, M.; Cord-Ruwisch, R. The effect of dissolved oxygen on PHB accumulation in activated sludge cultures. *Biotechnol. Bioeng.* **2003**, *82*, 238–250. [CrossRef] [PubMed]
25. Villadsen, J.; Nielsen, J.; Lidén, G. *Bioreaction Engineering Principles*, 3rd ed.; Springer: New York, NY, USA, 2011.
26. Klamt, S.; Saez-Rodriguez, J.; Gilles, E.D. Structural and functional analysis of cellular networks with cellnetanalyzer. *BMC Syst. Biol.* **2007**. [CrossRef] [PubMed]
27. Stephanopolulos, G.N.; Aristidou, A.A.; Nielsen, J. *Metabolic Engineering: Principles and Methodologies*; Academic Press: San Diego, CA, USA, 1998.
28. Escapa, I.F.; Garcia, J.L.; Buhler, B.; Blank, L.M.; Prieto, M.A. The polyhydroxyalkanoate metabolism controls carbon and energy spillage in *Pseudomonas putida*. *Environ. Microbiol.* **2012**, *14*, 1049–1063. [CrossRef] [PubMed]
29. Ren, Q.; de Roo, G.; Ruth, K.; Witholt, B.; Zinn, M.; Thony-Meyer, L. Simultaneous accumulation and degradation of polyhydroxyalkanoates: Futile cycle or clever regulation? *Biomacromolecules* **2009**, *10*, 916–922. [CrossRef] [PubMed]
30. Arias, S.; Bassas-Galia, M.; Molinari, G.; Timmis, K.N. Tight coupling of polymerization and depolymerization of polyhydroxyalkanoates ensures efficient management of carbon resources in *Pseudomonas putida*. *Microb. Biotechnol.* **2013**, *6*, 551–563. [CrossRef] [PubMed]
31. Lefebvre, G.; Rocher, M.; Braunegg, G. Effects of low dissolved-oxygen concentrations on poly(3-hydroxybutyrate-*co*-3-hydroxyvalerate) production by *Alcaligenes eutrophus*. *Appl. Environ. Microb.* **1997**, *63*, 827–833.
32. Lemos, P.C.; Serafim, L.S.; Reis, M.A.M. Synthesis of polyhydroxyalkanoates from different short-chain fatty acids by mixed cultures submitted to aerobic dynamic feeding. *J. Biotechnol.* **2006**, *122*, 226–238. [CrossRef] [PubMed]
33. Yu, J.; Si, Y.T. Metabolic carbon fluxes and biosynthesis of polyhydroxyalkanoates in *Ralstonia eutropha* on short chain fatty acids. *Biotechnol. Prog.* **2004**, *20*, 1015–1024. [CrossRef] [PubMed]
34. Kim, J.I.; Varner, J.D.; Ramkrishna, D. A hybrid model of anaerobic *E. Coli* GJT001: Combination of elementary flux modes and cybernetic variables. *Biotechnol. Prog.* **2008**, *24*, 993–1006. [CrossRef] [PubMed]
35. Zeng, A.P.; Ross, A.; Deckwer, W.D. A method to estimate the efficiency of oxidative-phosphorylation and biomass yield from ATP of a facultative anaerobe in continuous culture. *Biotechnol. Bioeng.* **1990**, *36*, 965–969. [CrossRef] [PubMed]
36. Shimizu, H.; Kozaki, Y.; Kodama, H.; Shioya, S. Maximum production strategy for biodegradable copolymer P(HB-*co*-HV) in fed-batch culture of *Alcaligenes eutrophus*. *Biotechnol. Bioeng.* **1999**, *62*, 518–525. [CrossRef]

bioengineering

MDPI

Article

Molecular Diagnostic for Prospecting Polyhydroxyalkanoate-Producing Bacteria

Eduarda Morgana da Silva Montenegro [1], Gabriela Scholante Delabary [1],
Marcus Adonai Castro da Silva [1], Fernando Dini Andreote [2] and André Oliveira de Souza Lima [1,*]

[1] Centro de Ciências Tecnológicas da Terra e do Mar, Universidade do Vale do Itajaí, R. Uruguai 458,
     88302-202 Itajaí-SC, Brazil; dudamorgana@gmail.com (E.M.S.M.); gabidelabary@hotmail.com (G.S.D.);
     marcus.silva@univali.br (M.A.C.S.)
[2] Department of Soil Science, "Luiz de Queiroz" College of Agriculture, University of São Paulo,
     Piracicaba-SP 13418-260, Brazil; fdandreo@gmail.com
*   Correspondence: andreolima@gmail.com; Tel.: +55-(47)-3341-7716; Fax: +55-(47)-3341-7715

Academic Editor: Martin Koller
Received: 24 April 2017; Accepted: 19 May 2017; Published: 25 May 2017

Abstract: The use of molecular diagnostic techniques for bioprospecting and microbial diversity study purposes has gained more attention thanks to their functionality, low cost and quick results. In this context, ten degenerate primers were designed for the amplification of polyhydroxyalkanoate synthase (*phaC*) gene, which is involved in the production of polyhydroxyalkanoate (PHA)—a biodegradable, renewable biopolymer. Primers were designed based on multiple alignments of *phaC* gene sequences from 218 species that have their genomes already analyzed and deposited at Biocyc databank. The combination of oligos *phaCF3/phaCR1* allowed the amplification of the expected product (PHA synthases families types I and IV) from reference organisms used as positive control (PHA producer). The method was also tested in a multiplex system with two combinations of initiators, using 16 colonies of marine bacteria (pre-characterized for PHA production) as a DNA template. All amplicon positive organisms ($n = 9$) were also PHA producers, thus no false positives were observed. Amplified DNA was sequenced ($n = 4$), allowing for the confirmation of the *phaC* gene identity as well its diversity among marine bacteria. Primers were also tested for screening purposes using 37 colonies from six different environments. Almost 30% of the organisms presented the target amplicon. Thus, the proposed primers are an efficient tool for screening bacteria with potential for the production of PHA as well to study PHA genetic diversity.

Keywords: bioprospecting; biopolymer; environmental diversity

1. Introduction

Bioplastics polymers have emerged as an alternative to the excessive use of polymers from petrochemical origin, which represent a problem in terms of waste management and environmental impact [1,2]. Polyhydroxyalkanoate (PHA) is among these polymers and has attracted increasing attention due to its properties and suitability for biodegradation, as well as its biocompatibility and thermoplastic characteristics [3,4]. These biopolymers accumulate in the cytoplasm of cells in the form of granules due to nutritional limitations that restrict growth [5]. They are generally associated with carbon reserves or excess in the medium, as well as reduced energy equivalents [6]. Although the procedure for the formation and accumulation of biopolymers is well-known, the main impediment to employing biopolymers is the large scale and the high cost of PHA production, which is nine times more expensive than the production of synthetic plastics [7,8]. In this sense, the bioprospection of bacteria capable of producing these biopolymers in greater quantities from the conversion of cheaper and renewable substrates is necessary, aiming a greater production and consequent reduction in cost [9–11].

The genes responsible for PHA synthesis can be classified into four different classes, according to the organization of gene locus and the structural and functional properties of enzymes PHA synthase [12,13]. Class I is represented by gene *phaC* of *Cupriavidus necator*, and class II by *Pseudomonas*, where PHA synthase is encoded by *phaC1* and *phaC2* [14]. Class III synthase is composed of the genes *phaC* and *phaE* and can be found in the model organism *Allochromatium vinosum* [15]. Class IV synthase is represented by *Bacillus megaterium*, in which the main genes are *phaC* and *phaR* [16]. Among the genes involved in PHA production, *phaC* is the most important since it encodes the key enzyme for PHA synthesis, thus justifying its choice as an indicative of possible producers of these biopolymers [11].

PHA-producing organisms can be identified and evaluated by different methods [17]. Among the traditional methods, the most frequently used are based on microscopy and specific dyes, such as the lipophilic dye Sudan Black B [18], the fluorescent dye from the Nile [19], and the Nile Red dye [20]. The traditional identification techniques require specific conditions for each bacterium and therefore become more laborious, and moreover offer no specificity and may indicate false positives [13,21,22]. In this context, molecular methods appear as an effective tool for the selection and diagnosis of PHA-producing bacteria, for agility of results, ease of handling, and low cost. Among these techniques, polymerase chain reaction (PCR) is simple and efficient for such a diagnosis [23], as it involves the use of specific primers for the locus of the gene responsible for PHA synthase, the biosynthesis of interest [13,21,22,24–27]. Thus, the present work aims to design primers capable of identifying bacteria that produce different classes of PHAs, as well as the prospection of environments for the pre-selection of the producing organisms and thus the analysis of the environmental diversity of such organisms.

2. Experimental Procedures

2.1. Bacteria Strains and Media

The reference bacteria strains *Bacillus pumilus* ATCC 14884, *Bacillus thurigiensis var. israelensis* 4Q2-72, *Bacillus megaterium* ATCC 14581, *Bacillus cereus* ATCC 14579, *Chromobacterium violaceum* (CV11), and *Cupriavidus necator* DSM 545, were used as positive controls for the presence of the gene *phaC* and a negative control was made by using DNA from *Escherichia coli* DH5α. Genomic DNA was isolated (DNeasy Blood & Tissue Kit, Qiagen, Hilden, Germany), quantified after agarose gel electrophoresis (Kodak 1D v.3.5.5b, Kodak, Rochester, NY, USA), and used as a DNA template for PCR (approximately 50 ng per reaction).

Genomic DNA from twelve marine bacteria, *Pseudomonas* sp. (LAMA 572), *Halomonas hydrothermalis* (LAMA 685), *Micrococcus luteus* (LAMA 702), *Brevibacterium* sp. (LAMA 758), *Halomonas* sp. (LAMA 761), LAMA 677, LAMA 726, LAMA 729, LAMA 737, LAMA 748, LAMA 760, LAMA 765, LAMA 790, and LAMA 896, previously recognized (tested by staining with Nile Red in seven culture media) as PHA producers, were used in the developed PHA PCR. Also, negative marine bacteria were applied; *Idiomarina loihiensis* and *Terribacillus saccharophilus*. To check the efficiency in pre-selection of PHA-producer bacteria isolated from the environment, 37 newly isolated bacteria from soils were tested. Genomic DNA of these bacteria was obtained as described, and similarly used for amplifications.

2.2. Design and Evaluation of Primers for the Amplification of the Gene phaC

The protocol for primer design was similar to that previously described Lima & Garcês [28]. A total of 218 sequences of the superfamily *phaC* gene were retrieved from the BioCyc Database Collection [29] and analyzed in Megan 4 program [30]. Sequences were aligned and phylogeny was inferred on the basis of neighbor-joining trees built from a similarity matrix determined by the Kimura-2 parameter. The sequences were also analyzed in the amino acid level, which was used to allocate them into the classes of PHA synthase. This was performed using the tool Conserved Domain search, available on NCBI [31]. The description of conserved regions was also evidenced by sequence alignment using the ClustalW algorithm [32] at Unipro UGENE 1.26.1 [33]. The best regions were selected for the primer design, using as parameters a high degree of identity, regions without gaps, and few degenerate bases.

The primer sequences were determined with the online program OligoAnalyzer version 3.1 (Integrate DNA Technologies, Coralville, IA, USA), in which important parameters for the efficiency of PCR reaction were defined, such as the melting temperature (Tm) and the percentage of C+G. The primers drawn were also evaluated at CLC Genomics Workbench 4.8 (CLC bio, Cambridge, MA, USA), enabling the visualization of which primer would produce more annealing results to all the gene sequences used during primer design. The parameter considered in this evaluation was the possibility of up to two degenerate bases for each primer.

2.3. Amplification of phaC Gene by PCR

The program used for the amplification of *phaC* gene fragments with all primer combinations was a cycle of 94 °C for 4 min, followed by 35 cycles of 94 °C for 45 s, 61 °C for 20 s, 72 °C for 10 s and a final extension of 72 °C for 2 min. As a template, extracted DNA or bacteria cells isolated from the environment were used for *phaC* amplification. Once the best set of primers was established, the program used for *phaC* gene amplification using reference strains was adjusted to one cycle of 94 °C for 4 min, followed by 10 cycles of 94 °C for 30 s, 68 °C for 20 s, as well as 25 cycles of 94 °C for 12 s, 65 °C for 10 s, 72 °C for 7 s and a final extension of 72 °C for 2 min. All reactions were held in the thermocycler Eppendorf Mastercycler Gradient, consisting of 20 μL containing 1× PCR amplification buffer (Invitrogen), 0.2 mM of each dNTP, 0.5 μM of each primer, 1U Taq DNA polymerase (Invitrogen), 2 mM MgCl$_2$, and the template DNA. PCR amplicons were observed by electrophoresis in 2% agarose gel further stained with ethidium bromide, and viewed under a UV transilluminator.

2.4. DNA Sequencing

The amplified fragments obtained from marine bacteria LAMA 677, LAMA 737, LAMA 748, LAMA 760, and the reference bacteria *C. violaceum* were purified (QIAquick PCR Purification Kit, Qiagen, Hilden, Germany) and sequenced in an ABI-Prism 3100 Genetic Analyzer at ACTGene (Alvorada, RS, Brazil). The identity of the sequences was evaluated through the Genomics Workbench 4.8 program accessing the tool of comparison BLASTX (Nacional Center for Biotechnology Information, Bethesda, MA, USA) [34]. The gene sequences retrieved by BLAST, in addition to the newly sequenced DNA, were pooled and analyzed for phylogenetic tree classification using multiple alignment calculated by the ClustalW algorithm [33] in Geneious v. 5.5.3 (Biomatters, Auckland, New Zealand).

3. Results

3.1. Design of Primers for Gene phaC Amplification

The phylogenetic classification of the 218 sequences of the gene *phaC* (1 sequence = 1 specie) used for primer design indicated a high percentage of organisms belonging to the phylum Proteobacteria (alpha, beta, and gamma) (Figure 1). The phyla Firmicutes and Spirochaetales were also presented, as well as organisms of the orders Chroococcales, Chloroflexales, and Actinomycelates. The domain Archea appears uniquely represented by organisms belonging to the order Halobacteriales (Figure 1). The analysis at the amino acid level showed that the regions of conserved domains were characteristic for three classes of PHA synthases; the classes I, II and III.

The application of the designed workflow resulted in the generation of 10 primers, as well as their determined characteristics (sequences, annealing temperatures, relative location to the consensus sequence) and compatibility for annealing with the 218 used sequences (Table 1). Among these, some sets were first selected to amplify the target gene *phaC*. For instance, primers *phaCF3* and *phaCR1* were selected due to their capacity to anneal with a large number of sequences. Also, primer *phaCF1* was picked due to its relative position to *phaCR1* and its ability to be used in multiplex PCR (Figure 2). These two combinations also resulted in the generation of small fragments; 304 bp for primers *phaCF1/phaCR1* and 239 bp for *phaCF3/phaCR*, which is desired when one is looking for fast detection and maximum amplification efficiency. For a shorter PCR period, amplicons are less likely to vary in size among distinct template sequences. Even so, it is important to consider that amplicon size may vary among different organisms, due to modifications that occurr during evolution. However, variation can be observed, for example, for amplifications with primers *phaCF1/phaCR1*, which resulted in fragments varying from 242 to 316 bp.

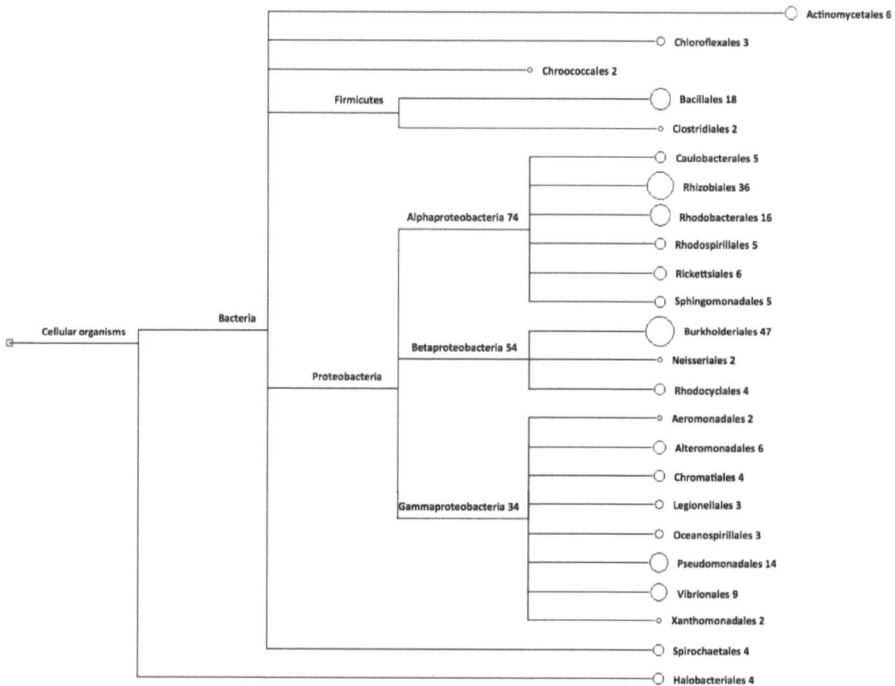

Figure 1. Analysis of the gene sequences used to design the primers, and their phylogenetic classification by the Megan 4 program.

Table 1. Primers designed showing: nomenclature, nucleotide sequences, melting temperatures, initial and final positions correspondent to the consensus sequence and the number of aligned sequences according to the database of organisms used in the study.

Primer ID	Sequence	Tm	Consensus Position	Aligned Sequences *
phaCF1	5′TGATSSAGCTGATCCAGTAC3′	53.9°	489–508	18
phaCF2	5′CCGCTGCTGATCGTBCCGCC3′	65.5°	539–558	41
phaCF3	5′CCGCCSTGGATCAACAAGT3′	58.0°	554–572	61
phaCF4	5′CTACATCCTCGACCTGMAGCCGGA3′	63.1°	574–597	24
phaCF5	5′GGCTACTGCRTCGGCGGCAC3′	65.1°	773–792	47
phaCF6	5′TGGAACDSCGACDCCACCAAC3′	61.6°	1078–1098	0
phaCF7	5′CGACRCCACCAACMTGCCGGG3′	65.8°	1086–1106	4
phaCR1	5′GTGCCGCCGAYGCAGTAGCC3′	65.1°	773–792	47
phaCR2	5′CCCGGCAKGTTGGTGGYGTCG3′	65.8°	1086–1106	4
phaCR3	5′CAGTSCGGCCACCAGSWGCC3′	66.3°	1432–1451	0

* Considering a maximum of two mismatches.

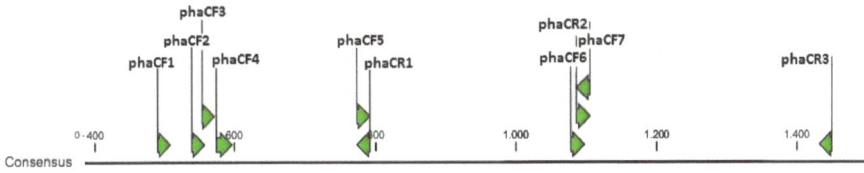

Figure 2. Positioning of the forward and reverse primers accordingly to the consensus sequence.

3.2. Partial Amplification of phaC Gene

Although a particular set of primers (*phaCF1*, *phaCF3*, and *phaCR1*) were revealed to be attractive for *phaC* amplification during in silico analysis, their efficiency and specificity has to be determined in vitro. Therefore, a total of seven pairs of primers were tested using LAMA 677 (PHA producer). Positive results were observed for three of these combinations (Figure 3A), with a remarkable match for functioning of sets previously elected by bioinformatics tools. This pair (*phaCF3/phaCR1*) was further tested using genomic DNA from already known *phaC* carrier species: the model organisms *B. pumilus* (ATCC 14884), *B. thuringiensis* var. israelensis (4Q2-72), *B. megaterium* (ATCC 14581), *B. cereus* (ATCC 14579), and *C. necator* (DSM 545). The expected fragments of 239bp were generated for all organisms tested (Figure 3B) and no amplicon was obtained from the negative control *E. coli* DH5α (data not shown).

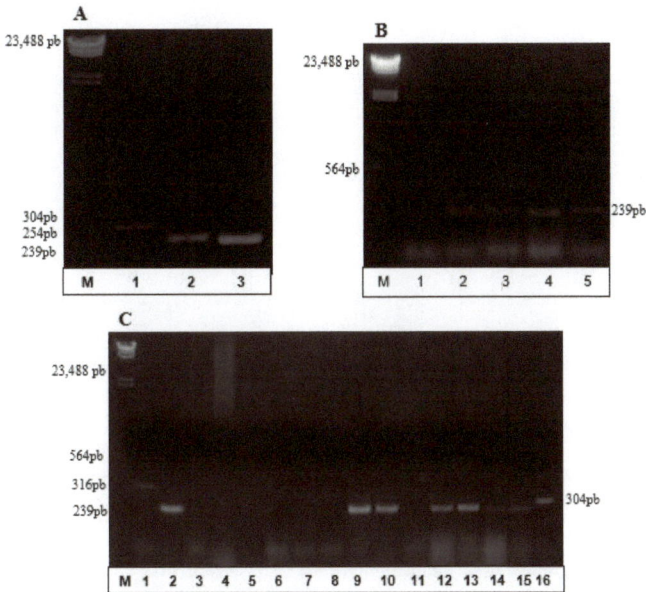

Figure 3. *phaC* gene amplification. Line M: ladder ʎ-hind III. (**A**) *phaC* amplicons for combinations of primers with positive results (organism LAMA 677): 1. *phaCF1/phaCR1*; 2. *phaCF2/phaCR1*; 3. *phaCF3/phaCR1*. (**B**) *phaC* amplicons in: 1. *Bacillus pumilus*; 2. *B. thurigiensis*; 3. *B. megaterium*; 4. *B. cereus* ATCC; 5. *C. necator*. (**C**) *phaC* amplicons in: 1. *Pseudomonas* sp. (LAMA 572); 2. LAMA 677; 3. *Halomonas hydrothermalis* (LAMA 685); 4. LAMA 691; 5. LAMA 694; 6. *Micrococcus luteus* (LAMA 702); 7. LAMA 726; 8. LAMA 729; 9. LAMA 737; 10. LAMA 748; 11. *Brevibacterium* sp. (LAMA 758); 12. LAMA 760; 13. LAMA 790; 14. *Halomonas* sp. (LAMA 761); 15. LAMA 765; 16. LAMA 896.

Once it was verified that primers were efficient in recognizing PHA-producing bacteria, they were also tested for the detection of potential new polymer producers isolated from environmental samples. For this purpose, two sets of primers (*phaCF1/phaCR1* and *phaCF3/phaCR1*) were used. The use of these primers in multiplex reactions allows for the increase of coverage for the detection of PHA producers. In this context, 16 marine organisms (14 positive and two negative PHA producers) and 37 environmental isolates (unknown PHA production) were screened. The proposed PCR protocol was able to detect *phaC* in nine marine isolates. No false positives were identified, highlighting the specificity of the primers designed. The positive control *C. violaceum* was amplified efficiently. When applying the *phaC* multiplex-PCR with the 37 environmental isolates, *phaC* amplicons were observed in approximately 30% of them (data not shown), revealing the great potential of this method for the screening of PHA producers.

3.3. DNA Sequencing and phaC Gene Identification

Amplicons from different reactions were sequenced and compared to the Genbank database. The sequences identities were compatible with *phaC* genes/proteins previously described. This indicates the specificity and efficiency of the proposed method. The originated sequences also allowed the taxonomic classification of organisms harboring the *phaC* gene (Figure 4). The differential allocation of positive isolates supports the inference that the developed tool is able to detect most of the *phaC* gene diversity that resides in bacterial cells belonging to distinct taxa.

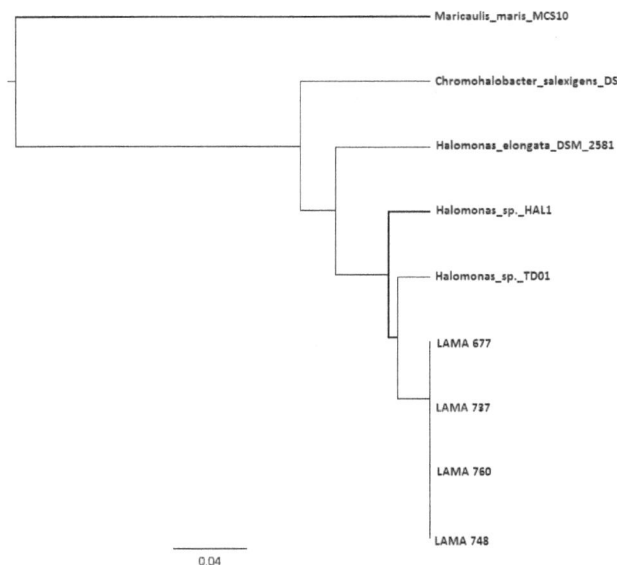

Figure 4. Sub-tree alignment of sequenced amplicons (LAMA 677, LAMA 737, LAMA 748, and LAMA 760), showing organisms with more genetic similarity. The analysis were through the software Geneious V.5.5.3.

4. Discussion

The use of molecular tools for the detection of organisms with particular features can aid in the field of biotechnology. These methodologies have been used for the determination of the microbial potential to produce PHA at the genetic level as well as to determine how much PHA can be produced by a given organism [17]. Here, we use the same approach to describe newly designed primers for the assessment of bacteria able to produce PHAs. Our approach is based on the growth of records for *phaC* gene sequences, which subsidizes our primer design [17,35]. An innovative aspect is the use of PCR to detect all distinct classes of PHA synthases; Shamala et al. [21] drew primers based only one gene sequence of *B. megaterium*, not obtaining a large breadth of results, and this approach was also employed by Solaiman & Ashby [13], as a result of the simplicity of the method used for primer design. Sheu et al. [24] used the *phaC* gene sequences from 13 gram-positive bacteria to design primers, and presented a greater breadth of results, which were capable of detecting organisms belonging PHA synthase class I and II. The present study used a wide variety of gene sequences belonging to different organisms, resulting in the generation of a precise tool for the detection of organisms with the potential for the production of PHAs from classes I and IV. This method was tested in reference organisms, and was also employed for the screening of new isolates, working in both systems for the detection of the targeted gene.

Colony PCR proved efficient, as this method used amplified regions of interest without necessary DNA extraction methods, which has also been suggested by Sheu et al. [24], making faster work of the screening of environmental organisms. Lane & Benton [23] obtained good results using the same method to determine if six cyanobacteria contained the *phaC* gene. Sasidharam et al. [22] identified the potential of *Vibrio azureus* BTKB33 isolated from marine sediments through PCR confirmation of PHA synthase class I. The use of the PCR technique considerably reduced the number of isolates and thus optimized the process. In addition to the traditional PCR, a multiplex PCR was performed. This methodology used more than one combination of primers to obtain a wider range of results and did not generate false positives, indicating that the use of specific primers for the samples and the chosen conditions were appropriate for the technique [17]. Castroverde et al. [36] showed the efficiency of identifying pathogens using three primers combined in a single PCR. The combination of primers used in the pre-selection of soil organisms in different environments was efficient, and the fragments showed the expected size in approximately 30% of the isolates. These results show the efficiency of using primers designed in the pre-selection of bacteria with the potential for PHA production in samples isolated from the environment. Tzu & Semblante [35] proved the efficiency of multiplex PCR by demonstrating that the primer set was more efficient than the primers tested individually, increasing the detection sensitivity of PHA synthases of classes I and II up to almost 90%. Class I and II PHA synthases were detected from alphaproteobacteria, betaproteobacteria, and gammaproteobacteria, indicating the wide diversity of PHA-accumulating bacteria in wastewater treatment from activated sludge.

Molecular detection of genes involved in PHA synthesis also allows for the prospection of PHA-producing organisms, as well as furthers the understanding and study of gene diversity and evolution [37,38]. Discrepancies in the phylogenetic trees for *phaA*, *phaB*, and *phaC* genes of the PHA biosynthesis have led to the suggestion that horizontal gene transfer may be a major contributor for their evolution [39]. In this way, the use of degenerate primers to study the genetic diversity of genes of biotechnological interest has been gaining prominence, as it aims to define the knowledge of conserved and variable regions of the gene, as well as the structural and functional organization of the enzyme. In the work described by Cheng et al. [40], degenerate primers are used to study the diversity of the subtilase gene with metagenomic DNA samples. This also indicates the potential use of primers in the study of environmental samples taken directly, through DNA metagenomics, which allows to access much of the genetic diversity present in the sample, since organisms that would not be cultivated in the laboratory can be studied directly by the DNA present in the sample. Tai et al. [41] successfully used a culture-independent approach for the detection of the presence of *phaC* genes in limestone soil using primers targeting the class I and II PHA synthases, reassuring the relevance of the approach used in our study.

The related sequences found in studies of diversity still have the potential to be used in genetic improvement programs by site-directed mutations, such as the DNA shuffling technique. Wang et al. [42] pointed out the efficiency of the variant technique called DNA family shuffling for metagenomic studies of homologous genes with specific primers, showing yet another possible application for the primers designed in the present study.

5. Conclusions

This study presents a powerful molecular tool for the identification and bioprospecting of bacteria that have the potential to produce PHAs. The tool also shows high potential for the identification of marine bacteria and pre-screening of environmental bacteria that have *pha*C gene, as well as for use in analyses of environmental diversity.

Acknowledgments: ICGEB/CNPq (Brazil, Process 577915/2008-8) and CNPq/INCT-Mar COI (Brazil, Process 565062/2010-7) supported this work. We also thank CNPq for the scholarship provided to A.O.S.L (Process 311010/2015-6) and G.S.D. (Process 400551/2014-4), as well as Santa Catarina State Govern for the E.M.S. scholarship.

Author Contributions: Eduarda Morgana da Silva Montenegro and Andre Oliveira de Souza Lima conceived and designed the study and experiments. Eduarda Morgana da Silva Montenegro performed experiments with Andre Oliveira de Souza Lima supervision. Marcus Adonai Castro da Silva and Fernando Dini Andreote contributed with analysis. Eduarda Morgana da Silva Montenegro, Andre Oliveira de Souza Lima and Gabriela Scholante Delabary wrote the paper with suggestions/corrections of Fernando Dini Andreote and Marcus Adonai Castro da Silva.

Conflicts of Interest: The authors declare no conflict of interest.

References

1. Song, J.H.; Murphy, R.J.; Narayan, R.; Davies, G.B.H. Biodegradable and compostable alternatives to conventional plastics. *Philos. Trans. R. Soc. Lond. B Biol. Sci.* **2009**, *364*, 2127–2139. [CrossRef] [PubMed]

2. Barnes, D.K.A.; Galgani, F.; Thompson, R.C.; Barlaz, M. Accumulation and fragmentation of plastic debris in global environments. *Philos. Trans. R. Soc. Lond. B Biol. Sci.* **2009**, *364*, 1985–1998. [CrossRef] [PubMed]

3. Steinbüchel, A.; Füchtenbush, B. Bacterial and other biological systems for polyester production. *Trends Biotechnol.* **1998**, *16*, 419–427. [CrossRef]

4. Rehm, B.H. Biogenesis of microbial polyhydroxyalkanoate granules: A platform technology for the production of tailor-made bioparticles. *Curr. Issues Mol. Biol.* **2007**, *9*, 41–62. [PubMed]

5. Bugnicourt, E.; Cinelli, P.; Lazzeri, A.; Alvarez, V.A. Polyhydroxyalkanoate (PHA): Review of synthesis, characteristics, processing and potential applications in packaging. *Express Polym. Lett.* **2014**, *8*, 791–808. [CrossRef]

6. Godbole, S. Methods for identification, quantification and characterization of polyhydroxyalkanoates. *Int. J. Bioassays* **2016**, *5*, 4977–4983. [CrossRef]

7. Khardenavis, A.A.; Kumar, M.S.; Mudliar, S.N.; Chakrabarti, T. Biotechnological conversion of agro industrial wastewaters into biodegradable plastic, poly-β-hydroxybutyrate. *Biosour. Technol.* **2007**, *98*, 3579–3584. [CrossRef] [PubMed]

8. Tan, G.Y.A.; Chen, C.L.; Li, L.; Ge, L.; Wang, L.; Razaad, I.M.N.; Li, Y.; Zhao, L.; Mo, Y.; Wang, J.-Y. Start are search on biopolymer polyhydroxyalkanoate (PHA): A review. *Polymers* **2014**, *6*, 706–754. [CrossRef]

9. Lee, S.Y.; Choi, J.; Wong, H.H. Recent advances in polyhydroxyalkanoate production by bacterial fermentation: Mini-review. *Int. J. Biol. Macromol.* **1999**, *25*, 31–36. [CrossRef]

10. Khanna, S.; Srivastava, A.K. Recent advances in microbial polyhydroxyalkanoates. *Process Biochem.* **2005**, *40*, 607–619. [CrossRef]

11. Silva, A.L.; dos Santosa, E.C.; dos Santosa, Í.A.; Lópeza, A.M. Seleção polifásica de microrganismos produtores de polihidroxialcanoatos. *Quim. Nova.* **2016**, *39*, 782–788.

12. Rehm, B.H. Polyester synthases: Natural catalysts for plastics. *Biochem. J.* **2003**, *376*, 15–33. [CrossRef] [PubMed]

13. Solaiman, D.K.; Ashby, R.D. Rapid genetic characterization of poly(hydroxyalkanoate) synthase and its applications. *Biomacromolecules* **2005**, *6*, 532–537. [CrossRef] [PubMed]

14. Hein, S.; Paletta, J.R.; Steinbüchel, A. Cloning, characterization and comparison of the *Pseudomonas mendocina* polyhydroxyalkanoate synthases *Pha*C1 and *Pha*C2. *Appl. Microbiol. Biotechnol.* **2002**, *58*, 229–236. [PubMed]

15. Yuan, W.; Jia, Y.; Tian, J.; Snell, K.D.; Muh, U.; Sinskey, A.J.; Lambalot, R.H.; Walsh, C.T.; Stubbe, J. Class I and III polyhydroxyalkanoate synthases from *Ralstonia eutropha* and *Allochromatium vinosum*: Characterization and substrate specificity studies. *Arch. Biochem. Biophys.* **2001**, *394*, 87–98. [CrossRef] [PubMed]

16. McCool, G.J.; Cannon, M.C. PhaC and PhaR are required for polyhydroxyalkanoic acid synthase activity in *Bacillus megaterium*. *J. Bacteriol.* **2001**, *183*, 4235–4243. [CrossRef] [PubMed]

17. Koller, M.; Rodríguez-Contreras, A. Techniques for tracing PHA-producing organisms and for qualitative and quantitative analysis of intra-and extracellular PHA. *Eng. Life Sci.* **2015**, *15*, 558–581. [CrossRef]

18. Murray, R.G.E.; Doetsch, R.N.; Robinow, C.F. Determinative and cytological light microscopy. *Am. Soc. Microbiol.* **1994**, *1*, 21–41.

19. Ostle, A.G.; Holt, J.G. Nile Blue A as a fluorescent stain for polybeta-hydroxybutyrate. *Appl. Environ. Microbiol.* **1982**, *44*, 238–241. [PubMed]

20. Spiekermann, P.; Rehm, B.H.; Kalscheuer, R.; Baumeister, D.; Steinbüchel, A. A sensitive, viable-colony staining method using Nile red for direct screening of bacteria that accumulate polyhydroxyalkanoic acids and other lipid storage compounds. *Arch. Microbiol.* **1999**, *171*, 73–80. [CrossRef] [PubMed]

21. Shamala, T.R.; Chandrashekar, A.; Vijayendra, S.V.; Kshama, L. Identification of polyhydroxyalkanoate (PHA)-producing Bacillus spp. using the polymerase chain reaction (PCR). *J. Appl. Microbiol.* **2003**, *94*, 369–374. [CrossRef] [PubMed]

22. Sasidharan, R.S.; Bhat, S.G.; Chandrasekaran, M. Amplification and sequence analysis of *phaC* gene of polyhydroxybutyrate producing *Vibrio azureus* BTKB33 isolated from marine sediments. *Ann. Microbiol.* **2016**, *66*, 299–306. [CrossRef]

23. Lane, C.E.; Benton, M.G. Detection of the enzymatically-active polyhydroxyalkanoate synthase subunit gene, *phaC*, in cyanobacteria via colony PCR. *Mol. Cell. Probes* **2015**, *29*, 454–460. [CrossRef] [PubMed]

24. Sheu, D.S.; Wang, Y.T.; Lee, C.Y. Rapid detection of polyhydroxyalkanoate accumulating bacteria isolated from the environment by colony PCR. *Microbiology* **2000**, *146*, 2019–2025. [CrossRef] [PubMed]

25. Solaiman, D.K.; Ashby, R.D.; Foglia, T.A. Rapid and specific identification of medium-chain-length polyhydroxyalkanoate synthase gene by polymerase chain reaction. *Appl. Microbiol. Biotechnol.* **2000**, *53*, 690–694. [CrossRef] [PubMed]

26. Kung, S.S.; Chuang, Y.C.; Chen, C.H.; Chien, C.C. Isolation of polyhydroxyalkanoates-producing bacteria using a combination of phenotypic and genotypic approach. *Lett. Appl. Microbiol.* **2007**, *44*, 364–371. [CrossRef] [PubMed]

27. Desetty, R.D.; Mahajan, V.S.; Khan, B.M.; Rawal, S.K. Isolation and heterologous expression of PHA synthesising genes from Bacillus thuringiensis R1. *World J. Microbiol. Biotechnol.* **2008**, *24*, 1769–1774. [CrossRef]

28. Lima, A.O.S.; Garcês, S.P.S. Intragenic Primer Design: Bringing Bioinformatics Tools to the Class. *Biochem. Mol. Biol. Educ.* **2006**, *34*, 332–337. [CrossRef] [PubMed]

29. Biocyc Database Collection. Available online: http://biocyc.org (accessed on 20 October 2016).

30. Huson, D.H.; Auch, A.; Qi, J.; Schuster, S.C. Megan Analysis of Metagenome Data. *Genome Res.* **2011**, *17*, 377–386. [CrossRef] [PubMed]

31. Nacional Center for Biotechnology Information. Available online: https://www.ncbi.nlm.nih.gov/Structure/cdd/wrpsb.cgi (accessed on 11 November 2016).

32. Larkin, M.A.; Blackshields, G.; Brown, N.P.; Chenna, R.; McGettigan, P.A.; McWilliam, H.; Valentin, F.; Wallace, I.M.; Wilm, A.; Lopez, R.; et al. ClustalW and ClustalX version 2. *Bioinformatics* **2007**, *23*, 2947–2948. [CrossRef] [PubMed]

33. Fursov, M.Y.; Oshchepkov, D.Y.; Novikova, O.S. UGENE: Interactive computational schemes for genome analysis. In Proceedings of the Fifth Moscow International Congress on Biotechnology, Moscow, Russia, 16–20 March 2009; Volume 3, pp. 14–15.

34. Altschul, S.F.; Madden, T.L.; Schaffer, A.A.; Zhang, J.; Zhang, Z.; Miller, W.; Lipman, D.J. Gapped BLAST and PSI-BLAST: A new generation of protein database search programs. *Nucleic Acids Res.* **1997**, *25*, 3389–3402. [CrossRef] [PubMed]

35. Tzu, H.Y.; Semblante, G.U. Detection of polyhydroxyalkanoate-accumulating bacteria from domestic wastewater treatment plant using highly sensitive PCR primers. *J. Microbiol. Biotechnol.* **2012**, *22*, 1141–1147.

36. Castroverde, C.D.M.; San Luis, B.B.; Monsalud, R.G.; Hedreyda, C.T. Differential detection of vibrios pathogenic to shrimp by multiplex PCR. *J. Gen. Appl. Microbiol.* **2006**, *52*, 273–280. [CrossRef] [PubMed]

37. Sujatha, K.; Mahalakshmi, A.; Shenbagarathai, R. Molecular characterization of *Pseudomonas* sp. LDC-5 involved in accumulation of poly 3 hydroxybutyrate and medium-chain-length poly 3-hydroxyalkanoates. *Arch. Microbiol.* **2007**, *188*, 451–462. [CrossRef] [PubMed]

38. Aneja, K.K.; Ashby, R.D.; Solaiman, D.K.Y. Altered composition of *Ralstonia eutropha* poly (hydroxyalkanoate) through expression of PHA synthase from *Allochromatium vinosum* ATCC 35206. *Biotechnol. Lett.* **2009**, *31*, 1601–1612. [CrossRef] [PubMed]

39. Kalia, V.C.; Lal, S.; Cheema, S. Insight in to the phylogeny of polyhydroxyalkanoate biosynthesis: Horizontal gene transfer. *Gene* **2007**, *389*, 19–26. [CrossRef] [PubMed]

40. Cheng, X.; Gao, M.; Wang, M.; Liu, H.; Sun, J.; Gao, J. Subtilase genes diversity in the biogas digester microbiota. *Curr. Microbiol.* **2011**, *62*, 1542–1547. [CrossRef] [PubMed]

41. Tai, Y.T.; Foong, C.P.; Najimudin, N.; Sudesh, K. Discovery of a new polyhydroxyalkanoate synthase from limestone soil through metagenomic approach. *J. Biosci. Bioeng.* **2016**, *121*, 355–364. [CrossRef] [PubMed]

42. Wang, Q.; Wu, H.; Wang, A.; Du, P.; Pei, X.; Li, H.; Yin, X.; Huang, L.; Xiong, X. Prospecting Metagenomic Enzyme Subfamily Genes for DNA Family Shuffling by a Novel PCR-based Approach. *J. Biol. Chem.* **2010**, *285*, 41509–41516. [CrossRef] [PubMed]

bioengineering

MDPI

Review

Cyanobacterial PHA Production—Review of Recent Advances and a Summary of Three Years' Working Experience Running a Pilot Plant

Clemens Troschl [1], Katharina Meixner [2,*] and Bernhard Drosg [2]

[1]   Institute of Environmental Biotechnology, Department of Agrobiotechnology, IFA-Tulln,
      University of Natural Resources and Life Sciences, Vienna, Tulln 3430, Austria; clemens.troschl@boku.ac.at
[2]   Bioenergy2020+ GmbH, Tulln 3430, Austria; bernhard.drosg@bioenergy2020.eu
*   Correspondence: katharina.meixner@bioenergy2020.eu; Tel.: +43-01-47654-97435

Academic Editor: Martin Koller
Received: 7 March 2017; Accepted: 16 March 2017; Published: 28 March 2017

Abstract: Cyanobacteria, as photoautotrophic organisms, provide the opportunity to convert CO_2 to biomass with light as the sole energy source. Like many other prokaryotes, especially under nutrient deprivation, most cyanobacteria are able to produce polyhydroxyalkanoates (PHAs) as intracellular energy and carbon storage compounds. In contrast to heterotrophic PHA producers, photoautotrophic cyanobacteria do not consume sugars and, therefore, do not depend on agricultural crops, which makes them a green alternative production system. This review summarizes the recent advances in cyanobacterial PHA production. Furthermore, this study reports the working experience with different strains and cultivating conditions in a 200 L pilot plant. The tubular photobioreactor was built at the coal power plant in Dürnrohr, Austria in 2013 for direct utilization of flue gases. The main challenges were the selection of robust production strains, process optimization, and automation, as well as the CO_2 availability.

Keywords: cyanobacteria; polyhydroxyalkanoates; CO_2 mitigation; flue gas utilization; photobioreactor

1. Introduction

Polyhydroxyalkanoates (PHAs) are considered as one of the most promising bioplastics. Their mechanical properties are similar to polypropylene and they can be processed in a similar way, including extrusion, injection molding, or fiber spinning [1]. One of the major advantages of PHAs are their biodegradability. They are degraded relatively rapidly by soil organisms, allowing easy composting of PHA waste material [2].

Currently, PHA is produced in large fermenters by heterotrophic bacteria, like *Cupriavidus necator* or recombinant *Escherichia coli* [3]. For these fermentation processes large amounts of organic carbon sources like glucose are necessary, accounting for approximately 50% of the total production costs [4]. An alternative way of producing PHA is the use of prokaryotic algae, better known as cyanobacteria. As part of the phytoplankton, they are global primary biomass producers using light as the sole energy source to bind atmospheric CO_2 [5]. Burning of fossil fuels has increased the atmospheric CO_2 concentration from approximately 300 ppm in 1900 to over 400 ppm today. The latest report of the intergovernmental panel on climate change (IPCC) clearly indicates anthropogenic CO_2 emissions as the main driver for climate change [6]. Given these facts, cultivation of cyanobacteria for PHA production could be a more sustainable way of producing bioplastics.

2. Cyanobacteria and Cyanobacterial Energy and Carbon Storage Compounds

Cyanobacteria are Gram-negative prokaryotes that perform oxygenic photosynthesis. They are abundant in illuminated aquatic ecosystems and contribute significantly to the world carbon and oxygen cycle [7]. According to current evidence, oxygen was nearly absent in the Earth's early atmosphere until 2.4 billion years ago [8]. Due to oxygenic photosynthesis of early cyanobacteria the CO_2-rich atmosphere gradually turned into an oxygen-rich atmosphere, providing the conditions for multicellular life [9,10]. Today there are an estimated 6000 species of cyanobacteria with great diversity, for example ranging in size from the 1 μm small unicellular *Synechocystis* sp. to the several millimeter-long multicellular filaments of *Oscillatoria* sp. [11]. The common feature of cyanobacteria is the presence of the pigment phycocyanin, which gives them their typical blue-green color. Figure 1 shows photographs of four different cyanobacterial species.

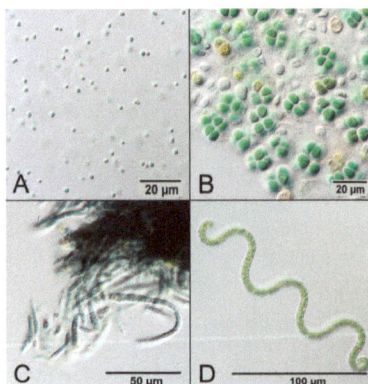

Figure 1. Microscopic photographs of different cyanobacterial species made in DIC (differential interference contrast). (**A**) *Synechocystis* sp.; (**B**) *Cyanosarcina* sp.; (**C**) *Calothrix* sp.; and (**D**) *Arthrospira* sp.

2.1. Cyanobacteria–Microalgae or Not?

For more than a century, cyanobacteria were considered as an algal group under the general name "blue-green algae". They were classified under the International Code of Botanical Nomenclature, nowadays called the International Code of Nomenclature for Algae, Fungi, and Plants (ICN). In 1980 the International Code of Nomenclature of Bacteria, nowadays called the International Code of Nomenclature of Prokaryotes (ICNP), was established. Stanier, one of the leading cyanobacteria researchers at that time, proposed the inclusion of cyanobacteria in the ICNP [12]. Nevertheless, the ICNP was not consistently applied for cyanobacteria and cyanobacteria are still covered by the ICN as well. The latest preamble of the ICN clarifies, that this code applies to all organisms traditionally treated as algae, fungi, or plants, including cyanobacteria [13]. Today cyanobacteria continue to be covered by both the Botanical Code (ICN) and Prokaryotic Code (ICNP). An effort to reconcile the status of this group of bacteria has been underway for several decades. Although some progress has been made, a final decision has not yet been reached [14]. From a phylogenetic point of view, there is a clear distinction between prokaryotic cyanobacteria and eukaryotic green algae. However, phycologists regard any organism with chlorophyll *a* and a thallus not differentiated into roots, stem, and leaves to be an alga. Therefore, in phycology, the term microalgae refers to both eukaryotic green algae and cyanobacteria, microscopic in size [15].

2.2. Cyanobacterial PHA

Polyhydroxyalkanoates (PHAs) can be classified into three groups: short-chain-length-PHA (scl-PHA), medium-chain-length-PHA (mcl-PHA), and long-chain-length-PHA (lcl-PHA). They differ in mechanical and thermal properties [16]. Among the different PHAs, polyhydroxybutyrate (PHB) is by far the most common and the only PHA produced under photoautotrophic conditions reported so far. Other scl-PHAs, like P[3HB-co-3HV], are only produced when adding organic carbon precursors, like valerate, to the medium. No mcl-PHA or lcl-PHA have been reported in cyanobacteria. Therefore, the term PHB is used in this study, if no other specific PHA is described.

PHB is frequently found in cyanobacteria as an energy and carbon storage compound. In the biosphere they often have to cope with unfavourable environmental conditions. One of the most important growth limiting factors is the absence of nutrients. Nitrogen limitation is the most important and best studied trigger for PHB production in cyanobacteria [17–19]. Non-diazotrophic strains are not able to bind molecular nitrogen and depend on nitrogen in the form of nitrate or ammonium. Nitrogen-depleted cells cannot synthesize the necessary proteins for reproduction and, therefore, start to accumulate storage compounds like PHB. Another important function of PHB synthesis is to compensate imbalanced metabolic situations, as it acts as an electron sink and delivers new reduction equivalents in the form of $NADP^+$ [18–20].

The model organism *Synechocystis* PCC6803 is the best-studied cyanobacterium, and its genome was fully sequenced in 1997 [21]. Most of the understanding of cyanobacterial PHB formation was gained by research done with *Synechocystis* PCC6803. Biosynthesis of PHB from the precursor acetyl-CoA takes place in three steps. Acetoacetyl-CoA is produced from two molecules of acetyl-CoA in a Claisen type condensation by β-ketothiolase. Next step is the reduction of acetoacetyl-CoA by the acetoacetyl-CoA reductase to form D-3-hydroxybutyryl-CoA. Ultimately, PHB is formed in a polymerization reaction by the PHA-synthase. The necessary three enzymes are encoded by the four genes phaA (slr1993), phaB (slr1994), phaC (slr1830), and phaE (slr1829). phaA and phaB are organized in one operon encoding for the β-ketothiolase and acetoacetyl-CoA reductase. phaC and phaE are also organized in one operon encoding the two subunits of the type III PHA synthase [22,23].

2.3. Cyanobacterial Glycogen

Regarding PHB synthesis, it should be kept in mind that cyanobacteria also produce glycogen as a second carbon and energy storage compound under nitrogen depletion. In fact, the glycogen content is most often higher than the PHB content and varies between 20% and 60% [24–28]. While PHB is produced in 3–8 larger granules, glycogen is stored in many small granules [18,29–33]. Glycogen is synthesized instantly after nitrogen depletion while PHB synthesis is slower [34]. Glycogen is also produced in non-depleted cells with lower content, aiding the cell to cope with short term energy deficits like the day-night cycle. Glycogen deficient mutants were shown to be highly sensitive to day-night cycles [35]. Glycogen synthesis is a highly-conserved feature abundant in all cyanobacterial genomes reported so far [36]. PHB synthesis on the other hand is common in many, but not all, cyanobacteria [37,38]. Glycogen shows similarities to starch in green algae, while PHB synthesis shows some similarities to triacylglycerol (TAG) synthesis in green algae, where TAG synthesis also serves as an electron sink and consumes excess NADPH [39,40].

In a recent study Damrow and colleagues compared PHB-deficient mutants to glycogen-deficient mutants of *Synechocystis* PCC6803. Glycogen-deficient mutants could not switch to a dormant metabolic state and could not recover from nitrogen depletion. Excess carbon was mostly secreted into the medium in the form of 2-oxoglutaric acid and pyruvate, although the PHB content also increased from 8% to 13%. PHB-deficient mutants, on the other hand, behaved very much like the wild-type with the same amount of glycogen accumulation and the same recovery capability. Only double-knockout mutants (glycogen and PHB deficient) were most sensitive and showed a reduced growth rate, signs for a very specific role of PHB in cyanobacteria, which is still not totally clear [41]. The reported studies show that inhibiting glycogen synthesis increases the PHB production, although cells suffer as glycogen plays an important role.

2.4. Nitrogen Chlorosis and Photosynthetic Activity

During nitrogen starvation the cells gradually change from a vegetative state to a dormant state. The most obvious feature of this is the change in colour from blue-green to brownish-yellow. This phenomenon is called "nitrogen chlorosis" and was described already at the begin of the 20th century [42]. It is caused by the degradation of the pigments phycocyanin and chlorophyll. When transferring *Synechococcus* PCC7942 to a nitrogen depleted medium, 95% phycocyanin was degraded within 24 h, and after 10 days 95% of the chlorophyll was also degraded [43]. Concomitantly, the activities of the photosystems (PS) I and II decrease strongly and are only about 0.1% compared to vegetative cells [44]. A recent and very interesting study examined the awakening of a dormant *Synechocystis* PCC6803 cell. After the addition of nitrate the yellow culture turned green again within 36 hours. Transmission electron microscopy revealed the rapid degradation of glycogen and PHB. During the first 24 h of this process the cells consumed oxygen. Transcriptome analysis showed the induction of RuBisCO and carboxysom associated RNAs, as well as the photosystem-related RNAs to prepare the cells for vegetative photoautotrophic growth [34]. The results indicate the decrease in photosynthetic activity during nitrogen starvation, which can be considered a significant challenge to photoautotrophic PHB production.

3. Different Cyanobacteria as PHA Producers

3.1. Synechocystis and Synechococcus

Synechocystis and *Synechococcus* are very small (0.5–2 μm) unicellular cyanobacteria abundant in almost all illuminated saltwater and freshwater ecosystems. One of the first detailed descriptions of PHB accumulation in *Synechocystis* PCC6803 was provided by Wu and colleagues. Nitrogen starved cells produced 4.1% PHB of cdw while under-balanced culturing conditions PHB content were under the detection limit [45]. The same strain was examined for PHB production some years later, where 9.5% PHB of cdw were produced under nitrogen limitation. Phosphorous-depleted cells showed 11.2% PHA of cdw. Interestingly, balanced cultivated control cultures already contained 4.5% PHB of cdw. Supplementation of acetate and fructose lead to a PHB content of 38% of cdw [46]. Recently, recombinant *Synechocystis* PCC6803 with overexpression of the native PHA genes were constructed. They showed a PHB content of 26% of cdw under nitrogen-depleted culturing conditions compared to 9.5% of cdw of the wild-type [47]. However, it must be considered that there are legal issues in most countries when cultivating recombinant strains outdoors. In another study the thermophilic strain *Synechococcus* MA19 showed a PHB content of 55% under phosphate-limited culturing conditions. This study was published in 2001 and still reports the highest PHB content under photoautotrophic conditions [48]. Table 1 shows reported PHA values of *Synechocystis* and *Synechococcus*.

Table 1. *Synechocystis* and *Synechococcus* as PHA producers. (cdw = cell dry weight, n.r. = not reported).

Carbon Source	Cyanobacterium	Culture Condition	%PHA of cdw	PHA Composition	Total cdw	Reference
Photoautotrophic	*Synechocystis* PCC6803	Photoautotrophic, nitrogen lim.	4.1%	PHB	0.65 g/L	[45]
	Synechocystis PCC6803	Photoautotrophic, nitrogen lim.	9.5%	PHB	n.r.	[46]
	Synechocystis PCC6803	Photoautotrophic, phosphate lim.	11.2%	PHB	n.r.	[46]
	Synechocystis PCC6803 (recombinant)	Photoautotrophic, nitrogen lim.	26%	PHB	n.r.	[47]
	Synechococcus MA19	Photoautotrophic, phosphate lim., 50 °C	55%	PHB	4.4 g/L	[48]
Heterotrophic	*Synechocystis* PCC6803	Acetate + Fructose supplementation	38%	PHB	n.r.	[46]
	Synechocystis PCC6803 (recombinant)	Acetate supplementation	35%	PHB	n.r.	[47]

3.2. Arthrospira (Spirulina)

Arthrospira (formally *Spirulina*) is a species of filamentous cyanobacteria that grows naturally in alkaline salt lakes. It has a high protein and vitamin content and is mainly grown as a food supplement. Recent studies have shown its antioxidant, immunomodulatory, and anti-inflammatory activities [49]. From all cyanobacterial species known, only *Arthrospira* sp. is produced at an industrial scale. The main reason for that is the possibility of cultivation in a highly alkaline environment that prevents contamination and enables the maintenance of a stable culture in open ponds. No exact data are available; however, we estimate the world annual production of around 5000–15,000 tons *Arthrospira* sp. dry weight per year [50–53].

The first description of PHB accumulation in *Arthrospira* was reported by Campbell and colleagues, who described a PHB content of 6% of cdw in a non-optimized mineral medium. Interestingly, the highest PHB content was measured at the end of exponential growth and decreased during stationary phase [54]. In a screening of 23 cyanobacterial strains, *Arthrospira platensis* had the lowest PHB concentration of only 0.5% in a non-optimized medium [37]. In a screening of several *Arthrospira* species the PHB amount never exceeded 1% of cdw in photoautotrophic growth. Addition of sodium acetate led to a PHB amount of 2.5% of cdw [55]. In another experiment *Arthrospira platensis* was grown under phosphate limitation and reached 3.5% PHB of cdw [56]. *Arthrospira subsalsa*, a strain isolated from the Gujarat coast, India, produced 14.7% PHB of cdw under increased salinity [57]. A detailed ultrastructural analysis of *Arthrospira* strain PCC8005 was conducted by Deschoenmaker and colleagues. Under nitrogen depleted conditions PHB granules were more abundant and larger. The nitrogen-starved cells showed an estimated four times higher PHB concentration [27]. Nitrogen starvation was performed in *Arthrospira maxima* and glycogen and PHB content was measured. While the glycogen content increased from around 10% to 60%–70% of cdw, PHB amount remained low at 0.7% of cdw. The addition of sodium acetate increased the PHB amount to 3% of cdw [26].

The performed studies support the idea, that PHB production in *Arthrospira* is highly strain-dependent. Most *Arthrospira* species produce PHB only in amounts of lower than 5%, even with the addition of sodium acetate. *Arthrospira* produces glycogen as storage compound, what has been shown in ultrastructural research, too [27]. Nevertheless, it must be emphasized that *Arthrospira*, at an industrial scale, is still one of the most promising candidates for PHB production with cyanobacteria. Indeed, PHB nanofibers were produced recently from *Arthrospira* PHB and showed highly favourable properties [58,59]. The biggest challenge for further research is to increase the relatively low PHB content of *Arthrospira*. Table 2 shows reported PHA values of *Arthrospira*.

Table 2. *Arthrospira* as a PHA producer. (cdw = cell dry weight, n.r. = not reported).

Carbon Source	Cyanobacterium	Culture Condition	%PHA of cdw	PHA Composition	Total cdw	Reference
Photoautotrophic	*Arthrospira platensis*	Photoautotrophic	6%	PHB	n.r.	[54]
	Arthrospira sp.	Photoautotrophic	<1%	PHB	n.r.	[55]
	Arthrospira platensis	Photoautotrophic, phosphate lim.	3.5%	PHB	0.3 g/L	[56]
	Arthrospira subsalsa	Photoautotrophic, nitrogen lim.	14.7%	PHB	1.97 g/L	[57]
	Arthrospira platensis	n.r.	22%	PHB	n.r.	[59]
Heterotrophic	*Arthrospira maxima*	Acetate + CO_2	5%	PHB	1.4 g/L	[26]
	Arthrospira sp.	Acetate + CO_2	2.5%	PHB	n.r.	[55]

3.3. Nostoc

Nostoc is a group of filamentous cyanobacteria very common in terrestrial and aquatic habitats. They are capable of fixing atmospheric nitrogen with specialized heterocysts and are suspected to maintain soil fertility in rice fields due to nitrogen fixation [60]. Ge-Xian-Mi, an edible *Nostoc* species, forms spherical colonies that have been collected in China for centuries [61]. The first reports found for PHB production in *Nosctoc muscorum* are from 2005, when Sharma and Mallick showed that *Nostoc muscroum* produced 8.6% PHB of cdw under phosphate and nitrogen limitation during the stationary phase. PHB content could be boosted to 35% of cdw with 0.2% acetate and seven days dark incubation [62]. Limited gas exchange and supply with 0.4% acetate increased the PHB content to 40% [63]. *Nostoc muscorum* was grown photoautotrophically without combined nitrogen sources and four days of phosphate deficiency increased PHB content from 4% to 22% [56]. The co-polymer P[3HB-co-3HV] could be produced by *Nostoc* in a propionate- and valerate-supplied medium. The 3HV fraction ranged from 10–40 mol% and showed desirable properties in terms of flexibility and toughness. Nitrogen and phosphate depletion led to a PHA content of 58%–60% of cdw, however, the total cdw did not exceed 1 g/L [64]. Further process optimization led to a PHA productivity of 110 mg/L/d and a P[3HB-co-3HV] content of 78% of cdw, the highest yield in heterotrophic grown cyanobacteria reported so far [65]. Recently, poultry litter was used for cultivation of *Nostoc muscorum agardh*. The poultry litter contained phosphate, ammonium, nitrate, and nitrite as nutrients for cyanobacterial growth. Optimized conditions, which included the addition of acetate, glucose, valerate, and CO_2-enriched air, led to a P[3HB-co-3HV] content of 70% cdw. However, total cdw remained relatively low at 0.68 g/L [66].

The reported studies show that PHB content in *Nostoc* can be significantly increased with organic carbon sources, especially in the form of acetate. However, those organic carbon sources lead to heterotrophic growth and may suppress CO_2 uptake by the cells, which is the most important argument for using cyanobacteria as PHA producers. All of the reported experiments of *Nostoc* were performed in shaking flasks or small reactors under sterile conditions. In mass cultivation *Nostoc* would have to be cultivated under non-sterile conditions and organic carbon sources could cause problems maintaining stable cultures. Although optimized conditions of several experiments lead to PHA contents of more than 50% of cdw, the total cdw remained mostly under 1 g/L and the overall productivity and growth rate of *Nostoc* is relatively low. Table 3 shows reported PHA values of *Nostoc*.

Table 3. *Nostoc* as a PHA producer. (cdw = cell dry weight, n.r. = not reported).

Carbon Source	Cyanobacterium	Culture Condition	%PHA of cdw	PHA Composition	Total cdw	Reference
Photoautotrophic	*Nostoc muscorum*	Photoautotrophic, nitrogen and phosphorous lim.	8.7%	PHB	n.r.	[62]
	Nostoc muscorum agardh	Photoautotrophic, 10% CO_2	22%	PHB	1.1 g/L	[66]
	Nostoc muscorum	Photoautotrophic, nitrogen and phosphorous lim.	22%	PHB	0.13 g/L	[56]
Heterotrophic	*Nostoc muscorum agardh*	Acetate, valerate, nitrogen lim.	58%	P[3HB-co-3HV]	0.29 g/L	[64]
	Nostoc muscorum	Acetate, limited gas exchange	40%	PHB	n.r.	[63]
	Nostoc muscorum agardh	Acetate, glucose, valerate, 10% CO_2	70%	P[3HB-co-3HV]	0.98 g/L	[66]
	Nostoc muscorum agardh	Acetate, glucose, valerate, nitrogen lim.	78%	P[3HB-co-3HV]	0.56 g/L	[65]
	Nostoc muscorum	Acetate, dark incubation, nitrogen and phosphorous lim.	35%	PHB	n.r.	[62]

3.4. Other Cyanobacteria

Recently, the PHB content of 137 cyanobacterial strains representing 88 species in 26 genera was determined under photoautotrophic conditions. High PHB content was highly strain-specific and was not associated with the genera. From the 137 tested strains, 134 produced PHB and the highest content was measured in *Calothrix scytonemicola* TISTR 8095 (Thailand Institute of Scientific and Technological Research). This strain produced 356.6 mg/L PHB in 44 days and reached a PHB content of 25% of cdw and a total biomass of 1.4 g/L. The PHB content of 25% was reached under nitrogen depletion, while cells with nitrogen supply reached a PHB content of only 0.4%. From the 19 tested *Calothrix* strains, only six produced more than 5% PHB of cdw. One of the greatest advantages of *Calothrix* is the relative ease of harvesting the dense flocs of algae, but cultivation of *Calothrix* is still at a very early stage [38].

The filamentous diazotroph cyanobacterium *Aulosira fertilissima* produced 10% PHB of cdw under photoautotrophic conditions and phosphate deficiency. The PHB content was boosted to 77% under phosphate deficiency with 0.5% acetate supplementation. This study also shows the positive effect of other carbon sources like citrate, glucose, fructose and maltose on PHB production [67]. Anabaena cylindrica, a filamentous cyanobacterium, was examined for PHB and P[3HB-co-3HV] production. Under nitrogen depletion with acetate supply, *Anabeana cylindrica* produced 2% PHB of cdw and a total biomass of 0.6 g/L. This organism was also able to produce the co-polymer P[3HB-co-3HV] when supplemented with valerate and propionate [68]. Table 4 shows reported PHA values of different cyanobacterial species.

Table 4. Different cyanobacterial species as PHA producers. (cdw = cell dry weight, n.r. = not reported).

Carbon Source	Cyanobacterium	Culture Condition	%PHA of cdw	PHA Composition	Total cdw	Reference
Photoautotrophic	*Phormidium* sp. TISTR 8462	Photoautotrophic, nitrogen lim.	14.8%	PHB	n.r.	[38]
	Oscillatoria jasorvensis TISTR 8980	Photoautotrophic, nitrogen lim.	15.7%	PHB	n.r.	[38]
	Calothrix scytonemicola TISTR 8095	Photoautotrophic, nitrogen lim.	25.2%	PHB	n.r.	[38]
	Anabaena sp.	Photoautotrophic	2.3%	PHB	n.r.	[69]
	Aulosira fertilissima	Photoautotrophic, phosphorous lim.	10%	PHB	n.r.	[67]
Heterotrophic	*Aulosira fertilissima*	Acetate, phosphorous lim.	77%	PHB	n.r.	[67]
	Aulosira fertilissima	Maltose, balanced	15.9%	PHB	2.3 g/L	[67]

4. CO_2 and Nutrient Supply for Mass Cultivation of Cyanobacteria

4.1. CO_2 Supply

Today, commercial microalgae production is still mainly taking place in open ponds. Here, the C source is normally sodium bicarbonate or atmospheric CO_2. In order to boost productivities in open systems, or if photobioreactor systems are employed, the use of commercial CO_2 from gas cylinders is common [70].

However, current production systems are used for the production of high value products (food, feed additives), where CO_2 price is not critical. If PHA is to be produced, which has a lower economic value, cheap CO_2 sources are of interest. Although there is considerable literature on various CO_2-sources (e.g., flue gases) and microalgae growth, there is very limited literature available on cyanobacteria and alternative CO_2-sources. Table 5 summarizes the literature on cyanobacterial growth on flue gases or fermentation gases.

Table 5. Growing cyanobacteria with alternative CO_2-sources.

Type of Gas	Cyanobacterium	CO_2 Source	Reference
Flue gases	*Phormidium valderianum*	Coal combustion flue gas	[71]
	Atrhrospira platensis	Coal combustion flue gas	[72]
	Arthrospira sp.	Synthetic flue gas	[73]
	Synechocystis sp.	Flue gas from natural gas combustion	[74]
CO_2 rich fermentation gases	*Arthrospira platensis*	CO_2-offgas from ethanol fermentation	[75]
	Arthrospira platensis	Biogas	[76]

4.2. Nutrient Supply

The cultivation of microalgae and cyanobacteria consume high amounts of nutrients, mainly nitrogen and phosphorous [77]. For research, and even cultivation, mainly synthetic nutrient sources are used [78]. By using alternative nutrient sources, like agro-industrial effluents, waste waters, or anaerobic digestate, questions concerning sustainability of cyanobacteria cultivation, which arise by using fertilizer as a synthetic nutrient source, can be answered [78]. The biomass concentration achieved in open, as well as in closed, cultivation systems are 0.5–1 g/L and 2–9 g/L, respectively [79]. Therefore, large amounts of water are needed. Recycling of process water is another important approach for a more sustainable microalgae cultivation.

In addition to their low costs, the advantages of using alternative nitrogen and phosphorous sources include the production of valuable biomass while removing nutrients from wastewaters, as well as the prevention of competition with food and feed production [78]. On the other hand, new challenges arise, including microbial contaminations, heavy metals and growth inhibitors, suspended solids, or dissolved organic compounds contained in wastewaters, as well as the seasonal composition and fluctuation in amounts [80]. To cope with these challenges recent research focused on cultivating cyanobacteria in anaerobic digestate and agro-industrial effluents or wastewaters for removing nutrients [81–86] (see Table 6) and on integrating cultivation processes into biorefinery systems [83].

Additionally, process water and nutrients after harvesting cyanobacterial biomass [79] and product extraction can be directly recycled. Biomass can also be anaerobically digested [87,88] or hydrothermally liquefied via HTL (mineralization of organic nutrients) [89,90] and then recycled. Recycling process water directly can increase the concentration of inhibitory substances and dissolved organic matter from the previous batch produced by cyanobacteria [91], which decrease the productivity of cyanobacteria. Furthermore, nutrient competition may arise by enhanced bacterial growth [79].

Although many publications deal with alternative nutrient sources for cultivating cyanobacteria, hardly any of them focus on cyanobacterial PHA production [66,92]. Reasons for that may be that PHA production requires nutrient limitation [93] and the balance between nutrient limitation, decreased growth and production rates is difficult. The colouring of the nutrient source must be respected as well [94].

Table 6. Overview of agro-industrial effluents and wastewaters and anaerobic digestates used as nutrient sources for cultivating cyanobacteria.

Nutrient Source		Cyanobacterium	Total cdw/Growth Rate	Product/Purpose	Reference
Agro-industrial effluents and waste waters	Raw cow manure	*Arthrospira maxima*	3.15 g/L	Biomass production	[80]
	Molasses	*Arthrospira platensis*	2.9 g/L	Biomass production	[95]
	Olive-oil mill wastewater	*Arthrospira platensis*	1.69 g/L	Nutrient removal	[84]
	Poultry litter	*Nostoc muscorum agardh*	0.62 g/L	PHA production	[66]
Anaerobic digestate	Waste from pig farm	*Arthrospira platensis*	20 g/m^2/d	Nutrient removal	[81]
	Digested sago effluent	*Arthrospira platensis*	0.52–0.61 g/L	Nutrient removal	[96]
	Digestate from municipal solid waste	*Arthrospira platensis*	Growth rate 0.04 d^{-1}	Nutrient removal	[97]
	Digestate from vegetable waste	*Arthrospira platensis*	Growth rate 0.20 d^{-1}	Nutrient removal	[97]
	Waste from pig farm	*Arthrospira* sp.	15 g/m^2/d	Nutrient removal	[85]
	Algal digestate	*Chroococcus* sp.	0.79 g/L	Nutrient removal	[86]
	Digestate sludges	*Lyngbya aestuarii*	0.28 g/L	Biomass production	[83]
	Digestates of *Scenedesmus* spp.	*Lyngbya aestuarii*	0.11 g/L	Biomass production	[83]
	Thin stillage digestate	*Synechocystis* cf. *salina* Wislouch	1.6 g/L	PHB production	[92]
	Anaerobic digester effluent	*Synechocystis* sp.	0.15 g/L	Lipid production	[98]

5. Three Years' Working Experience Running a Pilot Plant for Photoautotrophic PHB Production

5.1. Location and Reactor Description

The photobioreactor is situated in a glass house at the coal power station in Dürnrohr, Austria. It is a tubular system built from glass elements of Schott AG with an inner diameter of 60 mm, a total length of 80 m and a volume of 200 L (Figure 2). The main design of the photobioreactor is described elsewhere [99,100]. A central degassing unit serves to remove the oxygen as well as to compensate filling level. The medium is circulated with a 400 W centrifugal pump. pH value can be controlled through injection of pure CO_2 via a mass flow controller. Additional artificial light is provided by six 250 W gas-discharge lamps. Temperature is controlled with an air conditioner.

Figure 2. Two-hundred litre tubular photobioreactor with *Synechocystis salina* CCALA192. The central tower serves as a degasser. The centrifugal pump is situated at the lowest point of the reactor on the left side.

5.2. CO₂ Supply of the Reactor

The flue gases of the power plant at Dürnrohr usually contain between 11%–13% CO_2. Next to the chimney is a CO_2 separation plant (acronym SEPPL), providing the possibility to concentrate the CO_2 and fill it into gas bottles [101] though, for a more economic approach, the CO_2 should be used directly without prior compression. The SEPPL provides this option, as well as the possibility to wash the flue gases after the flue gas cleaning of the plant itself to remove residual NO_x and SO_x. Unfortunately for our research project, due to the current situation on the energy market, the power station is no longer run in full operation and only runs occasionally for balancing peak demands of the electrical grid. Therefore, a continuous cultivation with direct utilization of flue gas is not possible. Aspects like this must be respected when planning large industrial cultivation plants.

5.3. Automation and pH Control

The pH value is one of the most crucial parameters and needs to be controlled carefully. Due to CO_2 consumption, the pH value rises during photoautotrophic growth. This can be observed when turning on illumination. The tubular photobioreactor is equipped with a PI (Proportional–Integral) controller for pH setting which adjusts the mass-flow controller for CO_2 inlet. This allows an online control of the currently consumed CO_2, which is a suitable parameter for growth monitoring. Figure 3 shows a 24-h course of the pH value and the CO_2 mass flow. Lamps turned on at 02:00 and off at 22:00, causing a rise and decrease of the pH value, respectively. The setpoint of 8.5 is reached after first overshooting and held during the day. The decrease of CO_2 consumption at noon is caused by the shadow of the power plant's chimney that casts upon the greenhouse at this time.

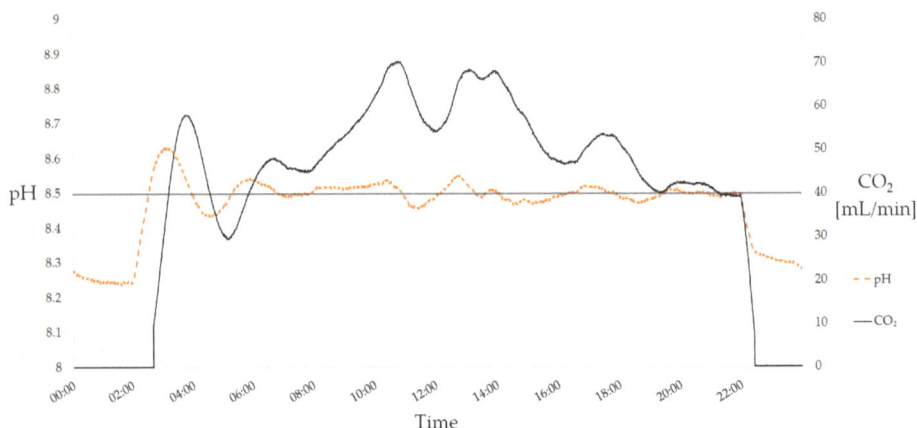

Figure 3. PI-controlled pH value. The setpoint of the pH is 8.5. Lamps turn on at 02:00 and turn off at 22:00, causing a rise and decrease of the pH value due to CO_2 consumption. In total, 59 L (118 g) of CO_2 were consumed on this day.

5.4. Overview of PHB Production Trials

Most of the trials (overview shown in Table 7) were performed using a modified BG 11 medium [102]. Modification in terms of PHB production is necessary, as normal BG 11 medium contains high amounts of nitrogen (1.5 g/L $NaNO_3$) and would not lead to nitrogen limitation. The modified BG 11 contains 0.45 g/L $NaNO_3$ and leads to a self-limitation of the culture. After 8–12 days nitrogen is consumed, PHB production starts and the color of the culture gradually turns yellow. This approach is necessary, as it is not possible to transfer large-scale cultures into a nitrogen-free medium [103].

Synechocystis salina CCALA192 was found to be a very suitable cyanobacterium. It is easy to handle and grows with small inoculation volumes of 1:50. Final biomass and PHB concentrations were in the range of 0.9–2.1 g/L and 4.8% to 9% of cdw, respectively. *Synechocystis salina* CCALA192 also grew with the addition of acetate, but no significant increase of biomass and PHB concentration was observed compared to photoautotrophic growth. When using acetate, contaminations with fungi were likely to occur and trials had to be stopped. Therefore, this approach was finally abandoned.

Digestate from a biogas reactor was successfully tested as an alternative nutrient source. The supernatant was produced by centrifugation with prior addition of precipitating agents. Before usage the supernatant was autoclaved and diluted 1:3 with water [92]. Figure 4 shows biomass and PHB production using digestate as nutrient source.

After one and a half years a new degassing system was installed, as the oxygen concentration was mostly above 250% saturation during the day. For an ideal cultivation of cyanobacteria the oxygen saturation should not exceed 200%. The new degasser led to a rise in biomass production with a maximum production rate of 0.25 g/L/d. Efficient degassing affected the cyanobacteria positively. However, during installation of the degasser dirt from the surrounding soil was brought into the reactor and from that time on culture crashes occurred due to ciliate contaminations (see Section 5.6).

The other tested cyanobacteria *Chlorogloeopsis fritschii* and *Arthrospira* sp. could not be successfully cultivated in the photobioreactor. It is assumed that these strains were sensitive to shear stress caused by the centrifugal pump [104].

Table 7. Overview of selected trials conducted in a tubular photobioreactor at pilot scale.

Trial	Strain	Nutrient Solution	Cultivation Time	Final Biomass Concentration	Final PHB-Concentration of cdw
1. Mineral medium	*Synechocystis salina* CCALA192	Optimized BG11	June 21 days	2.0 ± 0.12 g/L	$6.6\% \pm 0.5\%$
2. Acetate addition	*Synechocystis salina* CCALA192	Optimized BG11, 20 mM acetate	July 26 days	1.9 ± 0.02 g/L	$6.0\% \pm 0.1\%$
3. Acetate addition	*Synechocystis salina* CCALA192	Optimized BG11, 60 mM acetate	September 24 days	Trial cancelled, due to contaminations with fungi	
4. 24 h illumination	*Synechocystis salina* CCALA192	Optimized BG11	October 27 days	1.8 ± 0.02 g/L	$4.8\% \pm 0.0\%$
5. Alternative nutrient source	*Synechocystis salina* CCALA192	Digestate supernatant	November–December 40 days	1.6 ± 0.02 g/L	$5.5\% \pm 0.3\%$
6. Mineral medium	*Synechocystis salina* CCALA192	Optimized BG11	December–January 30 days	2.1 ± 0.03 g/L	$6.0\% \pm 0.02\%$
7. Optimal degassing	*Synechocystis salina* CCALA192	Optimized BG11	May 7 days	0.9 ± 0.03 g/L (Trial prematurely cancelled due to ciliates)	$9\% \pm 0.1\%$ (Trial prematurely cancelled due to ciliates)
8. Chlorogloeopsis fritschii CCALA39	*Chlorogloeopsis fritschii* CCALA39	Optimized BG11	February 11 days	Trial cancelled, due to lack of growth	
9. Arthrospira	*Arthrospira* sp.	Spirulina Medium	October 7 days	Trial cancelled, due to lack of growth	

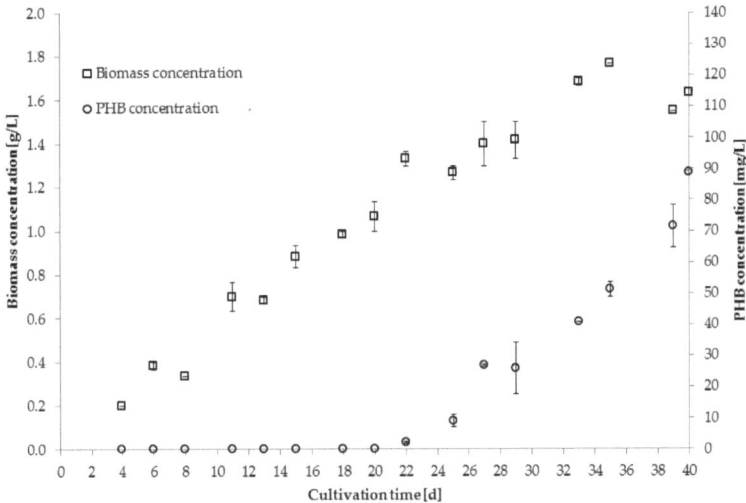

Figure 4. Biomass [g/L] and PHB [g/L] concentration of *Synechocystis salina* using digestate supernatant as nutrient source (Trial 5).

5.5. Downstream Processing of Cyanobacterial Biomass

Downstreaming of cyanobacterial cultures is particularly difficult, as cell densities are much lower compared to heterotrophic bacteria. Typical harvesting methods are sedimentation, filtration, or centrifugation [105]. The cyanobacterial biomass was harvested with a nozzle separator and stored at -20 °C. The biomass was then used to evaluate processing steps necessary to gain clean

PHB-samples for quality analysis. These downstream trials include (i) different cell disruption methods (milling, ultrasound, French press); (ii) different pigment extraction methods (with acetone and ethanol/methanol before or after extracting PHB); and (iii) different PHB extraction methods (soxhlet extraction with chloroform, biomass digestion with sodium hypochlorite) [106].

These trials showed that cell disruption with French press worked quite well but is very time consuming. Milling is assumed to decrease the molecular weight (polymer chain length). Pigment removal turned out to be necessary prior to PHB extraction, as pigments influenced the PHB properties negatively. This process step can be of advantage due to the generation of phycocyanin as a valuable side product [107]. A mixture of acetone and ethanol (70:30) was most suitable for this purpose. PHB extraction was performed with hot chloroform via a soxhlet extractor. Figure 5 compares the necessary processing steps of heterotrophs and cyanobacteria.

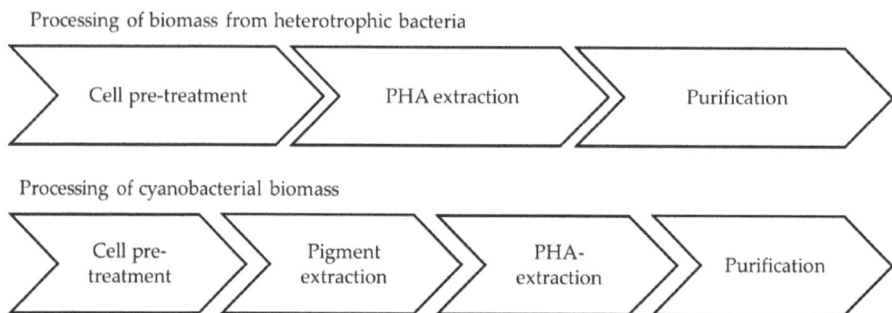

Processing of biomass from heterotrophic bacteria

Processing of cyanobacterial biomass

Figure 5. Comparison of processing steps needed to extract PHA from heterotrophic bacteria and cyanobacteria.

The PHB analysis showed that the polymers extracted from cyanobacterial biomass are comparable to commercially available PHB. Furthermore, it was shown that not only did the nutrient source, but also biomass pre-treatment and the method of polymer extraction influence the PHB properties. Pigment extraction and sample pre-treatment increased the average molecular weight (M_w) from 0.3 to 1.4 MD, but decreased degradation temperatures and crystallinity from 282 °C to 275 °C (T_{onset}) and from 296 °C to 287 °C (T_{max}), respectively. The M_w ranged from 5.8 to 8.0 MD, by using mineral medium and digestate, respectively. The thermal properties (T_{onset}: 283–282 °C; T_{max}: 293–292 °C), which are important for processing the polymer, are only slightly influenced by the nutrient source and are lower than, but comparable to, commercially available PHB. The crystallinity, responsible for higher final brittleness of the products, is about 17% lower than commercially available PHA.

5.6. Contaminations

Contaminations in non-sterile mass cultivation of microalgae are inevitable. It is only a matter of time before first contaminations appear, whether cultivation is done in open ponds or closed photobioreactors [108]. We observed certain bacterial and fungal contaminations with minor effects on *Synechocystis salina* CCALA192, when using CO_2 as the sole carbon source. Though, when adding acetate to the medium fungal contaminations were prevalent and difficult to control. After one and a half years the pilot plant was revised and a new degassing system was installed. From this moment on rapid culture crashes occurred. The microscopic image revealed a ciliated protozoa ingesting *Synechocystis* rapidly.

This ciliate forms highly resistant cysts and it is assumed that cysts from the surrounding soil were brought into the system during the revision work. Facts about contaminations in mass cultivation of *Synechocystis* are scarcely reported. Touloupakis and colleagues reported the grazing of *Synechocystis* PCC6803 by golden algae *Poterioochromonas* sp. [109]. High pH values of 10 and above helped to control the contaminant and maintain a stable culture. Unfortunately, the ciliate in our cultures survives those high pH values. Thoroughly cleaning and sanitizing the photobioreactor brought some success, but the ciliate is still occurring and leading to culture crashes. Due to the ciliate's capability to form cysts, it is very difficult to completely eliminate it from the reactor. Heat sterilization is not possible in tubular photobioreactors. Addressing further research, there is need for special cultivation methods for robustly growing *Synechocystis* in non-sterile environment.

6. Conclusions

Although not economical today, the idea of a sustainable PHB production with cyanobacteria, CO_2, and sunlight is still attractive and, more and more, researchers are working in this field. The main challenges today are similar to biofuel production with green algae: (i) realization of efficient low-cost cultivation systems at large scale; (ii) maintaining stable cultures under non-sterile conditions; (iii) increasing the total productivity and yield; and (iv) economic downstream processing and utilization of the residual biomass.

Looking for suitable production strains it must be considered that PHB production is a very common feature of many, but not all, cyanobacteria. The PHB content of cyanobacteria is highly strain specific, as strains of the same genus were reported with highly varying PHB contents. In addition to the PHB content the growth rate and robustness of a strain is particularly important. The only cyanobacterium cultivated in mass cultivation today is *Arthrospira* sp. and, therefore, one of the most promising candidates for photoautotrophic PHB production, although most *Arthrospira* sp. strains still show little PHB content.

Heterotrophic cultivation with acetate boosts the PHB content remarkably, as most reported values over 30% were achieved this way. However, it needs to be considered that using an organic carbon source impairs the most attractive feature of cyanobacteria, converting CO_2 to PHB. Using organic sources will also complicate non-sterile mass cultivation and could easily lead to contaminations and culture crashes. PHB production with organic carbon sources should be performed with heterotrophic bacteria, as their PHB productivity, as well as their cell density, are 10–100 times higher.

Nitrogen and phosphorous depletion are the most important factors to increase the PHB content and are often necessary to produce any PHB at all. Therefore, a two-stage cultivation with a self-limiting medium is necessary for large-scale photoautotrophic PHB production. With this strategy PHB was successfully produced in our 200 L photobioreactor. In tubular systems small unicellular organisms, like *Synechocystis* sp., are preferred over filamentous organisms, mainly because of the shear stress of the pump. Considering all of the difficulties to overcome, establishing a stable cyanobacterial culture is most important and most difficult to achieve.

Acknowledgments: The research project CO2USE was thankfully financed by the Austrian climate and energy fund and FFG (Austrian research promotion agency).

Author Contributions: Clemens Troschl designed and performed experiments, analyzed data, and wrote about 75% of the article. Katharina Meixner designed and performed experiments, analyzed data and wrote about 15% of the article. Bernhard Drosg supervised the work, analyzed data, and wrote about 10% of the article.

Conflicts of Interest: The authors declare no conflict of interest.

References

1.   Khosravi-Darani, K.; Mokhtari, Z.-B.B.; Amai, T.; Tanaka, K. Microbial production of poly(hydroxybutyrate) from C1 carbon sources. *Appl. Microbiol. Biotechnol.* **2013**, *97*, 1407–1424. [CrossRef] [PubMed]

2.   Dawes, E. Polyhydroxybutyrate: An intriguing biopolymer. *Biosci. Rep.* **1988**, *8*, 537–547. [CrossRef] [PubMed]

3.   Chen, G.-Q. A microbial polyhydroxyalkanoates (PHA) based bio- and materials industry. *Chem. Soc. Rev.* **2009**, *38*, 2434–2446. [CrossRef] [PubMed]

4.   Halami, P.M. Production of polyhydroxyalkanoate from starch by the native isolate Bacillus cereus CFR06. *World J. Microbiol. Biotechnol.* **2008**, *24*, 805–812. [CrossRef]

5. Ting, C.S.; Rocap, G.; King, J.; Chisholm, S.W. Cyanobacterial photosynthesis in the oceans: The origins and significance of divergent light-harvesting strategies. *Trends Microbiol.* **2002**, *10*, 134–142. [CrossRef]

6. IPCC. *2014: Climate Change 2014: Synthesis Report*; Core Writing Team, Pachauri, R.K., Meyer, L.A., Eds.; Contribution of Working Groups I, II and III to the Fifth Assessment Report of the Intergovernmental Panel on Climate Change; IPCC: Geneva, Switzerland, 2014; p. 151.

7. Stanier, R.J.; Cohen-Bazire, G. Phototrophic prokaryotes: The cyanobacteria. *Ann. Rev. Microbiol.* **1977**, *31*, 225–274. [CrossRef] [PubMed]

8. Bekker, A.; Holland, H.D.; Wang, P.-L.; Rumble, D.; Stein, H.J.; Hannah, J.L.; Coetzee, L.L.; Beukes, N.J. Dating the rise of atmospheric oxygen. *Nature* **2004**, *427*, 117–120. [CrossRef] [PubMed]

9. Schirrmeister, B.E.; Sanchez-Baracaldo, P.; Wacey, D. Cyanobacterial evolution during the Precambrian. *Int. J. Astrobiol.* **2016**, *15*, 1–18. [CrossRef]

10. Fischer, W.W.; Hemp, J.; Johnson, J.E. Evolution of Oxygenic Photosynthesis. *Annu. Rev. Earth Planet. Sci.* **2016**, *44*, 647–683. [CrossRef]

11. Nabout, J.C.; da Silva Rocha, B.; Carneiro, F.M.; Sant'Anna, C.L. How many species of Cyanobacteria are there? Using a discovery curve to predict the species number. *Biodivers. Conserv.* **2013**, *22*, 2907–2918. [CrossRef]

12. Stanier, R.Y.; Sistrom, W.R.; Hansen, T.A.; Whitton, B.A.; Castenholz, R.W.; Pfennig, N.; Gorlenko, V.N.; Kondratieva, E.N.; Eimhjellen, K.E.; Whittenbury, R.; et al. Proposal to Place the Nomenclature of the Cyanobacteria (Blue-Green Algae) Under the Rules of the International Code of Nomenclature of Bacteria. *Int. J. Syst. Bacteriol.* **1978**, *28*, 335–336. [CrossRef]

13. McNeill, J.; Barrie, F.R. *International Code of Nomenclature for Algae, Fungi, and Plants (Melbourne Code)*; Koeltz Scientific Books: Oberreifenberg, Germany, 2012.

14. Parker, C.T.; Tindall, B.J.; Garrity, G.M. International Code of Nomenclature of Prokaryotes. *Int. J. Syst. Evol. Microbiol.* **2015**. [CrossRef] [PubMed]

15. Richmond, A. *Handbook of Microalgal Culture*, 1st ed.; Blackwell Science Ltd.: Oxford, UK, 2004.

16. Tan, G.-Y.; Chen, C.-L.; Li, L.; Ge, L.; Wang, L.; Razaad, I.; Li, Y.; Zhao, L.; Mo, Y.; Wang, J.-Y. Start a Research on Biopolymer Polyhydroxyalkanoate (PHA): A Review. *Polymers (Basel.)* **2014**, *6*, 706–754. [CrossRef]

17. Nakaya, Y.; Iijima, H.; Takanobu, J.; Watanabe, A.; Hirai, M.Y.; Osanai, T. One day of nitrogen starvation reveals the effect of sigE and rre37 overexpression on the expression of genes related to carbon and nitrogen metabolism in *Synechocystis* sp. PCC 6803. *J. Biosci. Bioeng.* **2015**, 1–7. [CrossRef] [PubMed]

18. Schlebusch, M.; Forchhammer, K. Requirement of the nitrogen starvation-induced protein sll0783 for polyhydroxybutyrate accumulation in *Synechocystis* sp. strain PCC 6803. *Appl. Environ. Microbiol.* **2010**, *76*, 6101–6107. [CrossRef] [PubMed]

19. Hauf, W.; Schlebusch, M.; Hüge, J.; Kopka, J.; Hagemann, M.; Forchhammer, K. Metabolic Changes in Synechocystis PCC6803 upon Nitrogen-Starvation: Excess NADPH Sustains Polyhydroxybutyrate Accumulation. *Metabolites* **2013**, *3*, 101–118. [CrossRef] [PubMed]

20. Stal, L.J. Poly(hydroxyalkanoate) in cyanobacteria: An overview. *FEMS Microbiol. Lett.* **1992**, *103*, 169–180. [CrossRef]

21. Kaneko, T.; Sato, S.; Kotani, H.; Tanaka, A.; Asamizu, E.; Nakamura, Y.; Miyajima, N.; Hirosawa, M.; Sugiura, M.; Sasamoto, S.; et al. Sequence analysis of the genome of the unicellular cyanobacterium *Synechocystis* sp. strain PCC6803. II. Sequence determination of the entire genome and assignment of potential protein-coding regions. *DNA Res.* **1996**, *3*, 109–136. [CrossRef] [PubMed]

22. Hein, S.; Tran, H.; Steinbüchel, A. *Synechocystis* sp. PCC6803 possesses a two-component polyhydroxyalkanoic acid synthase similar to that of anoxygenic purple sulfur bacteria. *Arch. Microbiol.* **1998**, *170*, 162–170. [CrossRef] [PubMed]

23. Taroncher-Oldenburg, G.; Nishina, K.; Stephanopoulos, G. Identification and analysis of the polyhydroxyalkanoate-specific beta-ketothiolase and acetoacetyl coenzyme A reductase genes in the cyanobacterium *Synechocystis* sp. strain PCC6803. *Appl. Environ. Microbiol.* **2000**, *66*, 4440–4448. [CrossRef] [PubMed]

24. Aikawa, S.; Izumi, Y.; Matsuda, F.; Hasunuma, T.; Chang, J.S.; Kondo, A. Synergistic enhancement of glycogen production in *Arthrospira platensis* by optimization of light intensity and nitrate supply. *Bioresour. Technol.* **2012**, *108*, 211–215. [CrossRef] [PubMed]

25. Monshupanee, T.; Incharoensakdi, A. Enhanced accumulation of glycogen, lipids and polyhydroxybutyrate under optimal nutrients and light intensities in the cyanobacterium *Synechocystis* sp. PCC 6803. *J. Appl. Microbiol.* **2014**, *116*, 830–838. [CrossRef] [PubMed]

26. De Philippis, R.; Sili, C.; Vincenzini, M. Glycogen and poly-β-hydroxybutyrate synthesis in Spirulina maxima. *J. Gen. Microbiol.* **1992**, *138*, 1623–1628. [CrossRef]

27. Deschoenmaeker, F.; Facchini, R.; Carlos, J.; Pino, C.; Bayon-Vicente, G.; Sachdeva, N.; Flammang, P.; Wattiez, R. Nitrogen depletion in *Arthrospira* sp. PCC 8005, an ultrastructural point of view. *J. Struct. Biol.* **2016**, *196*, 385–393. [CrossRef] [PubMed]

28. Aikawa, S.; Nishida, A.; Ho, S.-H.; Chang, J.-S.; Hasunuma, T.; Kondo, A. Glycogen production for biofuels by the euryhaline cyanobacteria *Synechococcus* sp. strain PCC 7002 from an oceanic environment. *Biotechnol. Biofuels* **2014**, *7*, 88. [CrossRef] [PubMed]

29. Yoo, S.H.; Keppel, C.; Spalding, M.; Jane, J.L. Effects of growth condition on the structure of glycogen produced in cyanobacterium *Synechocystis* sp. PCC6803. *Int. J. Biol. Macromol.* **2007**, *40*, 498–504. [CrossRef] [PubMed]

30. Ostle, A.G.; Holt, J.G. Fluorescent Stain for Poly-3-Hydroxybutyrate. *Appl. Environ. Microbiol.* **1982**, *44*, 238–241. [PubMed]

31. Gorenflo, V.; Steinbüchel, A.; Marose, S.; Rieseberg, M.; Scheper, T. Quantification of bacterial polyhydroxyalkanoic acids by Nile red staining. *Appl. Microbiol. Biotechnol.* **1999**, *51*, 765–772. [CrossRef] [PubMed]

32. Tsang, T.K.; Roberson, R.W.; Vermaas, W.F.J. Polyhydroxybutyrate particles in *Synechocystis* sp. PCC 6803: Facts and fiction. *Photosynth. Res.* **2013**, *118*, 37–49. [CrossRef] [PubMed]

33. Hauf, W.; Watzer, B.; Roos, N.; Klotz, A.; Forchhammer, K. Photoautotrophic Polyhydroxybutyrate Granule Formation Is Regulated by Cyanobacterial Phasin PhaP in *Synechocystis* sp. Strain PCC 6803. *Appl. Environ. Microbiol.* **2015**, *81*, 4411–4422. [CrossRef] [PubMed]

34. Klotz, A.; Georg, J.; Bučínská, L.; Watanabe, S.; Reimann, V.; Januszewski, W.; Sobotka, R.; Jendrossek, D.; Hess, W.R.; Forchhammer, K. Awakening of a dormant cyanobacterium from nitrogen chlorosis reveals a genetically determined program. *Curr. Biol.* **2016**, *26*, 2862–2872. [CrossRef] [PubMed]

35. Gründel, M.; Scheunemann, R.; Lockau, W.; Zilliges, Y. Impaired glycogen synthesis causes metabolic overflow reactions and affects stress responses in the cyanobacterium *Synechocystis* sp. PCC 6803. *Microbiol. (UK)* **2012**, *158*, 3032–3043. [CrossRef] [PubMed]

36. Beck, C.; Knoop, H.; Axmann, I.M.; Steuer, R. The diversity of cyanobacterial metabolism: Genome analysis of multiple phototrophic microorganisms. *BMC Genom.* **2012**, *13*, 56. [CrossRef] [PubMed]

37. Ansari, S.; Fatma, T. Cyanobacterial polyhydroxybutyrate (PHB): Screening, optimization and characterization. *PLoS ONE* **2016**, *11*, 1–20. [CrossRef] [PubMed]

38. Kaewbai-ngam, A.; Incharoensakdi, A.; Monshupanee, T. Increased accumulation of polyhydroxybutyrate in divergent cyanobacteria under nutrient-deprived photoautotrophy: An efficient conversion of solar energy and carbon dioxide to polyhydroxybutyrate by Calothrix scytonemicola TISTR 8095. *Bioresour. Technol.* **2016**, *212*, 342–347. [CrossRef] [PubMed]

39. Hu, Q.; Sommerfeld, M.; Jarvis, E.; Ghirardi, M.; Posewitz, M.; Seibert, M.; Darzins, A. Microalgal triacylglycerols as feedstocks for biofuel production: Perspectives and advances. *Plant J.* **2008**, *54*, 621–639. [CrossRef] [PubMed]

40. de Jaeger, L.; Verbeek, R.E.; Draaisma, R.B.; Martens, D.E.; Springer, J.; Eggink, G.; Wijffels, R.H. Superior triacylglycerol (TAG) accumulation in starchless mutants of Scenedesmus obliquus: (I) Mutant generation and characterization. *Biotechnol. Biofuels* **2014**, *7*, 69. [CrossRef] [PubMed]

41. Damrow, R.; Maldener, I.; Zilliges, Y. The multiple functions of common microbial carbon polymers, glycogen and PHB, during stress responses in the non-diazotrophic cyanobacterium *Synechocystis* sp. PCC 6803. *Front. Microbiol.* **2016**, *7*, 1–10. [CrossRef] [PubMed]

42. Allen, M.M.; Smith, A.J. Nitrogen chlorosis in blue-green algae. *Arch. für Mikrobiol.* **1969**, *69*, 114–120. [CrossRef]

43. Gorl, M.; Sauer, J.; Baier, T.; Forchhammer, K. Nitrogen-starvation-induced chlorosis in Synechococcus PCC 7942: Adaptation to long-term survival. *Microbiology* **1998**, *144*, 2449–2458. [CrossRef] [PubMed]

44. Sauer, J.; Schreiber, U.; Schmid, R.; Völker, U.; Forchhammer, K. Nitrogen starvation-induced chlorosis in Synechococcus PCC 7942. Low-level photosynthesis as a mechanism of long-term survival. *Plant Physiol.* **2001**, *126*, 233–243. [CrossRef] [PubMed]

45. Wu, G.F.; Wu, Q.Y.; Shen, Z.Y. Accumulation of poly-beta-hydroxybutyrate in cyanobacterium *Synechocystis* sp. PCC6803. *Bioresour. Technol.* **2001**, *76*, 85–90. [CrossRef]

46. Panda, B.; Mallick, N. Enhanced poly-β-hydroxybutyrate accumulation in a unicellular cyanobacterium, *Synechocystis* sp. PCC 6803. *Lett. Appl. Microbiol.* **2007**, *44*, 194–198. [CrossRef] [PubMed]
47. Khetkorn, W.; Incharoensakdi, A.; Lindblad, P.; Jantaro, S. Enhancement of poly-3-hydroxybutyrate production in *Synechocystis* sp. PCC 6803 by overexpression of its native biosynthetic genes. *Bioresour. Technol.* **2016**, *214*, 761–768. [CrossRef] [PubMed]
48. Nishioka, M.; Nakai, K.; Miyake, M.; Asada, Y.; Taya, M. Production of poly-β-hydroxybutyrate by thermophilic cyanobacterium, *Synechococcus* sp. MA19, under phosphate-limited conditions. *Biotechnol. Lett.* **2001**, *23*, 1095–1099. [CrossRef]
49. Wu, Q.; Liu, L.; Miron, A.; Klimova, B.; Wan, D.; Kuca, K. The antioxidant, immunomodulatory, and anti-inflammatory activities of Spirulina: An overview. *Arch. Toxicol.* **2016**, *90*, 1817–1840. [CrossRef] [PubMed]
50. Spolaore, P.; Joannis-Cassan, C.; Duran, E.; Isambert, A. Commercial applications of microalgae. *J. Biosci. Bioeng.* **2006**, *101*, 87–96. [CrossRef] [PubMed]
51. Shimamatsu, H. Mass production of *Spirulina*, an edible microalga. *Hydrobiologia* **2004**, *512*, 39–44. [CrossRef]
52. Choi, S.-L.; Suh, I.S.; Lee, C.-G. Lumostatic operation of bubble column photobioreactors for Haematococcus pluvialis cultures using a specific light uptake rate as a control parameter. *Enzyme Microb. Technol.* **2003**, *33*, 403–409. [CrossRef]
53. Lu, Y.M.; Xiang, W.Z.; Wen, Y.H. Spirulina (Arthrospira) industry in Inner Mongolia of China: Current status and prospects. *J. Appl. Phycol.* **2011**, *23*, 265–269. [CrossRef] [PubMed]
54. Campbell, J.; Stevens, S.E.; Balkwill, D.L. Accumulation of poly-beta-hydroxybutyrate in Spirulina platensis. *J. Bacteriol.* **1982**, *149*, 361–363. [PubMed]
55. Vincenzini, M.; Sili, C.; de Philippis, R.; Ena, A.; Materassi, R. Occurrence of poly-beta-hydroxybutyrate in Spirulina species. *J. Bacteriol.* **1990**, *172*, 2791–2792. [CrossRef] [PubMed]
56. Panda, B.; Sharma, L.; Mallick, N. Poly-β-hydroxybutyrate accumulation in Nostoc muscorum and Spirulina platensis under phosphate limitation. *J. Plant Physiol.* **2005**, *162*, 1376–1379. [CrossRef] [PubMed]
57. Shrivastav, A.; Mishra, S.K.; Mishra, S. Polyhydroxyalkanoate (PHA) synthesis by Spirulina subsalsa from Gujarat coast of India. *Int. J. Biol. Macromol.* **2010**, *46*, 255–260. [CrossRef] [PubMed]
58. De Morais, M.G.; Stillings, C.; Roland, D.; Rudisile, M.; Pranke, P.; Costa, J.A.V.; Wendorff, J. Extraction of poly(3-hydroxybutyrate) from Spirulina LEB 18 for developing nanofibers. *Polímeros* **2015**, *25*, 161–167. [CrossRef]
59. De Morais, M.G.; Stillings, C.; Dersch, R.; Rudisile, M.; Pranke, P.; Costa, J.A.V.; Wendorff, J. Biofunctionalized nanofibers using *Arthrospira (spirulina)* biomass and biopolymer. *Biomed Res. Int.* **2015**, *2015*. [CrossRef] [PubMed]
60. Dodds, W.K.; Gudder, D.A.; Mollenhauer, D. Review the Ecology of Nostoc. *J. Phycol.* **1995**, *18*, 2–18. [CrossRef]
61. Qiu, B.; Liu, J.; Liu, Z.; Liu, S. Distribution and ecology of the edible cyanobacterium Ge-Xian-Mi (Nostoc) in rice fields of Hefeng County in China. *J. Appl. Phycol.* **2002**, *14*, 423–429. [CrossRef]
62. Sharma, L.; Mallick, N. Accumulation of poly-β-hydroxybutyrate in Nostoc muscorum: Regulation by pH, light-dark cycles, N and P status and carbon sources. *Bioresour. Technol.* **2005**, *96*, 1304–1310. [CrossRef] [PubMed]
63. Sharma, L.; Mallick, N. Enhancement of poly-β-hydroxybutyrate accumulation in Nostoc muscorum under mixotrophy, chemoheterotrophy and limitations of gas-exchange. *Biotechnol. Lett.* **2005**, *27*, 59–62. [CrossRef] [PubMed]
64. Bhati, R.; Mallick, N. Production and characterization of poly(3-hydroxybutyrate-co-3-hydroxyvalerate) co-polymer by a N 2-fixing cyanobacterium, Nostoc muscorum Agardh. *J. Chem. Technol. Biotechnol.* **2012**, *87*, 505–512. [CrossRef]
65. Bhati, R.; Mallick, N. Poly(3-hydroxybutyrate-co-3-hydroxyvalerate) copolymer production by the diazotrophic cyanobacterium Nostoc muscorum Agardh: Process optimization and polymer characterization. *Algal Res.* **2015**, *7*, 78–85. [CrossRef]
66. Bhati, R.; Mallick, N. Carbon dioxide and poultry waste utilization for production of polyhydroxyalkanoate biopolymers by Nostoc muscorum Agardh: A sustainable approach. *J. Appl. Phycol.* **2016**, *28*, 161–168. [CrossRef]
67. Samantaray, S.; Mallick, N. Production and characterization of poly-β-hydroxybutyrate (PHB) polymer from Aulosira fertilissima. *J. Appl. Phycol.* **2012**, *24*, 803–814. [CrossRef]

68. Lama, L.; Nicolaus, B.; Calandrelli, V.; Manca, M.C.; Romana, I.; Gambacorta, A. Effect of growth conditions on endo- and exopolymer biosynthesis in Anabaena cylindrical 10 C. *Phytochemistry* **1996**, *42*, 655–659. [CrossRef]
69. Gopi, K.; Balaji, S.; Muthuvelan, B. Isolation Purification and Screening of Biodegradable Polymer PHB Producing Cyanobacteria from Marine and Fresh Water Resources. *Iran. J. Energy Environ.* **2014**, *5*, 94–100. [CrossRef]
70. Benemann, J. Microalgae for biofuels and animal feeds. *Energies* **2013**, *6*, 5869–5886. [CrossRef]
71. Dineshbabu, G.; Uma, V.S.; Mathimani, T.; Deviram, G.; Arul Ananth, D.; Prabaharan, D.; Uma, L. On-site concurrent carbon dioxide sequestration from flue gas and calcite formation in ossein effluent by a marine cyanobacterium Phormidium valderianum BDU 20041. *Energy Convers. Manag.* **2016**, in press. [CrossRef]
72. Chen, H.W.; Yang, T.S.; Chen, M.J.; Chang, Y.C.; Lin, C.Y.; Wang, E.I.C.; Ho, C.L.; Huang, K.M.; Yu, C.C.; Yang, F.L.; et al. Application of power plant flue gas in a photobioreactor to grow Spirulina algae, and a bioactivity analysis of the algal water-soluble polysaccharides. *Bioresour. Technol.* **2012**, *120*, 256–263. [CrossRef] [PubMed]
73. Kumari, A.; Kumar, A.; Pathak, A.K.; Guria, C. Carbon dioxide assisted Spirulina platensis cultivation using NPK-10:26:26 complex fertilizer in sintered disk chromatographic glass bubble column. *J. CO$_2$ Util.* **2014**, *8*, 49–59. [CrossRef]
74. He, L.; Subramanian, V.R.; Tang, Y.J. Experimental analysis and model-based optimization of microalgae growth in photo-bioreactors using flue gas. *Biomass Bioenergy* **2012**, *41*, 131–138. [CrossRef]
75. Ferreira, L.S.; Rodrigues, M.S.; Converti, A.; Sato, S.; Carvalho, J.C.M. *Arthrospira (spirulina)* platensis cultivation in tubular photobioreactor: Use of no-cost CO$_2$ from ethanol fermentation. *Appl. Energy* **2012**, *92*, 379–385. [CrossRef]
76. Sumardiono, S.; Syaichurrozi, I.; Budi Sasongko, S. Utilization of Biogas as Carbon Dioxide Provider for Spirulina platensis Culture. *Curr. Res. J. Biol. Sci.* **2014**, *6*, 53–59.
77. Cuellar-Bermudez, S.P.; Aleman-Nava, G.S.; Chandra, R.; Garcia-Perez, J.S.; Contreras-Angulo, J.R.; Markou, G.; Muylaert, K.; Rittmann, B.E.; Parra-Saldivar, R. Nutrients utilization and contaminants removal. A review of two approaches of algae and cyanobacteria in wastewater. *Algal Res.* **2016**. [CrossRef]
78. Markou, G.; Vandamme, D.; Muylaert, K. Microalgal and cyanobacterial cultivation: The supply of nutrients. *Water Res.* **2014**, *65*, 186–202. [CrossRef] [PubMed]
79. Kumar, K.; Mishra, S.K.; Shrivastav, A.; Park, M.S.; Yang, J.W. Recent trends in the mass cultivation of algae in raceway ponds. *Renew. Sustain. Energy Rev.* **2015**, *51*, 875–885. [CrossRef]
80. Markou, G.; Georgakakis, D. Cultivation of filamentous cyanobacteria (blue-green algae) in agro-industrial wastes and wastewaters: A review. *Appl. Energy* **2011**, *88*, 3389–3401. [CrossRef]
81. Chaiklahan, R.; Chirasuwan, N.; Siangdung, W.; Paithoonrangsarid, K.; Bunnag, B. Cultivation of spirulina platensis using pig wastewater in a semi-continuous process. *J. Microbiol. Biotechnol.* **2010**, *20*, 609–614. [CrossRef] [PubMed]
82. Cicci, A.; Bravi, M. Production of the freshwater microalgae scenedesmus dimorphus and arthrospira platensis by using cattle digestate. *Chem. Eng. Trans.* **2014**, *38*, 85–90. [CrossRef]
83. Fouilland, E.; Vasseur, C.; Leboulanger, C.; Le Floc'h, E.; Carré, C.; Marty, B.; Steyer, J.-P.P.; Sialve, B. Coupling algal biomass production and anaerobic digestion: Production assessment of some native temperate and tropical microalgae. *Biomass Bioenergy* **2014**, *70*, 564–569. [CrossRef]
84. Markou, G.; Chatzipavlidis, I.; Georgakakis, D. Cultivation of *Arthrospira (Spirulina)* platensis in olive-oil mill wastewater treated with sodium hypochlorite. *Bioresour. Technol.* **2012**, *112*, 234–241. [CrossRef] [PubMed]
85. Olguín, E.J.; Galicia, S.; Mercado, G.; Pérez, T. Annual productivity of *Spirulina (Arthrospira)* and nutrient removal in a pig wastewater recycling process under tropical conditions. *J. Appl. Phycol.* **2003**, *15*, 249–257. [CrossRef]
86. Prajapati, S.K.; Kumar, P.; Malik, A.; Vijay, V.K. Bioconversion of algae to methane and subsequent utilization of digestate for algae cultivation: A closed loop bioenergy generation process. *Bioresour. Technol.* **2014**, *158*, 174–180. [CrossRef] [PubMed]
87. Daelman, M.R.J.; Sorokin, D.; Kruse, O.; van Loosdrecht, M.C.M.; Strous, M. Haloalkaline Bioconversions for Methane Production from Microalgae Grown on Sunlight. *Trends Biotechnol.* **2016**, *34*, 450–457. [CrossRef] [PubMed]

88. Nolla-Ardevol, V.; Strous, M.; Tegetmeyer, H.E.E. Anaerobic digestion of the microalga Spirulina at extreme alkaline conditions: Biogas production, metagenome and metatranscriptome. *Front. Microbiol.* **2015**, *6*, 1–21. [CrossRef] [PubMed]

89. Zhou, Y.; Schideman, L.; Yu, G.; Zhang, Y. A synergistic combination of algal wastewater treatment and hydrothermal biofuel production maximized by nutrient and carbon recycling. *Energy Environ. Sci.* **2013**, *6*, 3765. [CrossRef]

90. Zheng, M.; Schideman, L.C.; Tommaso, G.; Chen, W.-T.; Zhou, Y.; Nair, K.; Qian, W.; Zhang, Y.; Wang, K. Anaerobic digestion of wastewater generated from the hydrothermal liquefaction of Spirulina: Toxicity assessment and minimization. *Energy Convers. Manag.* **2016**, in press. [CrossRef]

91. Depraetere, O.; Pierre, G.; Noppe, W.; Vandamme, D.; Foubert, I.; Michaud, P.; Muylaert, K. Influence of culture medium recycling on the performance of Arthrospira platensis cultures. *Algal Res.* **2015**, *10*, 48–54. [CrossRef]

92. Meixner, K.; Fritz, I.; Daffert, C.; Markl, K.; Fuchs, W.; Drosg, B. Processing recommendations for using low-solids digestate as nutrient solution for production with *Synechocystis salina*. *J. Biotechnol.* **2016**, *240*, 1–22. [CrossRef] [PubMed]

93. Balaji, S.; Gopi, K.; Muthuvelan, B. A review on production of poly-b-hydroxybutyrates from cyanobacteria for the production of bio plastics. *Algal Res.* **2013**, *2*, 278–285. [CrossRef]

94. Marcilhac, C.; Sialve, B.; Pourcher, A.M.; Ziebal, C.; Bernet, N.; Béline, F. Digestate color and light intensity affect nutrient removal and competition phenomena in a microalgal-bacterial ecosystem. *Water Res.* **2014**, *64*, 278–287. [CrossRef] [PubMed]

95. Andrade, M.R.; Costa, J.A.V. Mixotrophic cultivation of microalga Spirulina platensis using molasses as organic substrate. *Aquaculture* **2007**, *264*, 130–134. [CrossRef]

96. Phang, S.M.; Miah, M.S.; Yeoh, B.G.; Hashim, M.A. Spirulina cultivation in digested sago starch factory wastewater. *J. Appl. Phycol.* **2000**, *12*, 395–400. [CrossRef]

97. Massa, M.; Buono, S.; Langellotti, A.L.; Castaldo, L.; Martello, A.; Paduano, A.; Sacchi, R.; Fogliano, V. Evaluation of anaerobic digestates from different feedstocks as growth media for Tetradesmus obliquus, Botryococcus braunii, Phaeodactylum tricornutum and Arthrospira maxima. *New Biotechnol.* **2017**, *36*, 8–16. [CrossRef] [PubMed]

98. Cai, T.; Ge, X.; Park, S.Y.; Li, Y. Comparison of *Synechocystis* sp. PCC6803 and Nannochloropsis salina for lipid production using artificial seawater and nutrients from anaerobic digestion effluent. *Bioresour. Technol.* **2013**, *144*, 255–260. [CrossRef] [PubMed]

99. Molina, E.; Fernández, J.; Acién, F.G.; Chisti, Y. Tubular photobioreactor design for algal cultures. *J. Biotechnol.* **2001**, *92*, 113–131. [CrossRef]

100. Acién Fernández, F.G.; Fernández Sevilla, J.M.; Sánchez Pérez, J.A.; Molina Grima, E.; Chisti, Y. Airlift-driven external-loop tubular photobioreactors for outdoor production of microalgae: Assessment of design and performance. *Chem. Eng. Sci.* **2001**, *56*, 2721–2732. [CrossRef]

101. Rabensteiner, M.; Kinger, G.; Koller, M.; Gronald, G.; Hochenauer, C. Pilot plant study of ethylenediamine as a solvent for post combustion carbon dioxide capture and comparison to monoethanolamine. *Int. J. Greenh. Gas Control* **2014**, *27*, 1–14. [CrossRef]

102. Rippka, R.; Deruelles, J.; Waterbury, J.B.; Herdman, M.; Stanier, R.Y. Generic Assignments, Strain Histories and Properties of Pure Cultures of Cyanobacteria. *J. Gen. Microbiol.* **1979**, *111*, 1–61. [CrossRef]

103. Drosg, B.; Fritz, I.; Gattermayr, F.; Silvestrini, L. Photo-autotrophic Production of Poly(hydroxyalkanoates) in Cyanobacteria. *Chem. Biochem. Eng. Q.* **2015**, *29*, 145–156. [CrossRef]

104. Michels, M.H.A.; van der Goot, A.J.; Vermuë, M.H.; Wijffels, R.H. Cultivation of shear stress sensitive and tolerant microalgal species in a tubular photobioreactor equipped with a centrifugal pump. *J. Appl. Phycol.* **2016**, *28*, 53–62. [CrossRef] [PubMed]

105. Milledge, J.J.; Heaven, S. A review of the harvesting of micro-algae for biofuel production. *Rev. Environ. Sci. Biotechnol.* **2013**, *12*, 165–178. [CrossRef]

106. Heinrich, D.; Madkour, M.H.; Al-Ghamdi, M.; Shabbaj, I.I.; Steinbüchel, A. Large scale extraction of poly(3-hydroxybutyrate) from Ralstonia eutropha H16 using sodium hypochlorite. *AMB Express* **2012**, *2*, 59. [CrossRef] [PubMed]

107. Ramos, A.; Acién, F.G.; Fernández-Sevilla, J.M.; González, C.V.; Bermejo, R. Development of a process for large-scale purification of C-phycocyanin from *Synechocystis aquatilis* using expanded bed adsorption chromatography. *J. Chromatogr. B Anal. Technol. Biomed. Life Sci.* **2011**, *879*, 511–519. [CrossRef] [PubMed]

108. Wang, H.; Zhang, W.; Chen, L.; Wang, J.; Liu, T. The contamination and control of biological pollutants in mass cultivation of microalgae. *Bioresour. Technol.* **2013**, *128*, 745–750. [CrossRef] [PubMed]

109. Touloupakis, E.; Cicchi, B.; Benavides, A.M.S.; Torzillo, G. Effect of high pH on growth of *Synechocystis* sp. PCC 6803 cultures and their contamination by golden algae (*Poterioochromonas* sp.). *Appl. Microbiol. Biotechnol.* **2015**, *100*, 1333–1341. [CrossRef] [PubMed]

bioengineering

MDPI

Article

Polyhydroxyalkanoate Production on Waste Water Treatment Plants: Process Scheme, Operating Conditions and Potential Analysis for German and European Municipal Waste Water Treatment Plants

Timo Pittmann [1,]* and Heidrun Steinmetz [2]

[1]   TBF + Partner AG, Herrenberger Strasse 14, 71032 Boeblingen, Germany
[2]   Department of Resource Efficient Wastewater Technology, University of Kaiserslautern, Paul-Ehrlich-Str. 14, 67663 Kaiserslautern, Germany; heidrun.steinmetz@bauing.uni-kl.de
*   Correspondence: pit@tbf.ch; Tel.: +49-7031-23806-60

Academic Editor: Martin Koller
Received: 28 April 2017; Accepted: 3 June 2017; Published: 6 June 2017

Abstract: This work describes the production of polyhydroxyalkanoates (PHA) as a side stream process on a municipal waste water treatment plant (WWTP) and a subsequent analysis of the production potential in Germany and the European Union (EU). Therefore, tests with different types of sludge from a WWTP were investigated regarding their volatile fatty acids (VFA) production-potential. Afterwards, primary sludge was used as substrate to test a series of operating conditions (temperature, pH, retention time (RT) and withdrawal (WD)) in order to find suitable settings for a high and stable VFA production. In a second step, various tests regarding a high PHA production and stable PHA composition to determine the influence of substrate concentration, temperature, pH and cycle time of an installed feast/famine-regime were conducted. Experiments with a semi-continuous reactor operation showed that a short RT of 4 days and a small WD of 25% at pH = 6 and around 30 °C is preferable for a high VFA production rate (PR) of 1913 mg$_{VFA}$/(L×d) and a stable VFA composition. A high PHA production up to 28.4% of cell dry weight (CDW) was reached at lower substrate concentration, 20 °C, neutral pH-value and a 24 h cycle time. A final step a potential analysis, based on the results and detailed data from German waste water treatment plants, showed that the theoretically possible production of biopolymers in Germany amounts to more than 19% of the 2016 worldwide biopolymer production. In addition, a profound estimation regarding the EU showed that in theory about 120% of the worldwide biopolymer production (in 2016) could be produced on European waste water treatment plants.

Keywords: biopolymer; municipal sewage plant; PHA; primary sludge; VFA

---

1. Introduction

Common plastic is derived from petrochemicals based on the limited natural resource petroleum. Besides the exploitation of natural resources, the use of plastic is responsible for major waste problems, as common plastic is non- or poor biodegradable [1].

Biopolymers present a possible alternative to common plastics. If they are fully biodegradable [2,3] their use not only allows the preservation of limited resources, but also suits the idea of sustainability.

The term "biopolymer" or "bioplastic" is not yet uniformly defined. Common definitions of the term "biopolymer" also include biodegradable plastics from fossil fuels and non-biodegradable plastics from renewable resources as seen in Figure 1. To eliminate the problems accompanied by polymer production from crude oil a more stringent definition is introduced by the authors:

"Biopolymers are made from renewable resources and/or biodegradable waste materials (e.g., waste water, sewage sludge, organic waste) and are fully biodegradable by naturally occurring microorganisms."

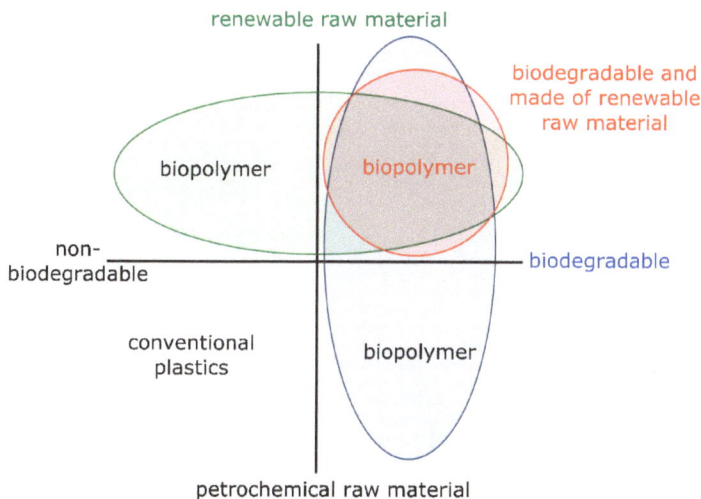

Figure 1. Definition for biopolymers, including the stringent definition on the upper right, modified after [4].

This definition ensures that polymers from fossil resources and non-biodegradable polymers, which cause at least one of the mentioned problems, are excluded and that the term biopolymer is just used for polymers, which allow the preservation of limited resources and also suit the idea of sustainability. This type of biopolymers is shown in the upper right of Figure 1.

Beside other polymers polyhydroxyalkanoates (PHA), which are biodegradable polyesters accumulated by bacteria under nutrient limited conditions [5] or under balanced growth, are a source for bioplastic production matching the above mentioned strict definition. More than 150 component parts of PHA have been identified so far [6]. The possibility for chemical modification of PHA provide a wide range of material properties and an even wider range of use [7,8]. However, so far the main raw material for the biopolymer production are starchy plants like maize [9], constituting the disadvantages of high land consumption, diminishing food resources as well as problems like leaching of nutrients, input of pesticide and soil erosion [10].

So far, municipal waste water treatment plants (WWTP) as alternative raw material and biomass source for the PHA production have not been widely investigated, although they offer the opportunity to compensate the disadvantages of the common PHA production using starchy plants.

PHA production in WWTP takes place in two steps, which composes the production of volatile fatty acids (VFA) in an anaerobic process and finally the PHA production in an aerobic process (see also Figure 2). In contrast to [11,12] the PHA production process described in this work is designed as a side stream process of a municipal WWTP and does not include the treatment of waste water. Therefore, the whole process can focus on polymer production only.

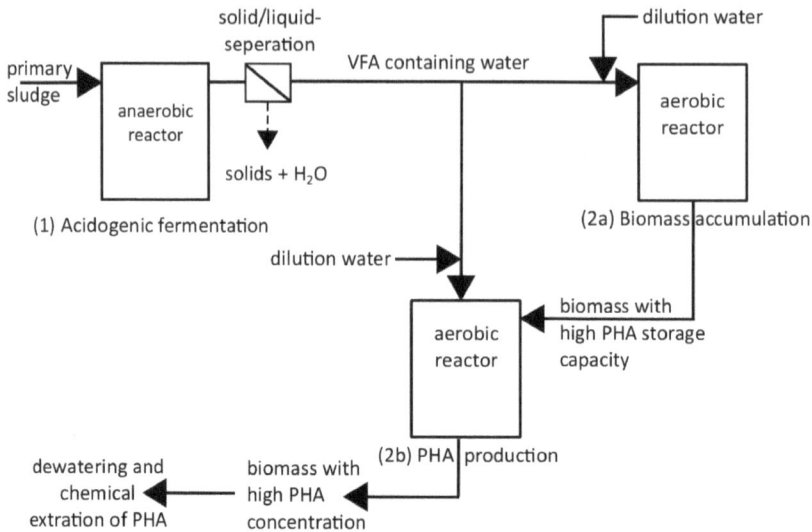

Figure 2. PHA production scheme.

The possibility to use ice-cream waste water as alternative source material for the VFA production was shown by [13] while [14] investigated the effect of pH, sludge-retention time (RT) and acetate concentration on the PHA production from municipal waste water. Diverse authors [15–18] stated that there is a general possibility to produce PHA from activated sludge.

In many of the research projects on PHA production, synthetic waste water was used to gain knowledge about one part of the PHA production or the production's operating conditions [19–25]. However, so far no research group has investigated the general possibility and all operating conditions of a biopolymer production using only material flows of a WWTP.

PHA production itself is based on a bacteria mixed culture selection from excess sludge via aerobic dynamic feeding. The installed feast/famine regime for enrichment of PHA producing bacteria is state of the art and tested by many authors [19,23,26,27]. The feast-phase is defined as a period of substrate availability and could be monitored via the reactors oxygen concentration. During the period of starvation (famine-phase) bacteria with the ability of polymer-storage gained a selection advantage as they are able to use the stored polymers as carbon and energy source.

The objective of this research project was to find the most suitable raw material and all operating conditions for the VFA and PHA production process using only material flows of a WWTP. At first the suitability of different raw materials of a municipal WWTP for VFA-production were investigated and afterwards the influence of operating conditions (temperature, pH, retention time (RT) and withdrawal (WD)) and reactor operation method. Another concern was, how the tested operating conditions or the diversity of the used material flows of a WWTP influence the VFA composition and the type of PHA produced. As there is a variation in the composition of the used material flows (different sludge) of a WWTP, it is of particular importance to observe their influence on VFA production and composition.

Then the possibility to produce PHA out of the VFA containing substrate was tested using a feast/famine regime (as shown in Figure 3). Subsequently, the influence of operating conditions (temperature, pH, cycle time (CT) and substrate concentration) on PHA production were investigated.

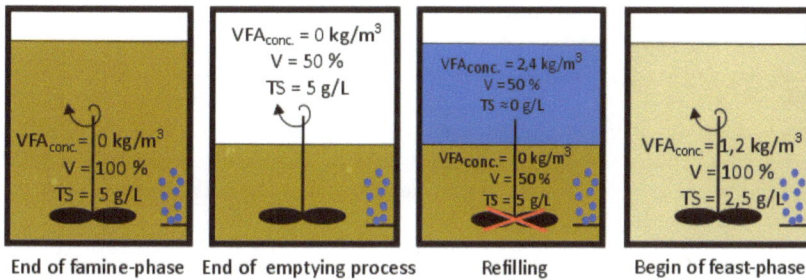

Figure 3. Emptying and refilling process during aerobic dynamic feeding (values shown are an example for investigated operation conditions), modified after [28].

The PHA potential on German WWTPs was calculated based on detailed data from operators of WWTPs [29] and the results of the PHA production experiments mentioned above. Finally, a profound estimation of the biopolymer potential of all WWTPs in the 28 member states of the European Union (EU) was made based on data provided by the EU [30] and the mentioned PHA production experiments.

2. Materials and Methods

2.1. VFA Production

2.1.1. Conception of the Experiments

For the VFA production ((1) Acidogenic fermentation in Figure 2) anaerobic reactors of different sizes (4 L, 15 L) were operated as batch reactors or as semi-continuous reactors. While the batch operation is defined as a one-time substrate filling at the beginning of the experiment with no withdrawal and refill during the test, a semi-continuously operation method allows to introduce and withdraw substrate to or from the reactor. Semi-continuously is defined as one-time substrate filling at the beginning of every cycle, e.g., daily within a test duration of one month.

All tests were conducted without sedimentation or biomass recirculation. Therefore, the hydraulic retention time equals the sludge age and both will be referred hereinafter as RT.

The raw material is the most important base for gaining high PHA production rates, so that the selection of suitable raw material has the number one priority. The improvement of the VFA production's operation conditions was examined afterwards with the most appropriate raw material found. A chronological test order was implemented as follows:

1. Selection of raw material
2. Investigation of the most suitable pH-level
3. Evaluation of a retention time (RT) range
4. Selection of a suitable combination of RT and withdrawal (WD)

2.1.2. Selection of Raw Material

For raw material selection continuously stirred batch reactors with a volume of 4 L were used. Four different types of sludge, namely primary sludge (average total solid $TS_a = 43$ g/L), excess sludge ($TS_a = 10$ g/L), a one to one mixture of primary-and digested sludge ($TS_a = 37.5$ g/L) and a one to one mixture of excess- and digested sludge ($TS_a = 21$ g/L) from a municipal WWTP were treated under anaerobic conditions. Thereby the digested sludge from the WWTP's digester was only used as inoculum for the anaerobic process in order to find out, if it could accelerate the process. All types of sludge were investigated under four different conditions: pH controlled at pH = 6, without pH-control and each at around 20 °C or around 30 °C reactor temperature. In summary 16 different tests were performed. The reactors were filled at the beginning of the experiments and samples of 50 mL were retrieved every day to determine the VFA concentration and composition. To achieve the selected temperature, the reactors were situated in temperature-controlled rooms. For pH-controlled tests, the pH-value was measured by a mobile pH meter (WTW pH340i) and adjusted with NaOH by hand twice a day. The test duration for all experiments was 18 days to 20 days. A sample of all tested types of sludge was taken before and after the tests to determine the chemical oxygen demand (COD), the total Kjeldahl nitrogen(TKN) and total P.

2.1.3. Evaluation of Operating Conditions

As the influence of fermentation temperature was already observed during the selection of the raw material, three additional operating conditions (pH, RT, WD) were investigated in continuous stirred tank reactors (CSTR) with a volume of 15 L. For all experiments primary sludge from a municipal WWTP was used as raw material and a sample was retrieved prior to the experiments to determine COD, TKN and total P. All reactors were placed in a temperature-controlled room at around 30 °C. pH was monitored at all times via a Metrom Profitrode pH probe and automatically adjusted with NaOH during all tests.

For the batch tests (pH, pre-RT) all reactors were filled with primary sludge at the beginning and samples of 100 mL were retrieved every day to determine the VFA concentration and composition. The batch test period was 18 days to 20 days long. Former studies show a high VFA production in a pH-range between 5 [11] and 9 [31] or 11 [32]. In consequence a range of pH-levels (pH = 6, 6.5, 7, 8, 10) were tested.

For the semi-continuously operated tests all reactors were filled with primary sludge at the beginning and operated under pH-controlled conditions at pH = 6. After a starting phase of 10 days, to accumulate VFAs, the semi-continuous operation phase began, for which a certain amount of the sludge in the reactor was exchanged. Samples of 100 mL were retrieved to analyse the VFA concentration and composition for about 40 days with a RT of 4 days, 6 days and 8 days, each with 25% and 50% WD. Additionally, a 75% WD was performed with a RT of 4 days. A RT of 2 days was also tested with a WD of 50%. RT and WD are related factors, e.g., a RT of 4 days was used when 25% of the sludge was exchanged every day, 50% every second day or 75% every third day.

2.1.4. Analytical Procedures

COD, TKN, total P and total solids (TS) were determined according to standard methods.

The concentration and composition of volatile fatty acids, namely formate (Fo), acetate (Ac), propionate (Pro) and butyrate (Bu), were detected by high performance liquid chromatography (HPLC). Therefore, the sample was acidified to pH = 2 and filtered through 0.45 μm membrane filter. Afterwards HPLC detection was performed using a HP1100 chromatographer equipped with an UV detector and a Varian Metacarb 87H column. Sulphuric acid (0.05 M) was used as eluent at a flow rate of 0.6 mL/min. The detection wavelength was 210 nm. Volatile fatty acid's concentration was calibrated using 4 nmol to 4000 nmol standards.

As the results of formic acid detection was below detection point for all except one test, formic acid is not shown in the VFA composition.

2.1.5. Conversion of Units

For a better comparability all results regarding VFA concentrations, COD, TKN and total P were converted into mg/L.

The concentration of VFA in terms of mg/L is defined as:

$$VFA = Ac + Pro + Bu \tag{1}$$

The degree of acidification (DA) was calculated according to [33] as shown in Equation (2). As VFA results are given in mg_{VFA}/L they have to be converted into COD units as shown in Equation (3)

$$DA = VFA/COD_S \text{ in } (mg_{COD}/L)/(mg_{COD}/L) \qquad (2)$$

with COD_S = COD of Substrate at the start.

$$COD_{VFA} = (\text{conc. } VFA_i/\text{molar mass } VFA_i) \text{ oxygen demand} \qquad (3)$$

with i = Ac, Pro, Bu.

For a better comparability regarding the different RT and WD, the average of the VFA concentration during the test period was calculated. In a second step, the average VFA concentration was used to calculate the average VFA production rate (PR_{VFA}). This step also eliminates the reactors size and can hence be considered as average VFA production rate per day and litre, which will be referred to production rate (PR) hereinafter (Equation (4)).

$$PR_{VFA} = \text{av. VFA conc.}/RT \text{ in } mg_{VFA}/(L \times day) \qquad (4)$$

This calculation helps to compare results from different reactor sizes and retention times.

2.2. PHA Production

2.2.1. Experimental Set-Up

The overall process describing the production of biopolymers from municipal waste water is displayed in Figure 2. For PHA production only "(2a): Biomass accumulation" and "(2b): PHA production" is considered. The first step, the volatile fatty acids (VFA) production, was already discussed in Chapter 2.1. The substrate produced in step 1 under anaerobic fermentation process at 30 °C, pH = 6, RT = 4 days and a withdrawal of 25% was frozen at −18 °C and defrosted about 24 h prior to its use as input material for phase 2 PHA production.

During all PHA production tests continuous stirred tank reactor (CSTR) with a volume of 15 L were used. All reactors were equipped with a pH- and oxygen-probe. If necessary, pH-value was adjusted via a dosing pump. pH levels were controlled by the pH probe and adjusted with NaOH or H_2SO_4. The reactor temperature was controlled via the pH-probe and manually adjusted using a heating bath (Haake DC30) and a heat exchanger, installed in the reactor. To maintain aerobic conditions an aerator was installed in the reactor. All reactors were operated in batch mode. The batch operation is defined as a one-time substrate filling at the beginning of the aerobic dynamic feeding-cycle with no withdrawal and refill during the cycle. There was no sedimentation or biomass recirculation in all tests. Therefore, the hydraulic retention time equals the sludge age and both will be referred hereinafter as RT. At the end of every cycle 7.5 L were withdrawn from reactor 2a and filled into Reactor 2b. Hence a RT of 2 days was implemented. Afterwards both reactors were filled with 7.5 L of substrate and fresh water to achieve the working volume of 15 L per reactor. Reactor 2b was emptied at the end of the feast-phase, and samples were taken to measure the PHA concentration and composition.

Both reactors were operated under similar conditions, just differing concerning nutrient availability. As the VFA enriched substrate showed nutrient limited conditions, with a Carbon:Nitrogen:Phosphorus (C:N:P)-ratio in a range of 100:2:0.5 to 100:3:0.8 [34] CH_4N_2O and KH_2PO_4 were added to Reactor 2a to create optimal conditions for the bacteria growth (C:N:P = 100:5:1) and selection process. Reactor 2b, however, was operated under the named nutrient limited conditions to reach a higher PHA concentration.

Samples to determine the PHA concentration were taken at the end of the feast-phase. The total solid (TS) concentration was measured at the end of each cycle and was about 5 ± 0.5 g/L in both reactors. This was done to figure out if the selection process in Reactor 2a was working correctly or if the biomass concentration was decreasing e.g., for lack of nutrients. As the biomass concentration was stable throughout all experiments no further tests regarding cell growth were conducted.

Experiments regarding the substrate concentration and temperature selection had the top priority. Afterwards the optimisation of all other operation conditions concerning the PHA production was examined with the most appropriate reactor temperature and substrate concentration found. A chronological test order was implemented as follows:

1. Selection of a suitable substrate concentration
2. Investigation of the reactor temperature
3. Evaluation of a suitable pH-level
4. Selection of a suitable cycle time

2.2.2. Investigation Concerning the Best Substrate Concentration

As there is a big variation in substrate concentration in literature, different substrate concentration of 1200 mg_{VFA}/L and 2000 mg_{VFA}/L were tested at 20 °C or 30 °C, pH = 7 or without pH control and a CT = 24 h. The named concentrations were chosen to avoid possible problems triggered by substrate inhibition.

2.2.3. Investigation Concerning the Best Temperature

The temperatures of material flows from a municipal WWTP are about 15 °C to 20 °C in temperate climates and might exceed 30 °C in hot climates or after mesophilic acidification. So tests with 15 °C, 20 °C and 30 °C were performed with a substrate concentration of 1200 mg_{VFA}/L, pH = 7 and a CT of 24 h or 48 h to find the best reactor temperature regarding PHA production. To avoid a substrate induced influence all tests were conducted with the same substrate batch.

2.2.4. Investigation Concerning the Best pH Level

As the best pH level to produce PHA depends on the used substrate, various pH levels (6, 7, 8, 9, without pH control) were tested in this study with a substrate concentration of 1200 mg_{VFA}/L, a temperature of 20 °C and a CT = 24 h.

2.2.5. Investigation Concerning the Best Cycle Time

To find the most suitable cycle time (CT) for the bacteria selection process, experiments with a substrate concentration of 1200 mg_{VFA}/L were conducted at a temperature of 20 °C and at pH = 7 or 8. As the feast/famine ratio is more important than the overall CT a constant substrate concentration should ensure that the feast phases of the cycles were constant and the variation in CT resulted in a different famine phase, only. [25] stated that the feast-phase should not last longer than 20% of the overall CT to create a selection pressure on non-PHA accumulating bacteria. Therefore, CTs of 24 h, 48 h and 72 h were tested.

2.2.6. Analytical Procedures

COD, TKN, total P and total solids (TS) were determined according to standard methods.

The concentration and composition of polyhydroxyalkanoates (PHA), namely polyhydroxybutyrate (PHB) and polyhydroxyvalerate (PHV), were detected by gas chromatography, according to [35] with some variations. Therefore, the biomass was separated via a centrifuge at 10,000 rpm for 20 min and dried at 105 °C. Afterwards the sample was pulverised with a ball mill and about 100 mg were digested and analysed. Detection was performed using a Perkin Elmer Autosystem XL chromatographer and a VF5ms 30 m × 0.25 column. Helium was used as carrier gas. PHA concentration was calibrated using 4 mL standards. The concentration and composition of volatile fatty acids, namely formate (Fo), acetate (Ac), propionate (Pro) and butyrate (Bu), were detected as described in Chapter 2.1.3.

2.2.7. Calculation of Parameters

For better comparability, all results regarding VFA concentrations, COD, TKN and total P were converted into mg/L. The concentration of PHA in terms of % cell dry weight (CDW) is defined as:

$$PHA = PHB + PHV \tag{5}$$

2.3. Potential Analysis

Calculations

Based on the PHA production results described in Chapter 3, a potential analysis was performed. The aim of the analysis was to determine the potential of biopolymer production (based on renewal resources and biodegradable, see definition in Chapter 1) on German and European waste water treatment plants (WWTP) by using sewage sludge as a substrate. All input data used for the calculations can be found in Table 1.

A plausibility analyses was performed to cross-check the most important input data like the amount of primary sludge (PS) per population equivalent (PE).

As detailed data about waste water and sewage sludge production are available in Germany, the first step of the potential analysis was calculated using these data together with results presented in Chapter 3. In a second calculation step, data provided by the European Union (EU) were used to create an in-depth estimation of the biopolymer potential considering all 28 member states.

Table 1. Input data used during the potential analysis.

Parameter	Unit	Value	Literature
Connected people equivalents (PE) on German WWTPs	Mio. PE	115.7	[29]
Proportion of PSP *-PEs regarding total PEs in Germany	%	92	[29]
PEs with PSP * in Germany	Mio. PE	106.56	
Amount of primary sludge per PE	L/(PE×d)	1.1	[36]
Total solid conc. of primary sludge/acidified material	g/L	35	[4]
VFA concentration	g_{VFA}/m^3	7,653	[4]
Retention time and withdrawal at the first production step	d; %/d	4; 25	[4]
Total solid concentration in the aerobic Reactors 2a/2b	g/L	5.0	[4]
Loading rate for PHA production	kg_{VFA}/m^3	1.2	[4]
Retention time and withdrawal at Reactor 2a	d and %/d	2 and 50	[4]
PHA proportion based on cell dry weight	CDW.-%	28.4	[36]
Yearly sewage sludge amount in the EU	t_{TS}/a	13,245,180	[30]
Yearly sewage sludge amount in Germany	t_{TS}/a	1,815,150	[30]

* PSP = German WWTPs with preliminary sedimentation potential (PSP = more than 10,000 PE).

3. Results and Discussion

3.1. VFA Production

3.1.1. Potential Analysis

Table 2 displays the results of the performed investigations ordered by degree of acidification. Primary sludge performed best under three out of four conditions and yielded by far the best degree of acidification (DA) with 31% at 30 °C under pH-controlled conditions [34]. The second best carbon source, a one to one mixture of primary and digested sludge at 20 °C under pH-uncontrolled conditions, achieved only a DA of 14%. In five out of eight experiments pH-uncontrolled conditions resulted in a higher DA. Therefore, a fermentation without pH control should be considered for all fermentation raw materials. However, primary sludge yielded better DAs with pH control at both investigated temperatures. At 30 °C the DA of primary sludge was twice as high than without pH control.

Table 2. Degree of acidification and VFA composition in dependence of substrate and operation conditions (batch-tests, 4 L).

Carbon Source	pH	Temperature	Max. Conc.	DA	Ac/Pro/Bu
		(°C)	(Day)	(%)	(%)
Primary sludge	6 *	30	9	31	52/48/0
Primary sludge	6 *	20	7	14	56/44/0
Primary sludge	4.6	30	10	14	41/59/0
Primary-/digested sludge	7	20	14	14	79/21/0
Primary sludge	4.5	20	15	13	42/58/0
Primary-/digested sludge	7.5	30	14	12	84/16/0
Excess sludge	7	30	5	10	59/20/20
Excess sludge	6.5	20	4	8	60/20/20
Primary-/digested sludge	6 *	30	5	7	75/25/0
Excess sludge	6 *	20	7	6	24/76/7
Excess sludge	6 *	30	5	6	67/33/0
Primary-/digested sludge	6 *	20	2	3	57/43/0
Excess-/digested sludge	6 *	30	4	3	100/0/0
Excess-/digested sludge	8	30	3	3	76/0/24
Excess-/digested sludge	7.5	20	7	2	100/0/0
Excess-/digested sludge	6 *	20	2	1	100/0/0

* Marks conditions pH-controlled.

Table 2 also shows the composition of the VFA. The results varied strongly between 24/76/7 (%Ac/%Pro/%Bu) and 100/0/0 depending on the used raw material. Primary sludge produced none butyric acid and acetic and propionic acid in nearly two equal sections. Excess sludge on the other hand produced up to 21% butyric acid, while the one to one mixture of primary and digested sludge produced the most acetic acid (up to 84%) of all tested raw materials. The results show that the raw material has a major influence on the VFA composition.

As the use of primary sludge resulted in highest DA and showed only small variations in VFA composition under the tested conditions, it was chosen as raw material and used in all further tests.

Beside the ability to produce VFAs, primary sludge has other advantages as raw material for the PHA production. As primary sludge is a mixture of organic material, water and fermenting microorganisms no longsome biological adaptation-phase or biomass recirculation for the fermentation process was necessary. During all experiments primary sludge showed nutrient limited conditions, as described in Chapter 2.2.1. This is of particular significance given that nutrient limited conditions are essential for the later PHA production [19,37].

3.1.2. Temperature

The aim of the investigations at two temperature levels was to ascertain if the VFA production at ambient temperature (20 °C) can reach the same VFA production compared with heating the sludge.

In six out of eight tested combinations a temperature increase from 20 to 30 °C caused a higher VFA production as shown in Table 2. The experiment confirmed the results of [32], who stated that the VFA concentration increases with higher fermentation temperature. Using primary sludge as raw material (under pH-controlled conditions) the temperature change from 20 to 30 °C caused a DA increase from 14% to 31%.

The general assumption that the acidification rate is higher at 30 °C than at 20 °C could not be confirmed. Only three out of eight tested combinations reached their VFA maximum at 30 °C in a shorter span of time than at 20 °C. Four out of them even reached their VFA maximum at 20 °C in a shorter span of time than at 30 °C. The results can be seen in Table 2. Primary sludge under pH-controlled conditions obtained its VFA maximum after 7 days at 20 °C and after 9 days at 30 °C. Nevertheless, the fact that the DA of primary sludge under pH-controlled conditions at 30 °C was twice as high as the DA at 20 °C is all the more important as the VFA production at 30 °C lasted only about 30% longer.

The variation of temperature has a wide range of influence on the VFA composition, depending on the used substrate. As primary sludge was already chosen as substrate for the optimisation tests, only its VFA composition was of interest for further tests. However, in the case of primary sludge the temperature change investigated resulted only in marginal changes in the VFA composition.

Consequently, a temperature around 30 °C for the further experiments was considered as reasonable.

3.1.3. pH

As illustrated in Figure 4, no big difference in the maximum VFA concentration between pH = 6 and pH = 8 was observed. A pH value of 7 yielded the highest result with 18,286 mg_{VFA}/L after a RT of 10 days. The fermentation at pH = 10 reached significantly worse results with a maximum of 10,050 mg_{VFA}/L only at 18 days retention time. This is in contrast to the results of [31] showing the best result at pH = 9 with excess sludge and food waste as source material and [32] yielding the highest result at pH = 11 with excess sludge as raw material.

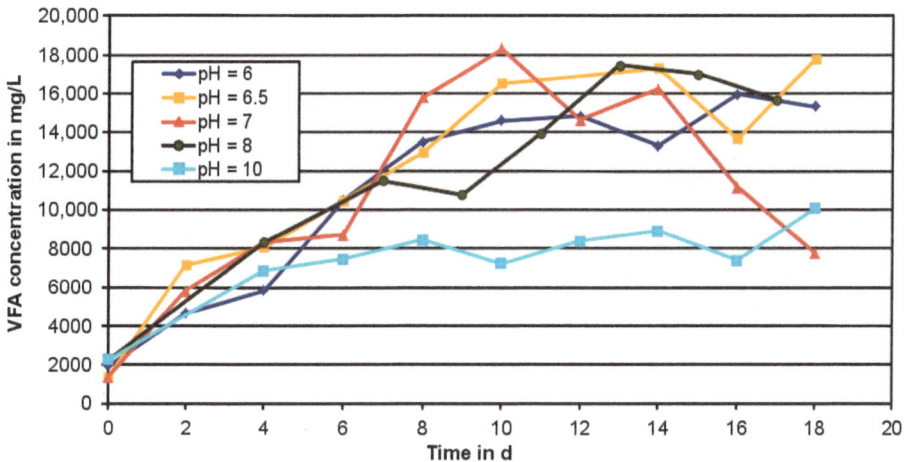

Figure 4. VFA concentrations during tests at different pH levels.

Although pH = 7 yielded the best result, methane production turned out to be an issue at this pH-value. After about 15 days the acetate concentration was falling rapidly and the overall VFA concentration after 18 days was less than 44% of the maximum (Figure 4). During this period more than 15 Vol.% methane was detected in the reactor via gas measurement. To prevent methanogenic conditions a pH-level of 6 has to be kept [38]. Consequently, further investigations were performed under pH = 6, although it produced about 12% less VFA within the batch experiments.

The variation of pH-level showed a strong influence on the VFA composition. Changing the pH from 6 to 7 within the batch experiments caused a constant decrease in the acetic acid ratio as shown in Table 3. In the same tests, the propionic acid ratio increased, while the butyric acid ratio decreased to zero. At pH = 8 conditions a reverse trend was observed. With 60% the maximum acetic acid ratio as well as the minimum propionic acid ratio (37%) was detected, while butyric acid was produced in small amounts (3%). In comparison to pH = 8, a reduction of acetic acid and propionic acid production was detected at the highest tested pH-level (pH = 10), while the butyric acid ratio increased to the highest level (8%) observed. In contrast to the other pH-values tested, formic acid was produced at pH = 10 with a ratio of 16.

Table 3. VFA composition and DA in dependence of pH or RT and WD.

	Batch			Semi Continuous		
pH	Ac/Pro/Bu	DA	pH	RT and WD	Ac/Pro/Bu	DA
	(%)	(%)		(day and %)	(%)	(%)
6	45/51/4	29	6	4 and 25	49/38/13	15
6.5	37/61/2	29	6	4 and 50	46/47/7	14
7	28/72/0	39	6	4 and 75	48/45/7	14
8	60/37/3	29	6	6 and 25	51/38/11	22
10	45/31/8*	14	6	6 and 50	46/48/6	17
			6	8 and 25	46/41/13	20
			6	8 and 50	43/49/8	19

* Missing to 100% is formate.

3.1.4. RT and WD

To get an idea about how much adapted bacteria are needed in the reactor to produce the most VFAs a wide range of RT and WD was tested. RT = 2 days and WD = 50% yielded poor results (VFA_{max} < 2000 mg_{VFA}/L) and after 10 days of semi-continuous operation the test was shut down. Therefore, these results are not shown. Further results ranged in a broad band between 5000 mg_{VFA}/L and 10,000 mg_{VFA}/L.

Obviously, the VFA production with short RTs and small WDs fluctuated less than using long RTs and large WDs, what can be explained by the changing composition of the introduced primary sludge. These changes of the used primary sludge are mostly due to weather events. A rainfall after a period of dry weather can transport a huge amount of organic matter to the WWTP. Daily changes in the waste water's composition or different contents during the week could be another reason for changing the primary sludge's composition. Smaller WD stabilises the fermentation process because only little material is turned over and the reactor is less sensitive to heterogeneous primary sludge input.

Figure 5 shows the average VFA concentration over a period of 40 days for all investigated combinations. Both, RT and WD influenced the VFA production. With higher WD the VFA concentration was decreasing at all tested RTs. The highest overall VFA concentration was reached at a RT of 6 days with a WD of 25%. Longer and shorter RTs (with a WD of 25%, too) resulted in lower VFA concentrations.

In order to have a high PHA-production in the second stage the VFA production rate (PR) is more important than the VFA concentration, which could, if it exceeds a certain value, lead to a substrate inhibition [23]. Therefore, the PR was calculated on Equation (4). A RT = 4 days and WD = 25% yielded the top production rate with PR = 1913 mg_{VFA}/(L×d) at a VFA concentration of 7653 mg_{VFA}/L on average [4].

The variation of RT influenced the VFA composition only slightly as shown in Table 3. Acetic acid and propionic acid were produced in a similar range, while butyric acid was always the smallest part. Nevertheless, the variation of WD had an effect. With a WD of 25% the fraction of propionic acid was about 20% smaller throughout all tests, compared to a WD of 50% and 75%, while the butyric acid ratio was nearly twice as much. The acetic acid ratio with a WD of 25% was slightly higher than for any other WDs.

As a stable VFA composition is necessary for high quality PHA production the possible fluctuation of the VFA composition during the whole test period is important. Figure 6 is exemplary for all semi-continuously operated tests and shows the concentrations of acetate, propionate and butyrate at RT = 4 days, WD = 50% during the whole test period of 44 days. After a starting phase of 10 days (not shown in the figure) the semi-continuous operation began. Due to the change in the operation method (from batch to semi-continuously) a transition phase with a decrease in VFA concentration was observed for the first six days of semi-continuous operation. Fluctuations in the VFA concentration between day six and day 44 were due to the changing concentration and composition of the introduced primary sludge. Although a fluctuation in VFA concentration after the fermentation step was observed, only small changes in the VFA composition were detected. Thus, it was possible to show that the variability of the raw material primary sludge did not affect the VFA composition significantly.

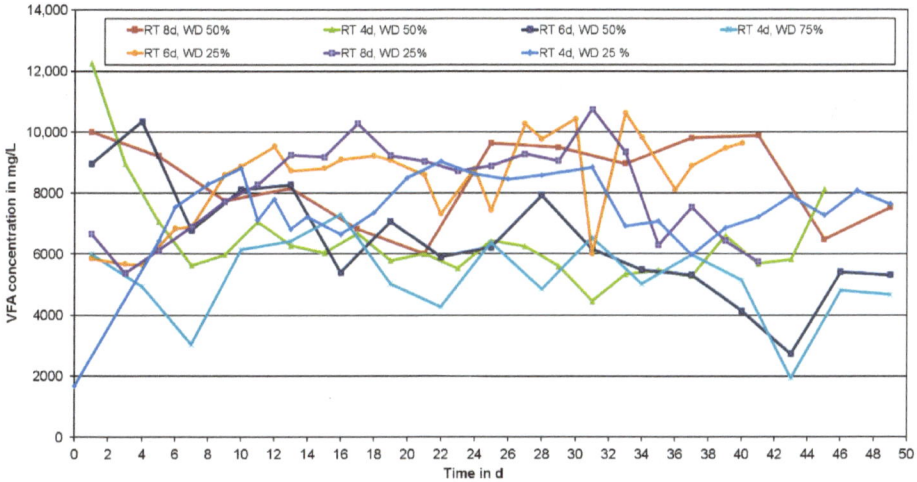

Figure 5. VFA concentration in dependence of RT and WD.

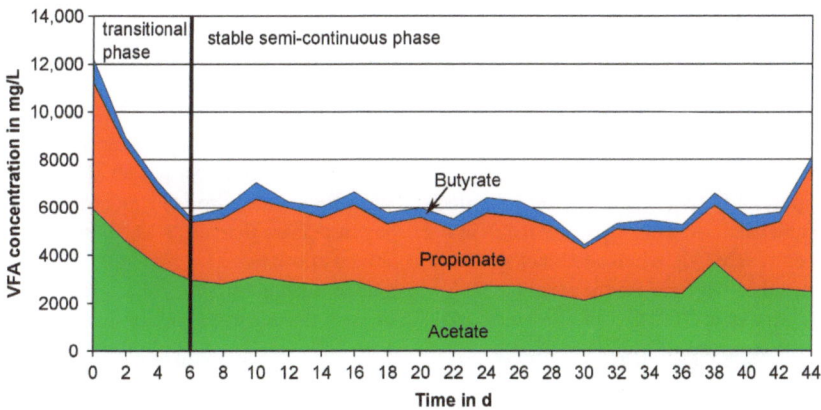

Figure 6. Development of the VFA concentration and composition at RT = 4 days, WD = 50%.

3.2. PHA Production

The acidified primary sludge from phase 1 was used for the production of PHA in step two where the influence of different operating conditions was investigated.

3.2.1. Substrate Concentration

The results of the conducted experiments are displayed in Table 4. It appears that higher PHA concentrations were gained at the lower substrate concentration tested. The maximum PHA concentration of 25.9% based on cell dry weight (CDW) was reached at a pH of 7 and a reactor temperature of 20 °C. With a VFA concentration of 2000 mg/L and a reactor temperature of 20 °C the highest PHA concentration achieved was 4.8% CDW, only and therefore much less than at a substrate concentration of 1200 mg/L at the same temperature. This confirms the observations of [8,23,39] that an increasing substrate concentration could result in a substrate inhibition.

Table 4. PHA production in dependence of two tested substrate concentrations.

Substrate Conc.	Temp.	pH	PHA	PHB/PHV
(mg$_{VFA}$/L)	(°C)		(% CDW)	(% CDW/% CDW)
1200	20	*	13.2	7.0/6.2
1200	20	7	25.9	13.2/12.7
1200	30	*	3.4	2.3/1.1
2000	20	*	4.8	3.3/1.5
2000	20	7	1.8	<2/1.8
2000	30	*	5.8	3.8/2.0

* Marks conditions without pH-control.

Furthermore, the experiment at the higher temperature and the lower substrate concentration (30 °C, 1200 mg/L) leads to a significant lower PHA production than at 20 °C and 1200 mg/L. This indicates that a reactor temperature of 20 °C may be preferable for a primary sludge based PHA production (see also Chapter 3.2.2).

PHA composition did not show any dependence regarding to substrate concentration. During all tests higher proportions of PHB than PHV were produced. About twice as much PHB (than PHV) was produced at all experiments at a substrate concentration of 2000 mg/L and at 30 °C with 1200 mg/L, while a nearly equal proportion of both PHAs was reached at 1200 mg/L and 20 °C.

3.2.2. Temperature

Table 5 displays the results of the PHA production at different temperatures. As the bacteria metabolism is slower at lower temperatures a very long feast-phase was observed at 15 °C. To ensure a sufficiently long famine-phase the cycle time of 24 h was doubled at all test with a reactor temperature of 15 °C. Figure 7 shows the length of the feast-phase based on the reactors oxygen concentrations. The very long feast-phase at 15 °C (around 22 h) is clearly visible. All other feast-phases at 20 °C or 30 °C were significantly shorter and not longer than 600 min. A somewhat surprising fact was that a faster bacteria metabolism at 30 °C did not result in a shorter feast-phase than at 20 °C. These observations are in contrast to [40], who stated that shorter fest-phases are observed at higher temperatures. At the same time, they confirm the results of [41], who gained the highest PHA concentration at reactor temperatures around 20 °C. This effect may origin in the fact, that the used sewage sludge was already adopted to temperature around 20 °C.

Table 5. PHA production in dependence of the reactor temperature.

Temp.	pH	CT	PHA	PHB/PHV
(°C)		(h)	(% CDW)	(% CDW/% CDW)
15	7	24/48	4.2	2.5/1.7
15	8	24/48	3.9	2.5/1.4
20	7	24	25.9	13.2/12.7
20	*	24	13.2	7.0/6.2
30	7	24	0.6	<2/0.6
30	*	24	3.4	2.3/1.1

* Marks conditions without pH-control.

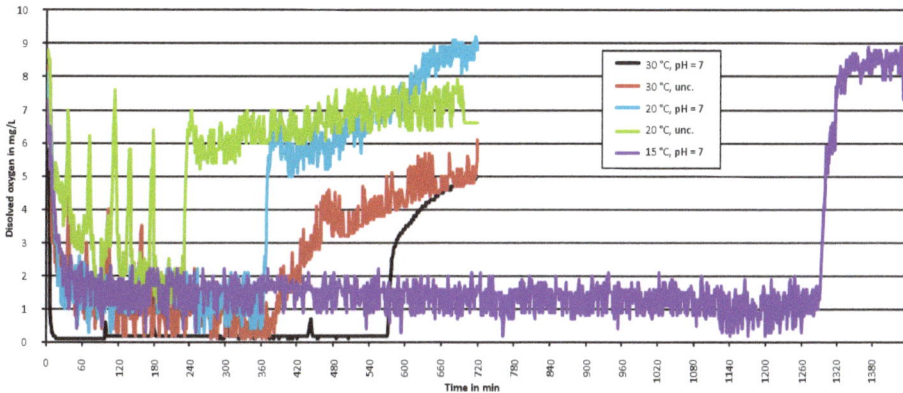

Figure 7. Reactor temperature influence on the feast/famine phase length.

There is no influence on PHA composition due to temperature changes. The results of Chapter 3.2.1 that the PHB/PHV ratio is about 2/1 at lower PHA concentrations produced, was confirmed (Table 5 at 15 °C and 30 °C), while both proportion are more or less equal at a higher PHA production (Table 5 at 20 °C).

As listed in Table 5 conditions of 20 °C yielded the highest PHA concentrations with 13.2% CDW and 25.9% CDW. At a reactor temperature of 15 °C or 30 °C a maximum PHA concentration of less than 4.5% CDW was reached, only. In consequence further experiments were conducted at 20 °C to achieve the best possible PHA concentration [42].

3.2.3. pH

Table 6 illustrates big differences in the maximum PHA concentration between the tested pH levels. All tests were operated in batch-mode with a cycle time of one day. The substrate's pH was adjusted before adding it into the reactor. An exception was the pH uncontrolled test. This experiment should clarify if a high PHA production is possible at fluctuating pH value. During the whole cycle the pH varied between 7.3 at the beginning of the feast-phase and 9.3 at the end of the famine-phase, with an average pH around 9.

Table 6. PHA production in dependence of the pH.

pH	PHA	PHB/PHV
	(% CDW)	(% CDW/% CDW)
unc. (av. 9)	13.2	7.0/6.2
6	—	—
7	25.9	13.2/12.7
8	28.4	14.7/13.7
9	4.4	3.0/1.4

The experiments with controlled pH values showed highly varying results. While there was no detectable PHA production at pH = 6, more than 25% CDW was produced between pH 7 and 8, with a maximum PHA production of 28.4% CDW at pH = 8. Then again, at the highest pH of 9 a low PHA production of 4.4% was reached only. The experiment without pH control produced around half as much PHA as the tests at pH 7 or 8. Nevertheless the results of the uncontrolled test were far better than at the same pH-level of 9 during the controlled test. Still a pH controlled PHA production is preferable, as the maximum PHA production was obtained at pH controlled operation [42].

pH changes did not influence the PHA composition. As described in Sections 3.2.2 and 3.2.3 the PHB/PHV ratio is 2/1 at low PHA concentrations and nearly 1/1 at a higher PHA production.

3.2.4. Cycle Time

When using a PHA production based on a bacteria mixed culture a feast/famine regime is crucial. This is the only way for PHA producing bacteria to gain a significant selection advantage over other microorganisms.

The produced amount of PHA in dependence of the length of the feast/famine-phase is displayed in Table 7. A correlation between the cycle time (CT) and the length of the feast-phase was observed, showing that a longer CT leads to a shorter feast-phase. This could be due to a longer and therefore harder famine-phase leading to a long period of starvation. It seems that after a bigger starvation, the bacteria's substrate uptake is higher and faster than at shorter cycle times. [39] highlights the cycle time must be such that the complete PHA is metabolised at the end of the famine-phase. This guideline was confirmed by testing samples, taken at the end of each famine-phase. No PHA was detected and therefore the required operating condition was kept during all tests. Thus, a cycle time of 24 h is sufficient.

Table 7. PHA production in dependence of the cycle time.

CT	Feast/Famine	Feast/Famine-Ratio	PHA	PHB/PHV
(h)	(min)	(%/%)	(% CDW)	(% CDW/% CDW)
24	524/916	36/64	28.4	14.7/13.7
48	500/2380	17/83	18.3	8.0/10.3
72	448/3872	10/90	21.4	8.3/13.1

A former study concluded that the proportion of the feast-phase should not exceed 20% of the CT, because a longer famine phase could lead to a lower selection pressure on non-PHA accumulating bacteria [26]. Table 7 shows a feast-phase proportion of 36% at a cycle time of 24 h and therefore nearly twice as high as suggested by [26]. Regardless, at this cycle time the highest PHA concentration (28.4% CDW) was reached. A feast-phase proportion of 17% at a CT of 48 h led to a PHA production of 18.3% CDW, while at the longest tested CT of 72 h a PHA production of 21.4% CDW was observed with a feast-phase proportion of 10%. This indicates that a fixed feast-phase proportion is unnecessary and the demanded operation condition by [39] is preferable. However, more experiments will be necessary to confirm these results.

An influence of the cycle time length on the PHA composition is shown in Table 7, also. While all other experiments reached a higher PHB than PHV proportion a cycle time of 48 h or longer led to a higher PHV production. This could be due to the fact that the PHA accumulation bacteria produced PHB at first while the PHV production started later. As no samples were taken at low oxygen levels during the feast-phase and no uptake rate was measured for the VFAs (acetate, butyrate, valerate) during the experiments this presumption could not be confirmed.

The preferable operating conditions for the VFA production described in Chapter 3.1 provide VFA every day. Regarding the pairing of both productions steps (VFA and PHA production) and having in mind that with a cycle time of one day the highest PHA concentration (28.4% CDW) was yielded, a CT = 24 h is favourable [42].

3.3. Potential Analysis

3.3.1. Calculation for German Waste Water Treatment Plants

Figure 8 shows the results for possible biopolymer production on German WWTPs. All material streams and reactor volumes in this figure are theoretical values showing the size of the flows, if the best substrate for acidification [34] of all German WWTPs with preliminary sedimentation potential (PSP = more than 10,000 PE) would be used for PHA production.

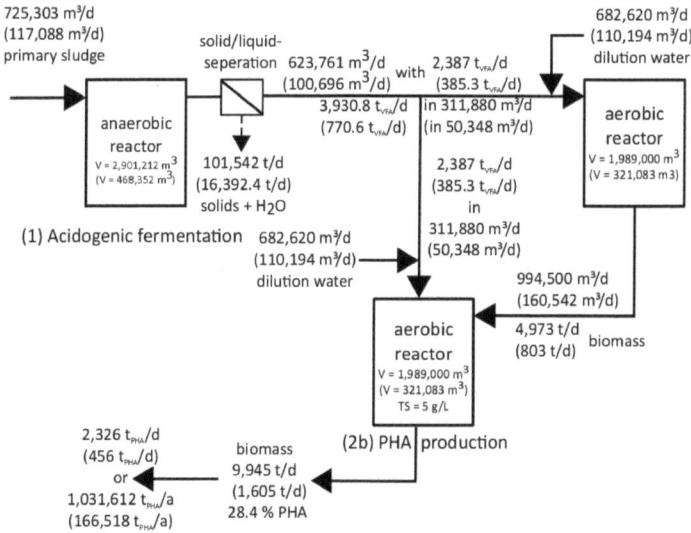

725,303 m³/d
(117,088 m³/d)
primary sludge

solid/liquid-
seperation 623,761 m³/d with 2,387 t$_{TM}$/d
(100,696 m³/d) (385.3 t$_{TM}$/d)
3,930.8 t$_{TM}$/d in 311,880 m³/d
(770.6 t$_{TM}$/d) (in 50,348 m³/d)

682,620 m³/d
(110,194 m³/d)
dilution water

anaerobic
reactor
V = 2,901,212 m³
(V = 468,352 m³)

101,542 t/d
(16,392.4 t/d)
solids + H₂O

2,387 t$_{TM}$/d
(385.3 t$_{TM}$/d)
in
311,880 m³/d
(50,348 m³/d)

aerobic
reactor
V = 1,989,000 m³
(V = 321,083 m3)

(1) Acidogenic fermentation 682,620 m³/d
(110,194 m³/d)
dilution water

994,500 m³/d
(160,542 m³/d)

aerobic
reactor
V = 1,989,000 m³
(V = 321,083 m³)
TS = 5 g/L

4,973 t/d biomass
(803 t/d)

2,326 t$_{PHA}$/d
(456 t$_{PHA}$/d)
or
1,031,612 t$_{PHA}$/a
(166,518 t$_{PHA}$/a)

biomass
9,945 t/d
(1,605 t/d)
28.4 % PHA

(2b) PHA production

Figure 8. Biopolymer potential scheme for European and German (figures in brackets) WWTPs.

First of all, the calculation for the amount of primary sludge in Germany regarding the PEs (Table 1) connected to German WWTPs is shown in Equation (6).

$$115,700,000 \text{ PE} \times 1.1 \frac{L_{PS}}{PE \times d} = 1,272,700,000 \frac{L_{PS}}{d} = 127,270 \frac{m^3_{PS}}{d} \tag{6}$$

Around 92% of PEs are coming from WWTPs with preliminary sedimentation potential (Table 1), on which a primary clarifier is installed or the construction of a primary clarifier would be preferable. Thus, the actual amount of primary sludge is calculated as follows:

$$127,270 \frac{m^3_{PS}}{d} \times 92\% = 117,088.4 \frac{m^3_{PS}}{d} \tag{7}$$

Using these data and the results of own experiments [34,42] and of Chapter 3.1 and 3.2 a calculation of the possible PHA production through various steps can be performed.

Implementing the best reactor operation method, using a retention time of 4 days and a daily withdrawal of 25% (Table 1) 117,088.4 $\frac{m^3}{d}$ acidified material could be used for PHA production every day (Equation (8)).

$$117,088.4 \frac{m^3_{PS}}{d} \times 4 \text{ d} \times 25 \frac{\%}{d} = 117,088.4 \frac{m^3}{d} \tag{8}$$

The total solid (TS) concentration of the acidified material of 35 $\frac{kg}{m^3}$ (Table 1) and the assumed residual moisture after de-watering via centrifuge of 75% leads to a daily acidified liquid production of 100,696 $\frac{m^3}{d}$ (Equations (9)–(11)).

$$117,088.4 \frac{m^3}{d} \times 35 \frac{kg_{TS}}{m^3} = 4,098,094 \frac{kg_{TS}}{d} = 4098.1 \frac{t_{TS}}{d} \tag{9}$$

The assumed residual moisture of 75% means that the calculated 4098.1 $\frac{t_{TS}}{d}$ biomass is 25% of the total mass separated by the centrifuge. Accordingly, 75% of the separated total mass is water. Assuming that the solid phase is completely separated it follows:

$$4098.1 \, \frac{t_{TS}}{d} + \frac{4098.1 \, \frac{t_{TS}}{d}}{25} \times 75\% \, \frac{t_{H_2O}}{d} = 16{,}392.4 \, \frac{t_{TS+H_2O}}{d} \tag{10}$$

With an assumed average density of $1 \, \frac{t}{m^3}$ the amount of

$$117{,}088.4 \, \frac{m^3}{d} - \frac{16{,}392.4 \, \frac{t}{d}}{1 \, \frac{t}{m^3}} = 100{,}696 \, \frac{m^3}{d} \tag{11}$$

of acidified liquid is available for the PHA production step.

Regarding the average VFA concentration of 7653 $\frac{mg_{VFA}}{L}$ (Table 1) the amount of VFA in the acidified water can be calculated (Equation (12)).

$$7653 \, \frac{g_{VFA}}{m^3} \times 100{,}696 \, \frac{m^3}{d} = 770.6 \, \frac{t_{VFA}}{d} \tag{12}$$

The substrate is divided into equal parts to both reactors (2a and 2b) of the second production step so that each reactor is supplied with 50,348 $\frac{m^3}{d}$ of acidified liquid (Equation (13)), containing 385.3 $\frac{t_{VFA}}{d}$ (Equation (14)).

$$\frac{100{,}696 \frac{m^3}{d}}{2} = 50{,}348 \, \frac{m^3}{d} \tag{13}$$

$$\frac{770.6 \, \frac{t_{VFA}}{d}}{2} = 385.3 \, \frac{t_{VFA}}{d} \tag{14}$$

In order to achieve the required loading rate of 1.2 $\frac{kg_{VFA}}{m^3 \times d}$ in both reactors of the second production step, the volume of Reactor 2a (sum over Germany) and 2b (sum over Germany) should be 321,083 m^3 each (Equation (15)).

$$\frac{385{,}300 \, \frac{kg_{VFA}}{d}}{1.2 \, \frac{kg_{VFA}}{m^3 \times d}} = 321{,}083.3 \, m^3 \tag{15}$$

Reactor 2a is operated with a retention time of 2 days, a daily withdrawal of 50% of the reactors volume and a total solid concentration of TS = 5 $\frac{g_{TS}}{L}$ (Table 1). Due to the fact that there is no biomass sedimentation before the removal of the material, the withdrawn material can be considered as fully mixed. Therefore, the amount of bacteria must double every cycle to achieve a constant concentration of total solids. During the own experiments it was shown that this necessary condition was fulfilled [42].

The removal of 50% of the reactor volume leads to the amount of substrate, which has to be filled in the Reactor 2b every day (Equation (16)).

$$321{,}083.3 \, \frac{m^3}{d} \times 50\% = 160{,}541.7 \, \frac{m^3}{d} \tag{16}$$

The emptying and refilling process at the beginning of every cycle is represented in Figure 3. In order to reach a VFA concentration of 1.2 $\frac{kg_{VFA}}{m^3 \times d}$ and regarding to the fact that the VFA concentration at the end of a cycle is zero, a substrate with a VFA concentration of 2.4 $\frac{kg_{VFA}}{m^3 \times d}$ is needed if half of the reactors volume is exchanged.

Equation (16) shows the required amount of substrate for one day for one reactor regarding the needed VFA concentration. Hence the amount of VFA rich liquid of 50,348 $\frac{m^3}{d}$ (Equation (13)) has to be diluted with 110,193.7 $\frac{m^3}{d}$ fresh water (Equation (17)).

$$160,541.7 \frac{m^3}{d} - 50,348 \frac{m^3}{d} = 110,193.7 \frac{m^3}{d} \tag{17}$$

When calculating the amount of dilution water it should be noted that the large amount of water is due to the reactors comparatively low total solid concentration of TS = 5 $\frac{g_{TS}}{L}$, which was installed during the experiments in [42]. This concentration is used for the potential analysis as well, to keep the calculation as close as possible to the operation conditions of the carried out experiments. Of course, a much higher solid concentration could be installed, leading to significantly smaller reactor volumes as well as less dilution water. As the amount of dilution water does not affect the result of the potential analysis it was kept.

As described, the biomass concentration in Reactor 2a was 5 $\frac{g_{TS}}{L}$ at the end of a cycle. Regarding the withdrawal of 50% of the reactor's volume a total of

$$321,083.3 \frac{m^3}{d} \times 5 \frac{kg_{TS}}{m^3} \times 5\% = 802,708.3 \frac{kg_{TS}}{d} = 802.71 \frac{t_{TS}}{d} \tag{18}$$

biomass is transferred into Reactor 2b. This reactor also had a dry matter content of 5 $\frac{kg_{TS}}{m^3}$ after the PHA production step (Table 1). Considering a cycle time of one day and the volume of 321,083 m³, the amount of biomass in Reactor 2b sums up to (Equation (19)):

$$\frac{321,083.3 \, m^3}{1 \, d} \times 5 \frac{kg_{TS}}{m^3} = 1,605,416.5 \frac{kg_{TS}}{d} = 1605.4 \frac{t_{TS}}{d} \tag{19}$$

With the reached PHA concentration of 28.4% of the cell dry weight (CDW) [42] the daily amount of biopolymer is calculated in Equation (20).

$$1605.4 \frac{t_{TS}}{d} \times 28.4\% \frac{PHA}{TS} = 455.9 \frac{t_{PHA}}{d} \tag{20}$$

Finally, the possible annual amount of PHA production on German waste water treatment plants can be calculated in Equation (21).

$$455.9 \frac{t_{PHA}}{d} \times 365.25 \frac{d}{a} = 166,517.5 \frac{t_{PHA}}{a} \tag{21}$$

Dividing the reactor volume (sum of all production stages) of 1,110,518 m³ by the people's equivalent with preliminary sedimentation potential of 106.56 Mio. PE (Table 1) results in a reactor volume (sum of all production stages) per capita of around 10.4 L/PE. Each PE can contribute to the production of 1.6 kg$_{PHA}$/(PE×a). Keeping in mind that the aeration tank volume on a WWTP with 100,000 PE sums up to 10,000 m³–15,000 m³ an additional reactor volume of approximately 1050 m³ would be needed for the biopolymer production, only. Thus, the extra volume would not be disproportional.

3.3.2. Estimation for European Waste Water Treatment Plants

The Biopolymer potential for European WWTPs are calculated similar to Section 3.3.1 and also shown in Figure 8.

As there are missing data about the amount of connected persons (PEs) for many EU member states as well as about the amount of municipal waste water, it is impossible to calculate the EU-wide production of primary sludge analogous to Equations (6) and (7). However, there are data for all 28 EU member states regarding the production of sewage sludge. These data show the dry weight of sewage sludge in $\frac{t_{TS}}{a}$ (Table 1) and hence their unit must be transferred into $\frac{m^3}{a}$ (Equation (22)) to compare them with the data used for German WWTP in Equation (6). Therefore, an average total solid concentration of 15 $\frac{g_{TS}}{L}$ or 66.67 $\frac{m^3}{t_{TS}}$ for European sewage sludge (primary and secondary) is assumed.

$$13,245,180 \ \frac{t_{TS}}{a} \times 66.67 \ \frac{m^3}{t_{TS}} = 883,056,150 \ \frac{m^3}{a} \tag{22}$$

On the assumption that the proportion of primary sludge in the amount of sewage sludge is more or less constant in all member states, the percentage can be calculated (Equation (25)) using the theoretical yearly amount of primary sludge produced in Germany (Equations (6) and (23)) and the yearly amount of German sewage sludge (Table 1) (Equation (24)).

$$115,700,000 \ PE \times 1.1 \ \frac{L_{PS}}{PE \times d} \times 365.25 \ \frac{d}{a} = 46,485,367.5 \ \frac{m^3_{PS}}{a} \tag{23}$$

$$1,815,150 \ \frac{t_{TS}}{a} \times 66.67 \ \frac{m^3}{t_{TS}} = 121,016,051 \ \frac{m^3}{a} \tag{24}$$

$$\frac{46,485,367.5 \ \frac{m^3_{PS}}{a}}{121,016,051 \ \frac{m^3}{a}} \times 100\% = 38.4\% \tag{25}$$

Assuming that not all European waste water treatment plants are equipped with a primary clarifier, the proportion of primary sludge is rounded off to 30%, so that the yearly amount of European primary sludge sums up to 265 Mio. $\frac{m^3_{PS}}{a}$ (Equation (26)) or 725,303 $\frac{m^3_{PS}}{d}$ (Equation (27)).

$$883,056,151 \ \frac{m^3_{PS}}{a} \times 30\% = 264,916,845.3 \ \frac{m^3_{PS}}{a} \tag{26}$$

$$\frac{264,916,845.3 \ \frac{m^3_{PS}}{a}}{365.25 \ \frac{d}{a}} = 725,302.8 \ \frac{m^3_{PS}}{d} \tag{27}$$

By now, the European biopolymer potential can be calculated analogous to Equations (8)–(21). The amount of acidified material is:

$$725,303 \ \frac{m^3_{PS}}{d} \times 4 \ d \times 25 \ \frac{\%}{d} = 725,303 \ \frac{m^3}{d} \tag{28}$$

Using Equations (9)–(11) the amount of acidified liquid can be calculated (Equation (31)):

$$725,303 \ \frac{m^3}{d} \times 35 \ \frac{kg_{TS}}{m^3} = 25,385,605 \ \frac{kg_{TS}}{d} = 25,385.6 \ \frac{t_{TS}}{d} \tag{29}$$

$$25,385.6 \ \frac{t_{TS}}{d} + \frac{25,385.6 \ \frac{t_{TS}}{d}}{25\%} \times 75\% \ \frac{t_{H_2O}}{d} = 101,542.4 \ \frac{t_{TS+H2O}}{d} \tag{30}$$

With an average density of 1 $\frac{t}{m^3}$

$$725,303 \ \frac{m^3}{d} - \frac{101,542.4 \ \frac{t}{d}}{1 \ \frac{t}{m^3}} = 623,760.6 \ \frac{m^3}{d} \tag{31}$$

of acidified liquid can be used within the second PHA production step. This leads to the amount of VFAs in the acidified water (Equation (32)):

$$7653 \ \frac{g_{VFA}}{m^3} \times 623,760.6 \ \frac{m^3}{d} = 4773.6 \ \frac{t_{VFA}}{d} \tag{32}$$

Analogous to Equations (13) and (14) both reactors of the second production step are supplied with 311,880 $\frac{m^3}{d}$ (Equation (33)) of acidified liquid containing 2386.8 $\frac{t_{VFA}}{d}$ (Equation (34)).

$$\frac{623,760.6 \frac{m^3}{d}}{2} = 311,880.3 \frac{m^3}{d} \tag{33}$$

$$\frac{4773.6 \frac{t_{VFA}}{d}}{2} = 2386.8 \frac{t_{VFA}}{d} \tag{34}$$

The reactor volumes (Equation (35)) can be calculated analogous to Equation (15):

$$\frac{2,386,800 \frac{kg_{VFA}}{d}}{1.2 \frac{kg_{VFA}}{m^3 \times d}} = 1,989,000 \ m^3 \tag{35}$$

The daily substrate amount for one reactor is (Equation (36)):

$$1,989,000 \frac{m^3}{d} \times 50\% = 994,500 \frac{m^3}{d} \tag{36}$$

Analogous to Equation (17) the amount of dilution water can be calculated (Equation (37)):

$$994,500 \frac{m^3}{d} - 311,880.3 \frac{m^3}{d} = 682,619.7 \frac{m^3}{d} \tag{37}$$

With a withdrawal of 50% a biomass transfer to Reactor 2b of 4094.58 $\frac{t_{TS}}{d}$ is necessary (Equation (38)).

$$1,989,000 \frac{m^3}{d} \times 5 \frac{kg_{TS}}{m^3} \times 50\% = 4,972,500 \frac{kg_{TS}}{d} = 4972.5 \frac{t_{TS}}{d} \tag{38}$$

With the described total solid concentration of 5 $\frac{kg_{TS}}{m^3}$ after the PHA production the amount of biomass in Reactor 2b is (Equation (39)):

$$1,989,000 \ m^3 \times 5 \frac{kg_{TS}}{m^3} = 9,945,000 \frac{kg_{TS}}{d} = 9945 \frac{t_{TS}}{d} \tag{39}$$

With a cycle time of one day and the reached PHA concentration of 28.4% CDW (Table 1) the daily amount of PHA sums up to (Equation (40)):

$$9945 \frac{t_{TS}}{d} \times 28.4\% \frac{PHA}{TS} = 2824.4 \frac{t_{PHA}}{d} \tag{40}$$

Finally, the possible annual amount of PHA production on European waste water treatment plants can be calculated in Equation (41).

$$2824.4 \frac{t_{PHA}}{d} \times 365.25 \frac{d}{a} = 1,031,612.1 \frac{t_{PHA}}{a} \tag{41}$$

3.3.3. Summary of the Results and Optimization Potential

The market for biopolymers is predicted to grow continuously [43]. In 2016, a worldwide biopolymer production of 4.16 Mio. $\frac{t}{a}$, of which 861,120 $\frac{t}{a}$ or 20.7% [43] suit the criteria of the stringent definition for biopolymers, introduced in Chapter 1, were achieved. Taking Equation (21) into account approximately 4.0% of the worldwide biopolymer production (bio- and non-biodegradable) could be produced just by using primary sludge from German WWTPs. Around 19.3% of the worldwide biopolymers could be produced on WWTPs in Germany considering the stringent definition, only.

For the biopolymer production on European WWTPs (Equation (41)) approximately 24.7% of 2016's worldwide biopolymer production (bio- and non-biodegradable) or around 119.8% of 2016's worldwide biopolymer production due to the stringent definition could be produced.

Assuming an improved PHA production with an achievable PHA concentration of 0.5 $\frac{g_{PHA}}{g_{VSS}}$ [44] or even around 60% CDW [45,46] a total amount of 2,179,446.8 $\frac{t_{PHA}}{a}$ (Equation (42)) could be produced on European WWTPs by using primary sludge, only.

$$9945 \frac{t_{TS}}{d} \times 60\% \times 36,525 \frac{d}{a} = 2,179,446.8 \frac{t_{PHA}}{a} \tag{42}$$

Thus, approximately 52.4% of 2016's worldwide biopolymer production (bio- and non-biodegradable) or 253.1% of 2016's worldwide biopolymer production due to the stringent definition could be produced in an improved production on WWTPs in the EU.

A large proportion of polymers (biopolymers and those from synthetic production) is used for packing materials. The PHAs feature similar characteristics like polypropylene (PP), which is the mostly sold plastic in the EU with 18.8% (around 8.6 Mio. $\frac{t}{a}$) market share in 2012 [47]. The potential analysis for Germany equates to approximately 1.9% of the EUs PP production. Using the calculation for the EU for PHA production from primary sludge around 12.0% of the conventional PP sold in the EU could be substituted which is a significant potential.

3.3.4. Plausibility Analysis

As some input parameters do have a strong effect to the calculations, a plausibility analysis was carried out. All critical parameters, like total solid concentration of the primary sludge, the daily amount of primary sludge per PE, or the daily amount of primary sludge per PE were analysed and considered plausible. A more detailed description of the plausibility analysis can be found in [28].

4. Summary and Conclusions

From the results, it could be concluded that the production of high amounts of VFAs with a stable VFA composition on a WWTP is possible. Using different raw materials shows a strong influences on degree of acidification and VFA composition. The VFA production and composition is strongly influenced by a pH-level change in the reactor. A semi-continuous operation method of the reactor with a short RT and small WD is preferable. With primary sludge as raw material no biomass recirculation is needed during the fermentation process.

The results showed that the produces VFA are suitable for PHA production in a second stage. This amount of PHA produced is strongly influenced by the reactors operating conditions (temperature, pH-level and substrate concentration), while the PHA composition is influenced by cycle time changes. At preferred conditions, a stable PHB/PHV composition was reached and both PHAs were produced in nearly the same proportion.

Nevertheless, further research is needed to couple both processes for constant and long term PHA production and for upscaling.

The results of the presented potential analysis clearly indicate the possibility to produce large amounts of PHAs on German and European WWTPs. It has been shown that municipal WWTPs could be used as a significant source for biopolymers and waste water is an important substituent for plant-based raw materials in the PHA production.

More than twice the amount of 2016's worldwide biopolymer production could be produced on European WWTPs with an upgraded operation. Thus, the production of biopolymers on waste water treatment plants contribute to a recycling of the organic material contained in waste water.

Acknowledgments: We thank the WILLY-HAGER-STIFTUNG, Stuttgart for funding the research project.

Author Contributions: Timo Pittmann and Heidrun Steinmetz conceived and designed the experiments; Timo Pittmann performed the experiments; Timo Pittmann and Heidrun Steinmetz analyzed the data; Timo Pittmann contributed reagents/materials/analysis tools; Timo Pittmann and Heidrun Steinmetz wrote the paper.

Conflicts of Interest: The authors declare no conflict of interest. The funding sponsor had no role in the design of the experiments; in the collection, analyses, or interpretation of date; in the writing of the manuscript, or the decision to publish the results.

References

1. United Nations Environment Programme (UNEP). *Marine Litter-A Global Challenge*; UNEP: Nairobi, Kenia, 2009.
2. Jendrossek, D.U.; Handrick, R. Microbial degradation of polyhydroxyalkanoates. *Annu. Rev. Microbiol.* **2002**, *56*, 403–432. [CrossRef] [PubMed]
3. Choi, G.G.; Kim, H.W.U.; Rhee, Y.H. Enzymatic and non-enzymatic degradation of poly(3-hydroxybutyrate-co-3-hydroxyvalerate) copolyesters produced by *Alcaligenes* sp. MT-16. *J. Microbiol.* **2004**, *42*, 346–352. [PubMed]
4. Pittmann, T. Herstellung von Biokunststoffen aus Stoffströmen einer kommunalen Kläranlage. Dissertation thesis, Institute for Sanitary Engineering, Water Quality and Solid Waste Management, Stuttgart, Germany, 2015.
5. Nikodinovic-Runic, J.U.; Guzik, M. Carbon-rich wastes as feedstocks for biodegradable polymer (polyhydroxyalkanoate) production using bacteria. *Adv. Appl. Microbiol.* **2013**, *84*, 139–200. [CrossRef] [PubMed]
6. Cavalheiro, J.M.B.T.; de Almeida, C.M.M.D.; Grandfils, C.; da Fonseca, M.M.R. Poly(3-hydroxybutyrate) production by *Cupriavidus necator* using waste glycerol. *Process Biochem.* **2009**, *44*, 509–515. [CrossRef]
7. Zinn, M.U.; Hany, R. Tailored material properties of polyhydroxyalkanoates through biosynthesis and chemical modification. *Adv. Eng. Mater.* **2005**, *7*, 408–411. [CrossRef]
8. Akaraonye, E.; Keshavarz, T.U.; Roy, I. Production of polyhydroxyalkanoates: The future green materials of choice. *J. Chem. Technol. Biotechnol.* **2010**, *85*, 732–743. [CrossRef]
9. Steinbuechel, A. *Angewandte Mikrobiologie, Biopolymere und Vorstufen*; Springer: Berlin/Heidelberg, Germany, 2005.
10. Faulstich, M.; Greiff, K.B. Klimaschutz durch biomasse, ergebnisse des SRU-sondergutachtens. *Umweltwissenschaften und Schadstoff-Forschung* **2007**. [CrossRef]
11. Bengtsson, S.; Werker, A.; Christensson, M.; Welander, T. Production of polyhydroxyalkanoates by activated sludge treating a paper mill wastewater. *Bioresour. Technol.* **2008**, *99*, 509–516. [CrossRef] [PubMed]
12. Morgan-Sagastume, F.; Valentino, F.; Hjort, M.; Cirne, D.; Karabegovic, L.; Geradin, F.; Dupont, O.; Johansson, P.; Karlsson, A.; Magnusson, P.; et al. Biopolymer production from sludge and municipal wastewater treatment. *Water Sci. Technol.* **2014**, *69*, 177–184. [CrossRef] [PubMed]
13. Chakravarty, P.; Mhaisalkar, V.; Chakrabarti, T. Study on poly-hydroxy-alkanoate (PHA) production in pilot scale continuous mode wastewater treatment system. *Bioresour. Technol.* **2010**, *101*, 2896–2899. [CrossRef] [PubMed]
14. Chua, A.; Takabatake, H.; Satoh, H.; Mino, T. Production of polyhydroxyalkanoates (PHA) by activated sludge treating municipal wastewater: Effect of pH, sludge retention time (SRT), and acetate concentration in influent. *Water Res.* **2003**, *37*, 3602–3611. [CrossRef]
15. Chua, H.; Yu, P. Production of biodegradable plastics from chemical wastewater-A novel method to reduce excess activated sludge generated from industrial wastewater treatment. *Water Sci. Technol.* **1999**, *39*, 273–280. [CrossRef]
16. Dionisi, D.; Majone, M.; Papa, V.; Beccari, M. Biodegradable polymers from organic acids by using activated sludge enriched by aerobic periodic feeding. *Biotechnol. Bioeng.* **2004**, *85*, 569–579. [CrossRef] [PubMed]
17. Reddy, S.V.; Thirumala, M.; Reddy, T.K.; Mahmood, S.K. Isolation of bacteria producing polyhydroxyalkanoates (PHA) from municipal sewage sludge. *World J. Microbiol. Biotechnol.* **2008**, *24*, 2949–2955. [CrossRef]
18. Lemos, P.C.; Serafim, L.S.; Reis, M. Synthesis of polyhydroxyalkanoates from different short-chain fatty acids by mixed cultures submitted to aerobic dynamic feeding. *J. Biotechnol.* **2006**, *122*, 226–238. [CrossRef] [PubMed]

19. Albuquerque, M.; Eiroa, M.; Torres, C.; Nunes, B.R.; Reis, M. Strategies for the development of a side stream process for polyhydroxyalkanoate (PHA) production from sugar cane molasses. *J. Biotechnol.* **2007**, *130*, 411–421. [CrossRef] [PubMed]

20. Albuquerque, M.G.E.; Martino, V.; Pollet, E.; Avérous, L.; Reis, M. Mixed culture polyhydroxyalkanoate (PHA) production from volatile fatty acid (VFA)-rich streams: Effect of substrate composition and feeding regime on PHA productivity, composition and properties. *J. Biotechnol.* **2011**, *151*, 66–76. [CrossRef] [PubMed]

21. Bengtsson, S. The utilization of glycogen accumulating organisms for mixed culture production of polyhydroxyalkanoates. *Biotechnol. Bioeng.* **2009**, *104*, 698–708. [CrossRef] [PubMed]

22. Bengtsson, S.; Pisco, A.R.; Reis, M.; Lemos, P.C. Production of polyhydroxyalkanoates from fermented sugar cane molasses by a mixed culture enriched in glycogen accumulating organisms. *J. Biotechnol.* **2010**, *145*, 253–263. [CrossRef] [PubMed]

23. Albuquerque, M.; Concas, S.; Bengtsoon, S.; Reis, M. Mixed culture polyhydroxyalkanoates production from sugar cane molasses: The use of a 2-stage CSTR system for culture selection. *Bioresour. Technol.* **2010**, *101*, 7112–7122. [CrossRef] [PubMed]

24. Choi, J.; Lee, S.Y. Process analysis and economic evaluation for Poly(3-hydroxybutyrate) production by fermentation. *Bioprocess Eng.* **1997**, *17*, 335. [CrossRef]

25. Dionisi, D.; Carucci, G.; Petrangeli Papini, M.; Riccardi, C.; Majone, M.; Carrasco, F. Olive oil mill effluents as a feedstock for production of biodegradable polymers. *Water Res.* **2005**, *39*, 2076–2084. [CrossRef] [PubMed]

26. Dionisi, D.; Majone, M.; Vallini, G.; Di Gregorio, S.; Beccari, M. Effect of the applied organic load rate on biodegradable polymer production by mixed microbial cultures in a sequencing batch reactor. *Biotechnol. Bioeng.* **2005**, *93*, 76–88. [CrossRef] [PubMed]

27. Johnson, K.; Jiang, Y.; Kleerebezem, R.; Muyzer, G.; van Loosdrecht, M. Enrichment of a mixed bacterial culture with a high polyhydroxyalkanoate storage capacity. *Biomacromolecules* **2009**, *10*, 670–676. [CrossRef] [PubMed]

28. Pittmann, T.; Steinmetz, H. Potential for polyhydroxyalkanoate production on German or European municipal waste water treatment plants. *Bioresour. Technol.* **2016**, *214*, 9–15. [CrossRef] [PubMed]

29. DWA: 28th Performance Comparison of Municipal Waste Water Treatment Plants. 2016. Available online: http://de.dwa.de/tl_files/_media/content/PDFs/1_Aktuelles/leistungsvergleich_2015.PDF (accessed on 2 June 2017).

30. Eurostat: Sewage Sludge Production and Disposal. Available online: http://ec.europa.eu/eurostat/data/database?nodecode=envwwspd (accessed on 2 June 2017).

31. Chen, H.; Meng, H.; Nie, Z.; Zhang, M. Polyhydroxyalkanoate production from fermented volatile fatty acids: Effect of pH and feeding regimes. *Bioresour. Technol.* **2013**, *128*, 533–538. [CrossRef] [PubMed]

32. Mengmeng, C.; Hong, C.; Qingliang, Z.; Shirley, S.N.; Jie, R. Optimal production of polyhydroxyalkanoates (PHA) in activated sludge fed by volatile fatty acids (VFAs) generated from alkaline excess sludge fermentation. *Bioresour. Technol.* **2009**, *100*, 1399–1405. [CrossRef] [PubMed]

33. Bengtsson, S.; Hallquist, J.; Werker, A.; Welander, T. Acidogenic fermentation of industrial wastewaters. Effects of chemostat retention time and pH on volatile fatty acids production. *Biochem. Eng. J.* **2008**, *40*, 492–499. [CrossRef]

34. Pittmann, T.; Steinmetz, H. Influence of operating conditions for volatile fatty acids enrichment as a first step for polyhydroxyalkanoate production on a municipal waste water treatment plant. *Bioresour. Technol.* **2013**, *148*, 270–276. [CrossRef] [PubMed]

35. Queirós, D.; Rossetti, S.; Serafim, L.S. PHA production by mixed cultures: A way to valorize wastes from pulp industry. *Bioresour. Technol.* **2014**, *157*, 197–205. [CrossRef] [PubMed]

36. *Dimensioning of a Single-Stage Biological Wastewater Tretment Plant*; Publishing Company of ATV-DVWK, Water, Wastewater, Waste: Hennef, Germany, 2000; Standard ATV-DVWK-A131. [CrossRef]

37. Serafim, L.S.; Lemos, P.C.; Oliveira, R.; Reis, M. Optimization of polyhydroxybutyrate production by mixed cultures submitted to aerobic dynamic feeding conditions. *Biotechnol. Bioeng.* **2004**, *87*, 145–160. [CrossRef] [PubMed]

38. Bischofsberger, W.; Dichtl, N.; Rosenwinkel, K.-H.; Seyfried, C.F.; Bohnke, B. *Anaerobtechnik*; Springer: Berlin/Heidelberg, Germany, 2015. [CrossRef]

39. Morgan-Sagastume, F.; Karlsson, A.; Johansson, P.; Pratt, S.; Boon, N.; Lant, P.; Werker, A. Production of polyhydroxyalkanoates in open, mixed cultures from a waste sludge stream containing high levels of soluble organics, nitrogen and phosphorus. *Water Res.* **2010**, *44*, 5196–5211. [CrossRef] [PubMed]

40. Johnson, K.; Kleerebezem, R.; van Loosdrecht, M.C.M. Influence of ammonium on the accumulation of polyhydroxybutyrate (PHB) in aerobic open mixed cultures. *J. Biotechnol.* **2010**, *147*, 73–79. [CrossRef] [PubMed]

41. Wen, Q.; Chen, Z.; Tian, T.; Chen, W. Effects of phosphorus and nitrogen limitation on PHA production in activated sludge. *J. Environ. Sci.* **2010**, *22*, 1602–1607. [CrossRef]

42. Pittmann, T.; Steinmetz, H. Polyhydroxyalkanoate production as a side stream process on a municipal waste water treatment plant. *Bioresour. Technol.* **2014**, *167*, 297–302. [CrossRef] [PubMed]

43. European Bioplastik e.V. Bioplastics. Bioplastics-Facts and Figures. 2016. Available online: http://docs. european-bioplastics.org/publications/EUBP_Facts_and_figures.pdf (accessed on 19 April 2017).

44. Morgan-Sagastume, F.; Hjort, M. Integrated production of polyhydroxyalkanoates (PHAs) with municipal wastewater and sludge treatment at pilot scale. *Bioresour. Technol.* **2015**, *181*, 78–89. [CrossRef] [PubMed]

45. Jia, Q.; Wang, H.; Wang, X. Dynamic synthesis of polyhydroxyalkanoates by bacterial consortium from simulated excess sludge fermentation liquid. *Bioresour. Technol.* **2013**, *140*, 328–336. [CrossRef] [PubMed]

46. Jia, Q.; Xiong, H. Production of polyhydroxyalkanoates (PHA) by bacterial consortium from excess sludge fermentation liquid at laboratory and pilot scales. *Bioresour. Technol.* **2014**, *171*, 159–167. [CrossRef] [PubMed]

47. Bonten, C. *Kunststofftechnik. Einführung und Grundlagen*; Hanser Carl Verlag: München, Germany, 2014.

Chapter 5:
PHA-Copolyester Production
and Application

bioengineering

MDPI

Article

Fed-Batch Synthesis of Poly(3-Hydroxybutyrate) and Poly(3-Hydroxybutyrate-*co*-4-Hydroxybutyrate) from Sucrose and 4-Hydroxybutyrate Precursors by *Burkholderia sacchari* Strain DSM 17165

Miguel Miranda de Sousa Dias [1], Martin Koller [2,3,]*, Dario Puppi [4], Andrea Morelli [4], Federica Chiellini [4] and Gerhart Braunegg [3]

[1] Université Pierre et Marie Curie UPMC, Institut national de la santé et de la recherche médicale INSERM, Centre national de la recherche scientifique CNRS, Institut de la Vision, Sorbonne Universités, 17 rue Moreau, 75012 Paris 06, France; migueldias670@hotmail.com

[2] Institute of Chemistry, University of Graz, NAWI Graz, Heinrichstrasse 28/III, 8010 Graz, Austria

[3] ARENA—Association for Resource Efficient and Sustainable Technologies, Inffeldgasse 21b, 8010 Graz, Austria; g.braunegg@tugraz.at

[4] BIOLab Research Group, Department of Chemistry & Industrial Chemistry, University of Pisa, UdR INSTM Pisa, Via Moruzzi, 13, 56124 Pisa, Italy; d.puppi@dcci.unipi.it (D.P.); a.morelli@dcci.unipi.it (A.M.); federica.chiellini@unipi.it (F.C.)

* Correspondence: martin.koller@uni-graz.at; Tel.: +43-316-380-5463

Academic Editor: Anthony Guiseppi-Elie

Received: 4 April 2017; Accepted: 19 April 2017; Published: 20 April 2017

Abstract: Based on direct sucrose conversion, the bacterium *Burkholderia sacchari* is an excellent producer of the microbial homopolyester poly(3-hydroxybutyrate) (PHB). Restrictions of the strain's wild type in metabolizing structurally related 3-hydroxyvalerate (3HV) precursors towards 3HV-containing polyhydroxyalkanoate (PHA) copolyester calls for alternatives. We demonstrate the highly productive biosynthesis of PHA copolyesters consisting of 3-hydroxybuytrate (3HB) and 4-hydroxybutyrate (4HB) monomers. Controlled bioreactor cultivations were carried out using saccharose from the Brazilian sugarcane industry as the main carbon source, with and without co-feeding with the 4HB-related precursor γ-butyrolactone (GBL). Without GBL co-feeding, the homopolyester PHB was produced at a volumetric productivity of 1.29 g/(L·h), a mass fraction of 0.52 g PHB per g biomass, and a final PHB concentration of 36.5 g/L; the maximum specific growth rate μ_{max} amounted to 0.15 1/h. Adding GBL, we obtained 3HB and 4HB monomers in the polyester at a volumetric productivity of 1.87 g/(L·h), a mass fraction of 0.72 g PHA per g biomass, a final PHA concentration of 53.7 g/L, and a μ_{max} of 0.18 1/h. Thermoanalysis revealed improved material properties of the second polyester in terms of reduced melting temperature T_m (161 °C vs. 178 °C) and decreased degree of crystallinity X_c (24% vs. 71%), indicating its enhanced suitability for polymer processing.

Keywords: 4-hydroxybutyrate; biopolymers; *Burkholderia sacchari*; copolyester; poly(3-hydroxybutyrate-*co*-4-hydroxybutyrate); polyhydroxyalkanoate (PHA); saccharose; sucrose; sugarcane

1. Introduction

Polyhydroxyalkanoates (PHA) are a versatile group of microbial biopolyesters with properties mimicking those of petrol-based plastics. A growing number of described bacterial and archaeal prokaryotic species accumulate PHA as refractive granular inclusion bodies in the cell's cytoplasm. PHA granules are surrounded by a complex membrane of proteins and lipids; these functional "carbonosomes" are typically accumulated under conditions of an excess exogenous carbon source in parallel with the limitation of a growth-essential component like the nitrogen source or phosphate [1–4]. Playing a major biological role, the presence of intracellular PHA supports bacterial survival under the conditions of carbon starvation. Moreover, PHA has pivotal functions in protecting cells against environmental stress conditions such as extreme temperature [5,6], exposure to oxidants [5,7], organic solvents [7], and UV-irradiation [6]. Depending on their composition, we distinguish homopolyesters, consisting of only one type of monomer, from heteropolyesters, composed of two or more types of monomers differing in their side chains (copolyesters) or both in their side chains and backbones (terpolyesters). In this context, the best known member of the PHA family, namely the homopolyester poly(3-hydroxybutyrate) (PHB), has restricted processability due to its high brittleness and crystallinity if compared to heteropolyesters consisting of different monomers such as 3-hydroxybutyrate (3HB), 3-hydroxyvalerate (3HV), 4-hydroxybutyrate (4HB), or 3-hydroxyhexanoate (3HHx) [8]. Changing PHA's composition on the monomeric level offers the possibility to fine-tune the polymer properties (melting temperature T_m, glass transition temperature T_g, degree of crystallinity X_c, degradability, elongation at break, or tensile strength) according to the customer's demands [9]. Apart from utilization in its crude form, PHA can be processed together with compatible organic or inorganic materials to make various composites and blends with tailored properties in terms of density, permeability, tensile strength, (bio)degradability, crystallinity, etc. [10–12]. To an increasing extent, the processing of PHA with nanoparticles is reported to generate novel designer bio-plastics especially useful for, *inter alia*, "smart packaging" [13,14].

Nowadays, there is an emerging trend of substituting petrol-based plastics with sustainable "bio-alternatives" with low environmental impact, that are biodegradable and bio-based in their nature [15,16]. Nevertheless, PHA production is still challenged by cost-decisive factors which make them considerably more expensive than their petrochemical counterparts; in order to optimize PHA production economically, all single process steps have to be taken into account [4,17]. Enhanced downstream processing to recover intracellular PHA from the biomass [18–21], bioreactor design and process regime [22–25], and in-depth understanding of the kinetics of the bioprocess [26] are crucial factors when designing a new PHA production process. Nevertheless, the selection of the most suitable carbonaceous raw materials to be used as feedstocks for PHA biosynthesis is the issue that is most difficult to solve. In this context, there is an increasing trend towards the application of carbon-rich (agro) industrial waste materials to produce the so called "2nd generation PHA" [4]. Among these materials, the current literature familiarizes us with PHA production based on surplus whey [27], abundant lignocelluloses [28–30], waste lipids from animal processing [31–33], used plant-and cooking oils [34–36], crude glycerol from biodiesel production [37–40], plant root hydrolysates [30], extracts and hydrolysates of spent coffee ground [41,42], and molasses [43]. Such waste materials already performed well as substrates on the laboratory scale, but are still awaiting their implementation in industrial-scale PHA production processes. This is mainly due to problems associated with upstream processing, insecure supply chains, presence of inhibitory compounds, or fluctuating composition of the industrial waste streams [4]. An emerging trend in using industrial waste streams is recognized in the direct conversion of CO_2 from industrial effluent gases [44]; here, cyanobacteria [45,46] or "Knallgasbacteria" [47] are potential cellular factories used to convert CO_2 to "3rd generation PHA" and additional valued products. Although also promising on the laboratory scale, development of these processes to industrial maturity has hitherto not been reached [44–47].

Apart from 2nd and 3rd generation PHA, the production of PHA based on materials relevant for food and feed purposes ("1st generation PHA") can also become economically viable given the integration of PHA-production facilities into existing production lines, where the raw material is generated [48]. This is successfully demonstrated at PHB Industrial SA (PHBISA), a company located in the Brazilian state of São Paolo. PHBISA is involved in the cane sugar business, predominantly fermenting hydrolyzed sucrose to bioethanol, and selling sucrose in its native form; a small part of sucrose is currently converted to PHA in a pilot plant with 100 ton annual capacity, and marketed under the trade mark Biocycle™ [48]. Remarkably, this bio-refinery process works energetically autarkic by the thermal conversion of surplus sugarcane bagasse to generate steam and electrical energy, which are used in the bioprocesses and the distillation for ethanol recovery. Moreover, distillative ethanol recovery generates a mixture of medium-chain-length alcohols (butanol, pentanol, etc.), which are used by the company for extractive PHA recovery from microbial biomass. This strategy saves expenses for the typically applied and often halogenated extraction solvents, which considerably contribute to the entire PHA production costs [48]. Currently, PHA production at PHBISA is carried out using the well-known production strain *Cupriavidus necator*, a eubacterial organism lacking the enzymatic activity for sucrose cleavage; hence, sucrose hydrolysis of the monomeric sugars (glucose and fructose) is a needed laborious operation step during upstream processing. For further optimization of this sucrose-based PHA production process, the assessment of alternative production strains appears reasonable. Such new whole-cell biocatalysts should fulfill some requirements: Growth rate and volumetric PHA productivity that are competitive with the data known for *C. necator*; direct sucrose conversion without the need for hydrolysis; temperature optima in the slightly thermophile range (in order to save cooling costs, a decisive cost factor under the climatic conditions prevailing in São Paolo); and last but not least, the strain should be able to produce copolyesters with advanced material properties.

A strain that appears promising in all these criteria is *Burkholderia sacchari* IPT 101 (DSM 17165), originally isolated from the soil of Brazilian sugarcane fields and investigated by Brämer and colleagues [49]. The strain is reported to accumulate high amounts of PHA inter alia from glucose [39,50], sucrose [49,50], glycerol [39,50], organic acids [51], pentose-rich substrate cocktails mimicking hydrolysates of bagasse [52], and hydrolyzed straw [53]. Aimed at the optimized utilization of lignocellulose hydrolysate, efforts are currently devoted to further improve the strain's substrate conversion ability in terms of xylose uptake [54]. PHA production by this organism and its mutant strains was demonstrated both in mechanically stirred tank bioreactors [52,53,55,56] and in airlift bioreactors [57]. As a drawback, the wild type strain displays insufficient ability for 3HV formation from structurally related precursors such as propionic acid, which is in contrast to pronounced 3HV formation by its mutant strain *B. sacchari* IPT 189 [54,56,58,59]. Formation of copolyesters consisting of 3HB and 4HB, hence P(3HB-*co*-4HB), was successfully demonstrated by co-feeding glucose or wheat straw hydrolysate (WSH) and the 4HB-related precursor compound γ-butyrolactone (GBL) [49]. Only recently has the production of copolyesters of 3HB and 3-hydroxyhexanoate (3HHx) by genetically engineered *B. sacchari* been reported [60]. In the present study, we demonstrate for the first time the feasibility of high-cell density production of PHB and P(3HB-*co*-4HB) by *B. sacchari* based on saccharose from PHBISA and the 4HB-precursor GBL, and for the first time, GBL's saponified form, 4-hydroxybutyrate sodium salt (Na-4HB). Furthermore, by addressing the contradictory literature information on the optimum temperature at which this organism thrives [50–54], we adapted the strain to an elevated cultivation temperature of 37 °C according to the requirements at the Brazilian production site [48,61]. Detailed kinetic data under controlled conditions in laboratory bioreactors, and an in-depth comparison of the polymer data of PHB and P(3HB-*co*-4HB), respectively, are provided.

2. Materials and Methods

2.1. Strain Maintenance and Adaptation to Elevated Temperature

Burkholderia sacchari DSM 17165 was purchased from DSMZ, Germany, and were grown on solid media plates (medium according to Küng [62] with 10 g/L of sucrose as the carbon source and 2 g/L ammonium sulfate as the nitrogen source). In two-week intervals, single colonies were transferred to new plates and incubated at 37 °C. All mineral components of the medium were purchased in p.a. quality (Company Roth, Graz, Austria), whereas sugarcane sucrose was obtained as unrefined saccharose directly from PHBISA.

2.2. Shaking Flask Cultivation to Assess Production of 4HB-Containing PHA

For preparation of pre-cultures, fresh single colonies from solid media were transferred to 100 mL of a liquid mineral medium containing the following components (g/L): KH_2PO_4, 9.0; $Na_2HPO_4 \cdot 2H_2O$,

3.0; $(NH_4)_2SO_4$, 2.0; $MgSO_4 \cdot 7H_2O$, 0.2 g; $CaCl_2 \cdot 2H_2O$, 0.02; $NH_4Fe(III)$citrate, 0.03; SL6, 1.0 (mL/L); sucrose, 15.0. These pre-cultures were incubated at 37 °C under continuous shaking; after 24 h, 5 mL of these pre-cultures were used for inoculation of four flasks each containing 100 mL of the minimal medium. The pH-value was adjusted to 7.0. After 8 h of incubation at 37 °C, 4HB-precursors were added to the cultures as follows: Two of the flasks were supplied with a solution of GBL, and two cultures with a solution of Na-4HB. Both solutions were added in a quantity to achieve a final precursor (GBL or the anion of 4HB, respectively) concentration of 1.5 g/L each. 15 h later, the re-feed of 4HB precursors was accomplished using the same quantity (1.5 g/L). After 47 h of cultivation, the experiment was stopped and the fermentation broth was analyzed for cell dry mass (CDM), PHA mass fraction in CDM, and PHA composition (fractions of 3HB and 4HB) (analytical methods *vide infra*).

2.3. Bioreactor Cultivations

2.3.1. PHB Production

Single colonies of *B. sacchari* were used to inoculate 100 mL (pre-cultures) of the medium according to Küng as described above. These pre-cultures were incubated (37 °C) for 36 h; then, 5 mL each of these pre-cultures were used for the inoculation of seven shaking flasks each containing 250 mL of the minimal medium. These cultures were incubated under continuous shaking at 37 °C for 36 h, until high cell densities (8–9 g/L) were reached, and two of them were used to inoculate a Labfors 3 bioreactor (Infors, CH) with an initial working volume of 1.5 L (1.0 L fresh medium with compounds calculated for 1.5 L plus 0.5 L inoculum). At the start of the cultivation, sucrose and $(NH_4)_2SO_4$ amounted to 15 g/L and 2.5 g/L, respectively. The set point for dissolved oxygen concentration (DOC) was 40% of the air saturation during the growth phase, and 20% during nitrogen-limited conditions; DOC was controlled by automatic adjustment of the stirrer speed and aeration rate. The pH-value was set to 7.0 and controlled automatically by the addition of H_2SO_4 (10%) to decrease the pH-value, and ammonia solution (25%) during the growth phase or NaOH (10%) during the accumulation phase to increase the pH-value. Hence, during the growth phase, the addition of the nitrogen source was coupled with pH-value correction. The cultivation was carried out at 37 °C. The time points of sugar addition (50% w/w aqueous solution of Brazilian sugarcane saccharose) are indicated in Figure 2 by arrows; the total amount of sucrose solution refeed amounted to 360 g.

2.3.2. P(3HB-*co*-4HB) Production:

This process was based on inoculum preparation according to the previous experiment. Cultivation in the bioreactor was performed using a minimal medium identical to the process at the company PHBISA (g/L): KH_2PO_4, 5.0; $(NH_4)_2SO_4$, 2.5; $MgSO_4 \cdot 7H_2O$, 0.8; NaCl; 1.0; $CaCl_2 \cdot 2H_2O$, 0.02; $NH_4Fe(III)$citrate, 0.05; trace element solution SL6 2.5 mL/L; sucrose 30; and the 4HB-precursor 4HB was provided by dropwise addition during the accumulation phase (total addition of GBL 15.5 g/L). Also in this case, a Labfors 3 bioreactor with an initial working volume of 1.5 L (1.0 L fresh medium with compounds calculated for 1.5 L plus 0.5 L inoculum) was used with the same basic parameters (DOC, T, pH-value) as described for the previous fermentation. The time points of sugar addition are indicated in Figure 7 by the arrows; the total amount of sucrose refeed amounted to 207 g of solution.

2.4. Cell Dry Mass (CDM) Determination

A gravimetric method was used to determine CDM in the fermentation samples. Five mL of the culture broth was centrifuged in pre-weighed glass screw-cap tubes for 10 min at 10 °C and 4000 rpm in a Heraeus Megafuge 1.0 R refrigerated centrifuge (Heraeus, Hanau, Germany). The supernatant was decanted, and subsequently used for substrate analysis. The cell pellets were washed with distilled water, re-centrifuged, frozen, and lyophilized (freeze-dryer Christ Alpha 1-4 B, Martin Christ Gefriertrocknungsanlagen GmbH, Osterode am Harz, Germany) to constant mass. CDM was expressed as the mass difference between the tubes containing cell pellets minus the mass of the empty tubes. The determination was done in duplicate. The lyophilized pellets were subsequently used for determination of intracellular PHA as described in the next paragraph.

2.5. Analysis of PHA Content in Biomass and Monomeric PHA Composition

For the analysis of PHA, standards of P(3HB-*co*-5.0%-3HV) (Biopol[TM], ICI, London, UK) were used for determination of the 3HB content; for determination of 4HB, "self-made" Na-4HB (next paragraph) was used as the reference material. Intracellular PHA in lyophilized biomass samples was transesterificated to volatile methyl esters of hydroxylkanoic acids via Braunegg's acidic methanolysis method [63]. Analyses were carried out with an Agilent Technologies 6850 gas chromatograph (30-m HP5 column, Hewlett-Packard, Palo Alto, CA, USA; Agilent 6850 Series Autosampler). The compounds were detected by a flame ionization detector; the split ratio was 1:10.

2.6. Preparation of Na-4HB

Na-4HB was synthesized by manually dropping a defined quantity of GBL into an equimolar aqueous solution of NaOH under continuous stirring and cooling. The obtained solution of Na-4HB was further frozen and lyophilized (freeze-dryer Christ Alpha 1–4 B) to obtain Na-4HB as a white powder. This powder was applied as a reference material for the analysis and as a co-substrate.

2.7. Substrate Analysis

The determination of carbon sources (sucrose and its hydrolysis products glucose, fructose, Na-4HB, and GBL) was accomplished by HPLC-RI using an Aminex HPX 87H column (thermostated at 75 °C, Biorad, Hercules, CA, USA), a LC-20AD pump, a SIC-20AC autosampler, a RID-10A refractive index detector, and a CTO-20AC column oven. Pure sucrose, glucose, fructose, Na-4HB, and GBL were used as standards for external calibration. Isocratic elution was carried out with 0.005 M H_2SO_4 at a flow rate of 0.6 mL/min.

2.8. Analysis of Nitrogen Source (NH_4^+)

The determination of the nitrogen source was done using an ammonium electrode (Orion) with ammonium sulfate solution standards (300–3000 ppm) as described previously [39].

2.9. PHA Recovery

After the end of the experiments, the fermentation broth was *in situ* pasteurized (80 °C, 30 min). Afterwards, the biomass was separated from the liquid supernatant via centrifugation (12,000 g; Sorvall® RC-5B Refrigerated Superspeed centrifuge, DuPont Instruments, Wilmington, NC, USA), frozen, and lyophilized (freeze-dryer Christ Alpha 1-4 B). Dry biomass was decreased by overnight stirring with a 10-fold mass of ethanol; after drying, PHA was extracted from the degreased, dried biomass by continuous overnight stirring in a 25-fold mass of chloroform in light-protected glass vessels. The solution containing the PHA was separated by vacuum-assisted filtration, and concentrated by evaporation of the major part of the solvent (Büchi Rotavapor® R-300). This concentrated PHA solution was dropped into permanently stirred ice-cooled ethanol. Precipitated PHA filaments of high purity were obtained by vacuum-assisted filtration, dried, and subjected to polymer characterization (*vide infra*).

2.10. Polymer Characterization

2.10.1. Molecular Mass Distribution

Gel Permeation Chromatography (GPC) analysis was carried out on a Waters 600 model (Waters Corporation, Milford, MA, USA) equipped with a Waters 410 Differential Refractometer and two PLgel 5 μm mixed-C columns (7.8 × 300 mm^2). The mobile phase constituted by chloroform (CHROMASOLV®

for HPLC amylene stabilized, Sigma-Aldrich, Milan, Italy) was eluted at a flow rate of 1 mL/min. Monodisperse polystyrene standards were used for calibration (range 500–1.800,000 g/mol). Samples were prepared at a concentration of ca. 0.5% (w/v).

2.10.2. Thermoanalysis

Differential Scanning Calorimetry (DSC) analysis was performed using a Mettler DSC-822E instrument (Mettler Toledo, Novate Milanese, Italy) under a nitrogen flow rate of 80 mL/min. The analysis was carried out in the range from -20 to 200 °C at a heating and cooling rate of 10 °C/min. By considering the second heating cycles in the thermograms, the glass transition temperature (T_g) was evaluated by analyzing the inflection point, while the melting temperature (T_m) and crystallinity (X_c) was evaluated by analyzing the endothermic peak. X_c was determined by considering the value of the melting enthalpy of 146 J/g for the 100% crystalline PHB. Both characterization tests were carried out on five replicates for each kind of sample and the data were presented as mean ± standard deviation. Statistical differences were analyzed using one-way analysis of variance (ANOVA), and a Tukey test was used for post hoc analysis. A p-value < 0.05 was considered statistically significant.

3. Results

3.1. Impact of 4HB-Precursors GBL and Na-4HB on Poly-(3-hydroxybutyrate-co-4-hydroxybutyrate) (P(3HB-co-4HB)) Biosynthesis by Burkholderia sacchari DSM 17165 on Sucrose

Figure 1 illustrates the outcomes of the shaking flask experiment comparing the effect of adding 4HB-precursors GBL and Na-4HB to *B. sacchari* cultivated on sucrose as main carbon source. After 47 h of incubation, the CDM concentration was in the range of 5 g/L in all experimental setups. Final PHA concentrations amounted to 1–2 g/L without significant differences between the individual cultivation setups. Using GBL as the 4HB-related precursor, PHA fractions in the CDM were slightly lower than in the case of using Na-4HB, but almost identical to the setups without precursor addition (ca. 30% vs. ca. 35%, respectively). The 4HB fractions in PHA (4HB/PHA) differ in dependence on the applied precursor; using GBL, this value amounted to 20.8%, while it was only 14.1% when using Na-4HB. As expected, the setups cultivated on sucrose as the sole carbon source (no addition of 4HB-related precursors) resulted in the generation of the PHB homopolyester. Here, it has to be emphasized that it is not clear from the available data if the generated polyester is definitely a P(3HB-*co*-4HB) copolyester with random distribution of the individual building blocks, a blend of homopolymers consisting of 3HB or 4HB, respectively, or a blend of different P(3HB-*co*-4HB) copolyesters with different 4HB fractions.

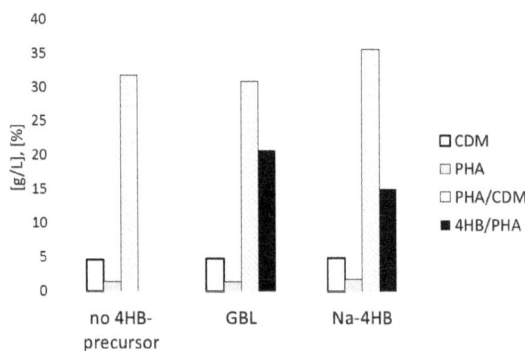

Figure 1. Cell dry mass (CDM) (g/L), polyhydroxyalkanoate (PHA) (g/L), mass fraction of PHA in CDM (%), and mass fraction of 4-hydroxybutyrate (4HB) in PHA (%): *B. sacchari* after 47 h of cultivation on 15 g/L sucrose and 4HB-precursors γ-butyrolactone (GBL) or Na-4HB (precursor addition: 1.5 g/L after 8 h, refeed of 1.5 g/L after 15 h).

3.2. Poly(3-hydroxybutyrate) (PHB) Production with Burkholderia sacchari on the Bioreactor Scale; Sucrose as the Sole Carbon Source

3.2.1. Bioprocess

This experiment aimed to test a medium similar to the one used at the industrial company PHBISA for sucrose-based PHA production by *C. necator*, and to study its influence on the kinetic data and on the polymer production (*cf.* Materials and Methods section). Of major importance, it was intended to considerably increase the concentration of the residual biomass and to achieve higher productivities for PHA. This was accomplished using an advanced strategy for adding the nitrogen source (NH_4^+) during the microbial growth phase by coupling the addition of the nitrogen source with the correction of the pH-value. Instead of a periodic re-feed of $(NH_4)_2SO_4$ solution to maintain the nitrogen concentration at the desired level, NH_4OH was used as a base for correction of the pH-value and, at the same time, to provide the nitrogen needed by the strain to grow. Hence, the addition of the nitrogen source was directly coupled to the excretion of acidic metabolites during the growth phase. After 12.5 h of fermentation, the NH_4OH solution as the pH-correction agent was replaced by NaOH solution (20%) in order to provoke a nutritional stress by limitation of the nitrogen source to stop the biomass formation and to enhance PHA production; this time point is marked by a full line in Figure 3. The depletion of the nitrogen source occurred after 19 h of cultivation.

Figure 2 illustrates the time curves of the sugar concentrations (sucrose, glucose, and fructose). It is easily seen that the strain possesses the metabolic ability to rapidly hydrolyze the disaccharide sucrose to its monomeric sugars by the excretion of an extracellular invertase enzyme. Immediately after inoculation, hydrolysis started, resulting in about 9 g/L sucrose and 6 g/L of monomers (glucose plus fructose) already present in the first sample taken at t = 0 h. The time points of sucrose additions are marked by arrows in Figure 2. Remarkably, the concentrations of the two monosaccharides do not follow the same trend with time, which might be due to the changing conversion rates of the individual monomers (glucose or fructose, respectively) with the changing environmental (nutritional) conditions during the cultivation. Mathematical modelling of the data to elucidate the metabolic processes should therefore be performed in follow-up experiments by specialists in the field of metabolic flux analysis. A total quantity of 360 g sucrose solution was added during the process. A total sugar consumption of 29.14 g/(L·h) was observed, and a conversion yield of sugar to CDM of 0.18 g/g (calculated for the entire sugar addition and also encompassing the not utilized sugar in the spent fermentation broth) (Table 1). Limitation of the carbon source was avoided during the entire cultivation period by permanent monitoring (HPLC) and re-feeding (Figure 2).

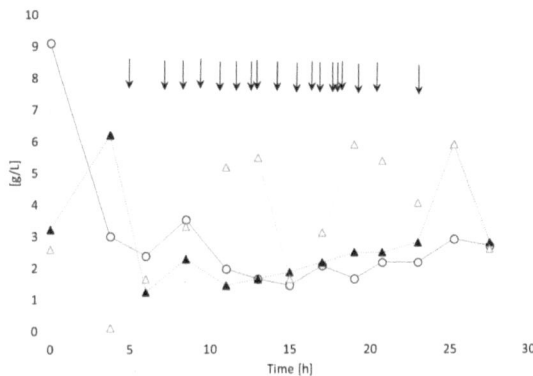

Figure 2. Substrate time curves: *B. sacchari* on sucrose without supplementation of 4HB-precursors. Open spheres: sucrose; black triangles: glucose; open triangles: fructose. Arrows indicate the time points of pulse feedings of the sucrose solution.

Table 1. Results of the bioreactor fermentations.

Kinetic Parameter	PHB Production Process (1st Bioreactor Cultivation)	P(3HB-*co*-4HB) Production Process (2nd Bioreactor Cultivation)
$\mu_{max.}$ (1/h)	0.41 (t = 3.75–6 h)	0.23 (t = 6–8 h)
max. CDM (g/L)	70.0 (t = 25.25 h)	78.6 (t = 32 h)
max. PHA concentration (g/L)	36.8 (t = 27.5 h)	55.8 (t = 29 h)
max. fraction of PHA in CDM (% w/w)	53.0 (t = 27.5 h)	72.6 (t = 29 h)
max. fraction of 4HB in PHA (% mol/mol)	-	1.6 (t = 39 h)
Volumetric productivity for PHA (g/L·h)	1.29 (t = 0–27.5 h)	1.87 (t = 0–39 h)
Yield$_{CDM/sucorse}$ (g/g)	0.18	0.38
Yield 4HB/GBL (g/g)	-	0.05
max. specific productivity q_P (g/(g·h))	0.19 (t = 7.25 h)	0.17 (t = 17.75 h)
Material Characterization		
Weight average molecular mass Mw (kDa)	627 ± 13	315 ± 24
Polydispersity P_i (Mw/Mn)	2.66 ± 0.13	2.51 ± 0.15
Glass transition temperature T_g (°C)	1.0 ± 0.6	1.8 ± 0.2
Melting point T_m (°C)	177.6 ± 0.6	160.9 ± 0.8
Degree of crystallinity X_c (%)	70.9 ± 0.9	24.0 ± 3.6

Figure 3 illustrates the time curves of the CDM, residual biomass, and PHA during the process. After the onset of nitrogen limitation after 19 h (indicated by a dashed line in Figure 3), the concentration of the residual biomass remained constant (35 g/L), whereas the PHA concentration increased, reaching a maximum concentration of 36.5 g/L at the end of the fermentation. This corresponds to a final CDM concentration of 70 g/L. Due to the fact that no 4HB-related precursors were supplied, homopolyester PHB was accumulated. The volumetric productivity for PHB, calculated for the entire process, amounted to 1.29 g/(L·h). For the entire process (t = 0 to 27.5 h), the yield for the conversion of sugars to CDM amounted to 0.18 g/g, whereas during the nitrogen-limited phase of cultivation, a conversion yield for sugars to PHB of 0.08 g/g was evidenced (Table 1).

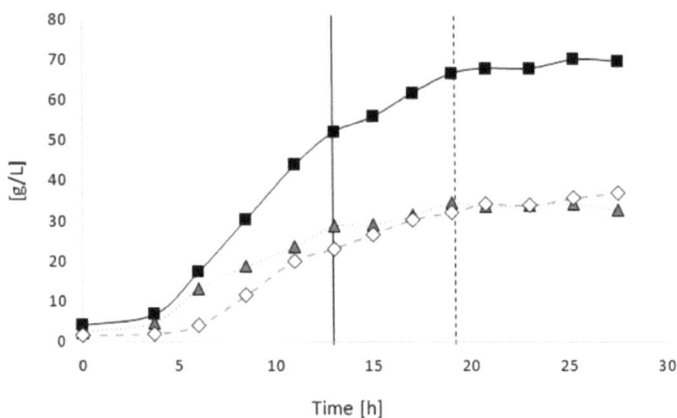

Time [h]

Figure 3. Time curves of CDM, residual biomass, and PHB concentration: *B. sacchari* on sucrose without supplementation of 4HB-precursors. Black squares: CDM; open rhombi: PHA; grey triangles: residual biomass. Full black line: Exchange of NH$_4$OH solution by NaOH solution as the pH-corrective agent, dashed line: start of nitrogen depletion in the medium.

Figure 4 illustrates the time curves of the specific growth rate μ and the specific product (PHB) formation rate q_P for the entire process. Here, it is visible that the maximum specific growth (μ_{max} = 0.41 1/h) was monitored at around 5 h of cultivation. For the entire growth phase (t = 3.75–13 h), μ_{max} was determined with 0.15 1/h by plotting the natural logarithm LN of the residual biomass concentration vs. time. After the exchange of NH$_4$OH by the NaOH solution and the resulting depletion of the nitrogen source, the specific growth tremendously decreased, and a slight decrease of the residual biomass concentration, indicated by the negative values for μ in Figure 4, was observed. The highest specific PHB production was observed starting from the onset of the exponential growth phase ($t = 5$ h) until the start of nitrogen depletion at $t = 12$ h; a q_P of about 0.19 g/(g·h) was measured for the period between the two subsequent samplings at $t = 6$ and 8.5 h. In later periods of the process, only a slight increase of PHB production, manifested by low values for q_P, was observed.

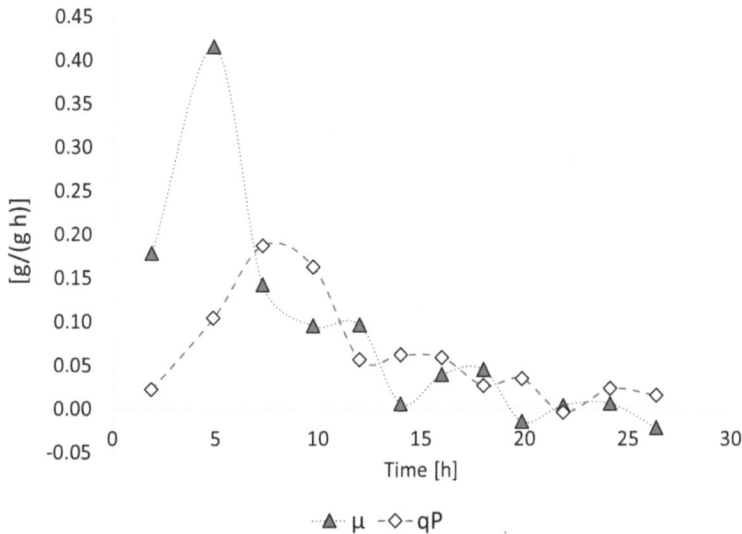

Figure 4. Time course of the specific growth rate μ and specific PHA production rate q_P: *B. sacchari* on sucrose without supplementation of 4HB-precursors. Full black line: Exchange of NH$_4$OH solution by NaOH solution as the pH-corrective agent; dashed line: start of nitrogen depletion in the medium.

3.2.2. Polymer Characterization:

After the end of the experiment, the biomass was separated from the liquid supernatant via centrifugation, and was frozen and lyophilized. The dry biomass was decreased with ethanol and the polymer was extracted using chloroform. The weight average molecular mass (Mw) and the polydispersity (P_i; Mw/Mn) values of the extracted homopolymer were determined by gel permeation chromatography (GPC). The Mw was 627 ± 13 kDa and P_i was 2.66 ± 0.13 kDa (Table 1). Differential scanning calorimetry (DSC) analysis was carried out to determine the glass transition temperature (T_g), melting temperature (T_m), and crystallinity (X_c) of the PHB samples. Analysis of the obtained data showed that the T_g of the produced PHB was 1.0 ± 0.6 °C and the T_m was 177.6 ± 0.6 °C, while X_c was 70.9% ± 0.9%.

3.3. Controlled Poly(3-Hydroxybutyrate-co-4-Hydroxybutyrate) (P(3HB-co-4HB)) Production with Burkholderia sacchari on the Bioreactor Scale: Sucrose plus GBL as Carbon Subsubstrates.

3.3.1. Bioprocess

Based on the results from the shaking flask scale reported in this study and previous findings which confirmed *B. sacchari*'s potential to produce PHA containing 4HB by co-feeding sucrose and 4HB-related precursor compounds, this material was produced under controlled conditions at the bioreactor scale. It was aimed at generating a residual biomass concentration of about 20 g/L and a

PHA mass fraction in CDM exceeding 60 g/L in order to be competitive with the *C. necator*—mediated sucrose-based PHA production process at PHBISA.

Figure 5 shows the time curves of the CDM, PHA, and residual biomass, whereas Figure 6 illustrates the corresponding time curves of the sugar concentrations; again, arrows mark the time points of sucrose addition. Also in this cultivation, the nitrogen source (NH_4^+) served as the growth-limiting factor. NH_4^+ was added continuously during the growth phase as aqueous NH_4OH solution (25%) according to the response of the pH-electrode. The maximum specific growth rate μ_{max} measured between two subsequent samplings (t = 6–8 h) amounted to 0.23 1/h for the entire growth phase (t = 0–10 h), and the μ_{max} for the entire exponential growth phase was determined to be 0.18 1/h. About 21 g/L of catalytically active residual biomass was produced until the onset of nitrogen depletion. Figure 7 shows the time curve of the main carbon source sucrose and its hydrolysis products glucose and fructose, which are produced by the extracellular invertase excreted by the organism; again, the rapid hydrolysis of sucrose is evident.

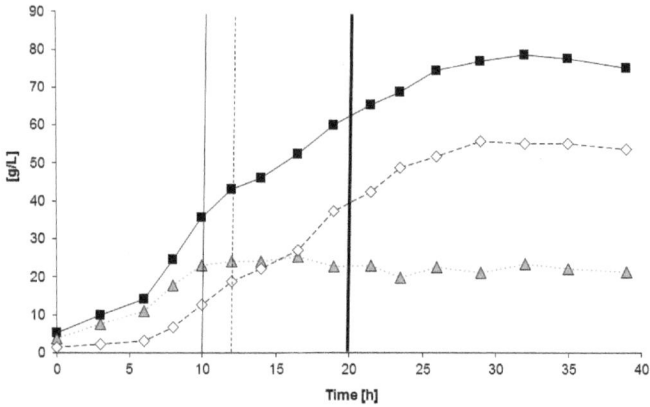

Figure 5. Time curves of product concentrations: *B. sacchari* on sucrose and the addition of γ-butyrolactone (GBL) as 4HB precursor. Black squares: CDM; open rhombi: PHA; grey triangles: residual biomass. Thin black line: Exchange of NH_4OH solution by NaOH solution as the pH-corrective agent; dash line: start of nitrogen depletion in the medium; bold black line: start of GBL feed.

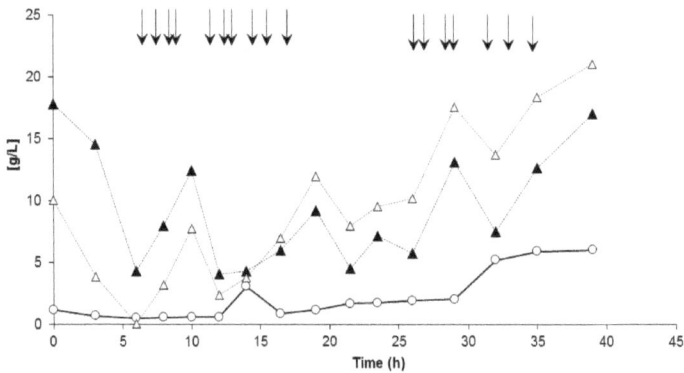

Figure 6. Actual concentrations of sugars: *B. sacchari* on sucrose and the addition of GBL as 4HB precursor. Arrows indicate the refeed with sucrose solution. Open spheres: sucrose; black triangles: glucose; open triangles: fructose.

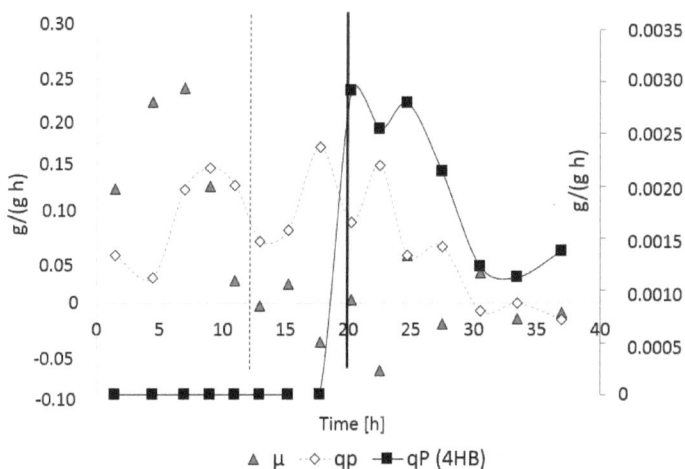

Figure 7. Time course of the specific growth rate μ and specific PHA production rate q_P (left axis): *B. sacchari* on sucrose and the addition of GBL as 4HB precursor. Thin black line: Exchange of NH_4OH solution by NaOH solution as the pH-corrective agent; dashed line: start of nitrogen depletion in the medium; bold black line: start of GBL feed.

After 10 h of fermentation, the nitrogen source supply was stopped by exchanging NH_4OH with NaOH as the pH-value correction agent; now, the second phase of the process was initiated (accumulation phase). During this phase, the time curve of the residual biomass was constant and the increase of CDM until the end of the experiment was only due to the increasing intracellular concentration of PHA (see Figure 5). It is visible that already during the exponential phase of the microbial growth (t = 7–10 h) that considerable amounts of PHA were produced ("growth associated product formation"). During the phase of product formation, GBL was added dropwise in order to not move into inhibiting concentration ranges. The actual GBL concentration was always below the detection limit when analyzing the samples; hence, GBL was completely converted by the cells. During the process, a total of 15.5 g/L GBL was added to the culture, distributed to a total of ten pulses of the substrate feed.

At the end of the process, the final concentrations of CDM and PHA of 75.1 g/L and 53.7 g/L, respectively, were achieved, corresponding to a PHA mass fraction in CDM of 71.5%. The total PHA concentration remained constant from t = 27.5 h. The volumetric productivity of PHA for the entire process and the conversion yield of sugar to CDM were calculated as 1.87 g/(L·h) and 0.38 g/g, respectively, which signifies an enormous enhancement in comparison to the previous experiment (Table 1).

Figure 7 illustrates the time curves of the specific growth rate μ, the specific PHA production rate q_P, and the specific 4HB production rate for the entire process. Again, starting with nitrogen limitation at about t = 12 h, the values for μ drastically decreased, whereas the specific PHA productivity q_P reached its highest values under nitrogen limited conditions; the maximum value for q_P was reached between t = 16.5 and 19 h, and amounted to 0.17 g/(g·h). Maximum specific 4HB production occurred between t = 20 and 35 h, and was calculated with about 0.003 g/(g h).

Co-feeding of GBL started after 20 h; until this time, the PHB homopolyester was produced (Figures 7 and 8). Starting with the sample taken at t = 23.5 h, 4HB-building blocks were detected in the polymer. The achieved 4HB fraction in PHA at the end of the fermentation was determined with 1.6% (mol/mol). The time curve of the polyester composition is illustrated in Figure 8. The essential process results are collected in Table 1 and directly compared with the outcomes of the previous process for the PHB production.

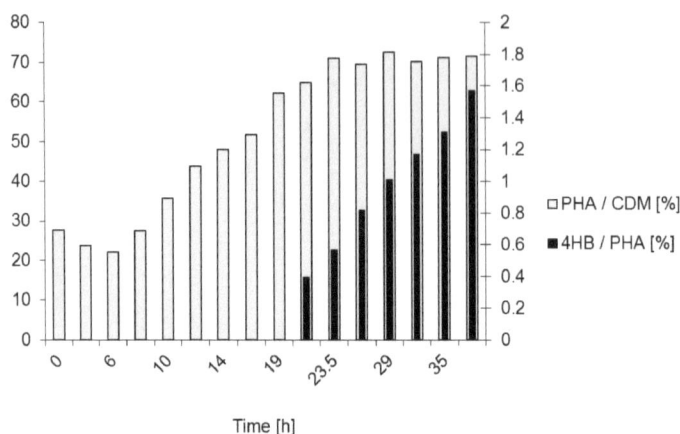

Figure 8. *B. sacchari* on sucrose and the addition of GBL as 4HB precursor. Composition of PHA during the process. Grey bars: Mass fraction of PHA in CDM (left axis). Black bars: Molar 4HB fraction on PHA (right axis). GBL addition started at *t* = 20 h.

3.3.2. Polymer Characterization:

After the end of the experiment, the biomass was separated from the liquid supernatant via centrifugation, and was frozen and lyophilized. The dry biomass was decreased with ethanol and the polymer was extracted using chloroform. The Mw and P_i values of the extracted copolymer, determined by GPC, were 315 ± 24 kDa and 2.51 ± 0.15 kDa, respectively (Table 1). Statistical differences analyses showed that the Mw of P(3HB-*co*-4HB) was significantly lower than that of PHB. In addition, analysis of the DSC data showed that P(3HB-*co*-4HB) had significantly lower X_c (24.0% ± 3.6%) and T_m (160.9 ± 0.8 °C) than PHB, while T_g was in the same range (1.8 ± 0.2 °C).

Table 1 compares both kinetic data and data from polymer characterization of both bioprocesses on the bioreactor scale.

4. Discussion

4.1. Bioprocess

The organism *B. sacchari* DSM 17165 possesses the desired ability to produce 4HB-containing PHA from sucrose plus both investigated 4HB precursors GBL and Na-4HB. The successful conversion of GBL towards 4HB building blocks is in agreement to previous findings reported by Cesário, who used glucose or WSH plus GBL for P(3HB-*co*-4HB) biosynthesis by this strain. These authors also tested P(3HB-*co*-4HB) production by this organism by using 1,4-butanediole as the 4HB-related precursor, revealing the incorporation of 4HB by GBL supplementation and the strain's inability to utilize 1,4-butanediole. No reports were previously available on the utilization of Na-4HB by this strain. The results reported by Cesário et al. show varying PHA fractions in CDM for the fed-batch cultivation of *B. sacchari* on glucose/GBL mixtures, dependent on the ratio of glucose/GBL. Cultivation on pure glucose resulted in 49.2% PHB in CDM; this value decreased with increasing GBL portions in the feed stream to only 7.1% using GBL as the sole carbon source [28]. In our shaking flask setups, the rather modest precursor supplementation of 1.5 g/L neither significantly impacted the CDM production or the PHA fraction in CDM compared to the precursor-free setups (sucrose as the sole carbon source). Remarkably, the application of the GBLs saponified from Na-4HB resulted in considerably lower 4HB fractions in PHA than observed when using the annular lactone (GBL) (21% vs. 14%). As assumed for *C. necator* [64] and *Hydrogenophaga pseudoflava* [65], GBL is imported into the cells as an intact lactone ring, which is opened only intracellularly. According to Valentin et al., only a part of 4HB is converted to 4-hydroxybutyryl-CoA (4HB-CoA) in the cells, whereas 4HB's major share is converted to succinic acid semialdehyde and succinic acid, which finally undergo conversion to the 3-hydroxybutyryl-CoA (3HB-CoA) precursor acetyl-CoA. PHA synthase in turn polymerizes 3HB-CoA and 4HB-CoA to P(3HB-*co*-4HB) [66].

As shown previously [39,52,53,55,57] and confirmed by the present work, nitrogen limitation is a suitable approach to boost PHA biosynthesis by *B. sacchari*. Generally, the strategy to constantly supply a nitrogen source by coupling the NH_4OH supply to microbial growth by automatically responding to the signal of the pH-electrode was performed successfully to rapidly generate a high concentration of catalytically active biomass at a high specific growth rate. Only about 9 h (PHB production) or 12 h (production of 4HB-containing PHA) were needed to boost the concentration of the residual biomass above 20 g/L. This shows significant progress to comparable experiments carried out by Rocha and colleagues, who used the same strategy and achieved a maximum residual biomass of about 16 g/L after 24 h of cultivation using the mutant *B. sacchari* IPT 189 [55]. The maximum growth rates μ_{max} obtained in our experiments (0.15 1/h for the first, 0.18 1/h for the second bioreactor cultivation; calculated for the entire growth phase; 0.41 and 0.23 1/h maximum valued between two subsequent samplings) can be compared to related reports found in the literature; for shaking flask cultivations of *B. sacchari* LFM 101 on sucrose, Nascimento et al. report a μ_{max} of 0.544 and 0.546 1/h at 30 and 35 °C, respectively [50]. At the bioreactor scale, Rocha et al. obtained a μ_{max} of 0.4 1/h for the first 10 hours of continuous cultivation of *B. sacchari* IPT 189 [55]; this value was also obtained by da Cruz Pradella with *B. sacchari* IPT 189 by using a fedbatch feeding regime in an airlift reactor [57]. Reliable μ_{max} values from the bioreactor scale cultivations of our production strain *B. sacchari* IPT 101 (DSM 17165) are available for xylose-based experiments, where μ_{max} amounted to 0.07–0.21 1/h with dependence on the initial xylose concentration [52]. Using glucose during the growth phase, Rodriguez-Contreras obtained a μ_{max} of 0.42 1/h [39]. Testing the effect of GBL on the growth of *B. sacchari* in shaking flask setups, Cesário et al. noticed a continued decrease of μ_{max} from 0.32 to 0.19 1/h with GBL concentrations increasing from 5 to 40 g/L, with 40 g/L glucose as the main carbon source. In this study, μ_{max} was unfortunately not reported for the fedbatch cultivations in the bioreactors for the production of PHB and P(3HB-*co*-4HB) [53].

Furthermore, we demonstrated that the organism can successfully be cultivated at an elevated temperature of 37 °C, which is beneficial for large scale operation in reactors integrated into the production facilities of the Brazilian sugarcane industry [48,61]. The cultivation temperature of 37 °C is in contrast to previous literature reports for this organism and its close relatives. Generally, 30 °C is reported as the optimum temperature to efficiently thrive most *B. sacchari* sp. [50]. In a mechanically stirred tank bioreactor, Raposo and colleagues cultivated the same strain for the production of PHB, xylitol, and xylonic acid at a temperature of 32 °C [61], whereas 30–32 °C was used by da Cruz Pradella et al. to culture its mutant strain *B. sacchari* IPT 189 for PHB biosynthesis in an airlift reactor [57], or by Rocha and colleagues in continuously operated bioreactor cultivations [55]. *B. sacchari* LFM 101, a strain that is most likely closely related to our production strain, was only recently tested by Nascimento et al. for PHA production on sucrose, glucose, and glycerol at both 30 and 35 °C. These authors report higher volumetric productivity and PHA fractions in CDM, and unaltered specific growth rates for cultivations carried out on glucose or sucrose at 35 °C or 30 °C, respectively. When using glycerol as the carbon source, no biomass formation or significant substrate consumption was observed, probably due to the lack of energy needed to convert the glycerol molecules [50]. As demonstrated by Rodriguez-Contreras et al. who operated a *B. sacchari*-mediated PHB production process at 37 °C, this problem can be overcome by feeding the cells with energy-rich carbohydrates like glucose or sucrose in the first stage (growth phase), and subsequently switching to glycerol feeding in the second phase (PHA accumulation) [39].

Values of 1.29 g/(L·h) (PHB) and 1.87 g/(L·h) (4HB-containing PHA) were achieved for the volumetric PHA productivity in the two conducted bioreactor experiments. These values are considerably higher than that reported for comparable experiments by Rodriguez-Contreras et al., who reported a volumetric productivity of 0.08 g/(L·h) for a two-stage process based on the co-feeding of *B. sacchari* with glucose and glycerol [39], and by Cesário and colleagues, who obtained 0.7 g/(L·h) for fed-batch cultures supplied with glucose and GBL, and 0.5 g/(L·h) when using WSH plus GBL for fed-batch P(3HB-*co*-4HB) production [53]. Here, it has to be emphasized that Cesário et al. [53] used considerably higher GBL dosage than we did in the study at hand; this, on the one hand, resulted in tripling the molar fractions of 4HB in PHA in comparison to our results, but, on the other hand, negatively influenced the overall volumetric PHA productivity as the fundamental economic parameter in PHA production. Regarding the obtained PHA contents in the biomass, our results show final PHA fractions in CDM of 52.4% for PHB, and 71.5% for P(3HB-*co*-4HB), respectively. The results by Cesário and colleagues report 73% PHB in CDM in fed-batch cultures with glucose as the sole carbon source, and 45% P(3HB-*co*-4HB) in CDM with pulse feeding 8 g/L GBL in the accumulation phase followed by continuously feeding GBL at a rate of 2.3 g/h. Fed-batch cultures of *B. sacchari* on WSH plus GBL reported in the same study resulted in a P(3HB-*co*-4HB) fraction in CDM of 27%. Interestingly, the authors found that in *B. sacchari*, the conversion yield of GBL towards 4HB can considerably be improved by supplementing acetate or propionate as additional "stimulants" for the 4HB biosynthesis [53]. Based on the works carried out by Lee et al. with *C. necator*, it was known previously that an increased acetyl-CoA pool from acetate conversion or from propionate ketolysis, respectively, inhibits the conversion of 4HB-CoA to acetyl-CoA, thus preserving a high 4HB-CoA pool available for the P(3HB-*co*-4HB) biosynthesis [67]. Using the mutant strain *B. sacchari* IPT 189, PHA copolyesters consisting of 3HB and 3HV were produced by Rocha et al. by co-feeding sucrose and propionic acid in two-stage bioreactor setups at a volumetric productivity of 1 g/(L·h); in these experiments, the biomass contained a PHA mass fraction of up to 60%, which is higher than in our PHB production process (52.4%), but lower than the value obtained in the present study for P(3HB-*co*-4HB) production (71.5%) [55]. The two-stage co-feeding experiments with *B. sacchari* carried out by Rodriguez-Contreras et al. on glucose and glycerol generated a PHA fraction in CDM that hardly exceeded 10% [39]. Using mixtures of xylose and glucose to mimic differently composed lignocellulosic hydrolysates, Raposo and associates produced PHB by fed-batch cultivations of *B. sacchari* in laboratory bioreactors. Changing the pulse size, feeding rate, and glucose/xylose ratio, the volumetric productivities decreased from 2.7 g/(L·h) (73% PHB in CDM) for pure glucose feeding to 0.07 (11% PHB in CDM) for xylose as the sole carbon source, indicating the inhibitory effect of this pentose sugar [52].

4.2. Polymer Characterization:

The obtained data for polymer characterization were in the same range as the results provided by Cesário and colleagues, who extracted PHB and P(3HB-*co*-4HB) from *B. sacchari* biomass, cultivated on WSH, via the same method used in the present study. These authors describe a Mw for PHB of 790 kDa, and between 450 and 590 kDa for P(3HB-*co*-4HB); higher 4HB fractions gradually decreased the Mw values [29]. Our results report a M_W of 627 kDa for PHB, and 315 kDa for P(3HB-*co*-4HB). The P_i of our sucrose-based polyester samples was higher than the values reported for WSH-based PH. For PHB, we obtained a P_i of 2.66, which is similar to the value obtained for the P(3HB-*co*-4HB) sample (2.51). For comparison, the PHB and P(3HB-*co*-4HB) samples produced by Cesário and colleagues had significantly lower P_i, ranging from 1.4 to 1.7 [29]. Other comparable results were provided by Rosengart et al., who reported a P_i of 2.33 for a *B. sacchari*-based PHB [68]. A considerably lower Mw (200 kDa) was described by Rodriguez-Contreras et al. for PHB obtained by co-feeding *B. sacchari* with glucose and glycerol; in this study, a P_i of 2.5 was reported [39]. Here, it should be noted that glycerol feeding generally results in low molecular mass PHA if compared to sugar-based PHA production, as reported elsewhere [37,68]. This is due to the "endcapping effect", a phenomenon describing the termination of the *in vivo* PHA chain propagation in the presence of glycerol and other polyols [69]. The melting temperature T_m reported by Cesário and colleagues amounted to 171.7 °C for PHB, and to 158.8 and 164.3 °C for P(3HB-*co*-4HB) with 7.6 or 4.6 mol% of 4HB, respectively [29]. In our case, the T_m for PHB amounted to 177.6 °C, whereas for P(3HB-*co*-4HB) was only 160.9 °C, which matches well with the cited literature data. Our PHB displayed an X_c of 70.9 °C, which is slightly higher than that reported for the WSH-based material (64.8%) [29]. A remarkably low X_c of 24.0% was measured for our P(3HB-*co*-4HB), which is considerably lower than the value reported for P(3HB-*co*-4HB) based on WSH (between 47.2% and 52.3%) [29]. The PHB produced by Rodriguez-Contreras et al. on glucose plus glycerol displayed an X_c of 72.8% and a T_m of 163.3°C [39]. Using PHB-rich biomass from a cultivation of *B. sacchari* on glucose, Rosengart et al. [21] compared the extraction performance of unusual extraction solvents (anisol, phenetole, and cyclohexanone) with the performance of classical chloroform extraction as used in our study, and by Cesário and colleagues [53]. As an outcome, the thermal properties (T_m, T_g, X_c) and molecular mass were fully comparable to the values obtained via chloroform extraction, thus demonstrating the feasibility of switching to sustainable, non-chlorinated alternatives to chloroform [21].

5. Conclusions

The highest (up to now) reported productivity for *B. sacchari*-mediated biosynthesis of PHA with building blocks differing from 3HB is described in the present work. Adaptation of the production strain to an elevated temperature optimum of 37 °C makes it a feasible candidate for cost-efficient on-site PHB and P(3HB-*co*-4HB) production starting from cane sugar on the industrial scale. In any case, PHA production facilities should also in future be integrated into the existing production lines for sucrose-based bioethanol production in order to profit from reduced transportation costs, energetic autarky, and in-house availability of extraction solvents for PHA recovery from the biomass. Further efforts should be devoted to high-throughput continuous PHA production by this organism in a chemostat ("chemical environment is static") process regime. Similar to the results recently obtained by other production strains [70], the application of a multistep-continuous production in a bioreactor cascade displays a viable process-engineering tool to further increase volumetric productivity, and to trigger the distribution of 3HB and 4HB monomers in tailor-made copolyesters. Moreover, the highly effective invertase enzyme excreted by this strain deserves in-depth characterization and might be of interest for applications in food technology. Together with PHA production and other metabolites generated by this strain, such as xylitol or xylonic acid [52], this might open the door to implementing *B. sacchari* as a versatile platform to catalyze a bio-refinery plant starting from inexpensive feedstocks.

Acknowledgments: ARENA is grateful for the funding received by PHBISA for the industrial project "Production of the copolyester Poly(3-hydroxybutyrate-*co*-3-hydroxyvalerate) from cane sugar by fermentation with special regard to new strains adapted to high temperature and direct use of sucrose".

Author Contributions: Miguel Miranda de Sousa Dias, Gerhart Brauneg, and Martin Koller conceived and designed the biotechnological experiments; Miguel Miranda de Sousa Dias performed the biotechnological experiments; Dario Puppi, Andrea Morelli, and Federica Chiellini analyzed and interpreted the polymer data; Martin Koller wrote the predominant portion of the paper. All authors read, edited, and approved the final manuscript.

Conflicts of Interest: The authors declare no conflict of interest.

References

1. Jendrossek, D.; Pfeiffer, D. New insights in the formation of polyhydroxyalkanoate granules (carbonosomes) and novel functions of poly(3-hydroxybutyrate). *Environ. Microbiol.* **2014**, *16*, 2357–2373. [CrossRef] [PubMed]
2. Chen, G.Q.; Hajnal, I. The 'PHAome'. *Trends Biotechnol.* **2015**, *33*, 559–564. [CrossRef] [PubMed]
3. Tan, G.Y.A.; Chen, C.L.; Li, L.; Ge, L.; Wang, L.; Razaad, I.M.N.; Li, Y.; Zhao, L.; Mo, Y.; Wang, J.Y. Start a research on biopolymer polyhydroxyalkanoate (PHA): A review. *Polymers* **2014**, *6*, 706–754. [CrossRef]
4. Koller, M.; Maršálek, L.; de Sousa Dias, M.; Braunegg, G. Producing microbial polyhydroxyalkanoate (PHA) biopolyesters in a sustainable manner. *New Biotechnol.* **2017**, *37*, 24–38. [CrossRef] [PubMed]

5. Obruca, S.; Sedlacek, P.; Mravec, F.; Samek, O.; Marova, I. Evaluation of 3-hydroxybutyrate as an enzyme-protective agent against heating and oxidative damage and its potential role in stress response of poly(3-hydroxybutyrate) accumulating cells. *Appl. Microbiol. Biotechnol.* **2016**, *100*, 1365–1376. [CrossRef] [PubMed]

6. Ayub, N.D.; Pettinari, M.J.; Ruiz, J.A.; López, N.I. A polyhydroxybutyrate-producing *Pseudomonas* sp. isolated from Antarctic environments with high stress resistance. *Curr. Microbiol.* **2004**, *49*, 170–174. [CrossRef] [PubMed]

7. Obruca, S.; Marova, I.; Stankova, M.; Mravcova, L.; Svoboda, Z. Effect of ethanol and hydrogen peroxide on poly(3-hydroxybutyrate) biosynthetic pathway in *Cupriavidus necator* H16. *World J. Microbiol. Biotechnol.* **2010**, *26*, 1261–1267. [CrossRef] [PubMed]

8. Steinbüchel, A.; Valentin, H.E. Diversity of bacterial polyhydroxyalkanoic acids. *FEMS Microbiol. Lett.* **1995**, *128*, 219–228. [CrossRef]

9. Zinn, M.; Witholt, B.; Egli, T. Occurrence, synthesis and medical application of bacterial polyhydroxyalkanoate. *Adv. Drug Deliv. Rev.* **2001**, *53*, 5–21. [CrossRef]

10. Pérez Amaro, L.; Chen, H.; Barghini, A.; Corti, A.; Chiellini, E. High performance compostable biocomposites based on bacterial polyesters suitable for injection molding and blow extrusion. *Chem. Biochem. Eng. Q.* **2015**, *29*, 261–274. [CrossRef]

11. Kovalcik, A.; Machovsky, M.; Kozakova, Z.; Koller, M. Designing packaging materials with viscoelastic and gas barrier properties by optimized processing of poly(3-hydroxybutyrate-*co*-3-hydroxyvalerate) with lignin. *React. Funct. Polym.* **2015**, *94*, 25–34. [CrossRef]

12. Koller, M. Poly(hydroxyalkanoates) for food packaging: Application and attempts towards implementation. *Appl. Food Biotechnol.* **2014**, *1*, 3–15.

13. Khosravi-Darani, K.; Bucci, D.Z. Application of poly(hydroxyalkanoate) in food packaging: Improvements by nanotechnology. *Chem. Biochem. Eng. Q.* **2015**, *29*, 275–285. [CrossRef]

14. Martínez-Sanz, M.; Villano, M.; Oliveira, C.; Albuquerque, M.G.; Majone, M.; Reis, M.A.M.; Lopez-Rubio, A.; Lagaron, J.M. Characterization of polyhydroxyalkanoates synthesized from microbial mixed cultures and of their nanobiocomposites with bacterial cellulose nanowhiskers. *New Biotechnol.* **2014**, *31*, 364–376. [CrossRef] [PubMed]

15. Narodoslawsky, M.; Shazad, K.; Kollmann, R.; Schnitzer, H. LCA of PHA production–Identifying the ecological potential of bio-plastic. *Chem. Biochem. Eng. Q.* **2015**, *29*, 299–305. [CrossRef]

16. Dietrich, K.; Dumont, M.J.; Del Rio, L.F.; Orsat, V. Producing PHAs in the bioeconomy—Towards a sustainable bioplastic. *Sustain. Prod. Consum.* **2017**, *9*, 58–70. [CrossRef]

17. Koller, M. Poly(hydroxyalkanoate) (PHA) biopolyesters: Production, Performance and processing aspects. *Chem. Biochem. Eng. Q.* **2015**, *29*, 261.

18. Koller, M.; Niebelschütz, H.; Braunegg, G. Strategies for recovery and purification of poly[(*R*)-3-hydroxyalkanoates] (PHA) biopolyesters from surrounding biomass. *Eng. Life Sci.* **2013**, *13*, 549–562. [CrossRef]

19. Madkour, M.H.; Heinrich, D.; Alghamdi, M.A.; Shabbaj, I.I.; Steinbüchel, A. PHA recovery from biomass. *Biomacromolecules* **2013**, *14*, 2963–2972. [CrossRef] [PubMed]

20. Murugan, P.; Han, L.; Gan, C.Y.; Maurer, F.H.; Sudesh, K. A new biological recovery approach for PHA using mealworm, *Tenebrio molitor*. *J. Biotechnol.* **2016**, *239*, 98–105. [CrossRef] [PubMed]

21. Rosengart, A.; Cesário, M.T.; de Almeida, M.C.M.; Raposo, R.S.; Espert, A.; de Apodaca, E.D.; da Fonseca, M.M.R. Efficient P(3HB) extraction from *Burkholderia sacchari* cells using non-chlorinated solvents. *Biochem. Eng. J.* **2015**, *103*, 39–46. [CrossRef]

22. Kaur, G.; Roy, I. Strategies for large-scale production of polyhydroxyalkanoates. *Chem. Biochem. Eng. Q.* **2015**, *29*, 157–172. [CrossRef]

23. Koller, M.; Muhr, A. Continuous production mode as a viable process-engineering tool for efficient poly(hydroxyalkanoate) (PHA) bio-production. *Chem. Biochem. Eng. Q.* **2014**, *28*, 65–77.

24. Sindhu, R.; Pandey, A.; Binod, P. Solid-state fermentation for the production of poly(hydroxyalkanoates). *Chem. Biochem. Eng. Q.* **2015**, *29*, 173–181. [CrossRef]

25. Haas, C.; El-Najjar, T.; Virgolini, N.; Smerilli, M.; Neureiter, M. High cell-density production of poly(3-hydroxybutyrate) in a membrane bioreactor. *New Biotechnol.* **2017**, *37*, 117–122. [CrossRef] [PubMed]

26. Novak, M.; Koller, M.; Braunegg, M.; Horvat, P. Mathematical modelling as a tool for optimized PHA production. *Chem. Biochem. Eng. Q.* **2015**, *29*, 183–220. [CrossRef]

27. Koller, M.; Bona, R.; Chiellini, E.; Fernandes, E.G.; Horvat, P.; Kutschera, C.; Hesse, P.; Braunegg, G. Polyhydroxyalkanoate production from whey by *Pseudomonas hydrogenovora*. *Bioresour. Technol.* **2008**, *99*, 4854–4863. [CrossRef] [PubMed]

28. Obruca, S.; Benesova, P.; Marsalek, L.; Marova, I. Use of lignocellulosic materials for PHA production. *Chem. Biochem. Eng. Q.* **2015**, *29*, 135–144. [CrossRef]

29. Cesário, M.T.; Raposo, R.S.; de Almeida, M.C.M.; Van Keulen, F.; Ferreira, B.S.; Telo, J.P.; da Fonseca, M.M.R. Production of poly(3-hydroxybutyrate-*co*-4-hydroxybutyrate) by *Burkholderia sacchari* using wheat straw hydrolysates and gamma-butyrolactone. *Int. J. Biol. Macromol.* **2014**, *71*, 59–67. [CrossRef] [PubMed]

30. Haas, C.; Steinwandter, V.; Diaz De Apodaca, E.; Maestro Madurga, B.; Smerilli, M.; Dietrich, T.; Neureiter, M. Production of PHB from chicory roots—Comparison of three *Cupriavidus necator* strains. *Chem. Biochem. Eng. Q.* **2015**, *29*, 99–112. [CrossRef]

31. Muhr, A.; Rechberger, E.M.; Salerno, A.; Reiterer, A.; Schiller, M.; Kwiecień, M.; Adamus, G.; Kowalczuk, M.; Strohmeier, K.; Schober, S.; et al. Biodegradable latexes from animal-derived waste: Biosynthesis and characterization of mcl-PHA accumulated by *Ps. citronellolis*. *React. Funct. Polym.* **2013**, *73*, 1391–1398. [CrossRef]

32. Muhr, A.; Rechberger, E.M.; Salerno, A.; Reiterer, A.; Malli, K.; Strohmeier, K.; Schober, S; Mittelbach, M.; Koller, M. Novel description of mcl-PHA biosynthesis by *Pseudomonas chlororaphis* from animal-derived waste. *J. Biotechnol.* **2013**, *165*, 45–51. [CrossRef] [PubMed]

33. Titz, M.; Kettl, K.H.; Shahzad, K.; Koller, M.; Schnitzer, H.; Narodoslawsky, M. Process optimization for efficient biomediated PHA production from animal-based waste streams. *Clean Technol. Environ. Policy* **2012**, *14*, 495–503. [CrossRef]

34. Obruca, S.; Marova, I.; Snajdar, O.; Mravcova, L.; Svoboda, Z. Production of poly(3-hydroxybutyrate-*co*-3-hydroxyvalerate) by *Cupriavidus necator* from waste rapeseed oil using propanol as a precursor of 3-hydroxyvalerate. *Biotechnol. Lett.* **2010**, *32*, 1925–1932. [CrossRef] [PubMed]

35. Obruca, S.; Snajdar, O.; Svoboda, Z.; Marova, I. Application of random mutagenesis to enhance the production of polyhydroxyalkanoates by *Cupriavidus necator* H16 on waste frying oil. *World J. Microbiol. Biotechnol.* **2013**, *29*, 2417–2428. [CrossRef] [PubMed]

36. Walsh, M.; O'Connor, K.; Babu, R.; Woods, T.; Kenny, S. Plant oils and products of their hydrolysis as substrates for polyhydroxyalkanoate synthesis. *Chem. Biochem. Eng. Q.* **2015**, *29*, 123–133. [CrossRef]

37. Hermann-Krauss, C.; Koller, M.; Muhr, A.; Fasl, H.; Stelzer, F.; Braunegg, G. Archaeal production of polyhydroxyalkanoate (PHA) co-and terpolyesters from biodiesel industry-derived by-products. *Archaea* **2013**, *2013*. [CrossRef] [PubMed]

38. Cavalheiro, J.M.; Raposo, R.S.; de Almeida, M.C.M.; Cesário, M.T.; Sevrin, C.; Grandfils, C.; Da Fonseca, M.M.R. Effect of cultivation parameters on the production of poly(3-hydroxybutyrate-*co*-4-hydroxybutyrate) and poly(3-hydroxybutyrate-4-hydroxybutyrate-3-hydroxyvalerate) by *Cupriavidus necator* using waste glycerol. *Bioresour. Technol.* **2012**, *111*, 391–397. [CrossRef] [PubMed]

39. Rodríguez-Contreras, A.; Koller, M.; Miranda-de Sousa Dias, M.; Calafell-Monfort, M.; Braunegg, G.; Marqués-Calvo, M.S. Influence of glycerol on poly(3-hydroxybutyrate) production by *Cupriavidus necator* and *Burkholderia sacchari*. *Biochem. Eng. J.* **2015**, *94*, 50–57. [CrossRef]

40. Koller, M.; Marsalek, L. Potential of diverse prokaryotic organisms for glycerol-based polyhydroxyalkanoate production. *Appl. Food Biotechnol.* **2015**, *2*, 3–15.

41. Obruca, S.; Petrik, S.; Benesova, P.; Svoboda, Z.; Eremka, L.; Marova, I. Utilization of oil extracted from spent coffee grounds for sustainable production of polyhydroxyalkanoates. *Appl. Microbiol. Biotechnol.* **2014**, *98*, 5883–5890. [CrossRef] [PubMed]

42. Obruca, S.; Benesova, P.; Petrik, S.; Oborna, J.; Prikryl, R.; Marova, I. Production of polyhydroxyalkanoates using hydrolysate of spent coffee grounds. *Process Biochem.* **2014**, *49*, 1409–1414. [CrossRef]

43. Carvalho, G.; Oehmen, A.; Albuquerque, M.G.; Reis, M.A. The relationship between mixed microbial culture composition and PHA production performance from fermented molasses. *New Biotechnol.* **2014**, *31*, 257–263. [CrossRef] [PubMed]

44. Khosravi-Darani, K.; Mokhtari, Z.B.; Amai, T.; Tanaka, K. Microbial production of poly(hydroxybutyrate) from C1 carbon sources. *Appl. Microbiol. Biotechnol.* **2013**, *97*, 1407–1424. [CrossRef] [PubMed]

45. Drosg, B.; Fritz, I.; Gattermayr, F.; Silvestrini, L. Photo-autotrophic production of poly(hydroxyalkanoates) in cyanobacteria. *Chem. Biochem. Eng. Q.* **2015**, *29*, 145–156. [CrossRef]

46. Koller, M.; Marsalek, L. Cyanobacterial Polyhydroxyalkanoate Production: *Status Quo* and *Quo Vadis*? *Curr. Biotechnol.* **2015**, *4*, 464–480. [CrossRef]
47. Tanaka, K.; Miyawaki, K.; Yamaguchi, A.; Khosravi-Darani, K.; Matsusaki, H. Cell growth and P(3HB) accumulation from CO_2 of a carbon monoxide-tolerant hydrogen-oxidizing bacterium, *Ideonella* sp. O-1. *Appl. Microbiol. Biotechnol.* **2011**, *92*, 1161–1169. [CrossRef] [PubMed]
48. Nonato, R.; Mantelatto, P.; Rossell, C. Integrated production of biodegradable plastic, sugar and ethanol. *Appl. Microbiol. Biotechnol.* **2001**, *57*, 1–5. [PubMed]
49. Brämer, C.O.; Vandamme, P.; da Silva, L.F.; Gomez, J.G.; Steinbüchel, A. Polyhydroxyalkanoate-accumulating bacterium isolated from soil of a sugar-cane plantation in Brazil. *Int. J. Syst. Evol. Microbiol.* **2001**, *51*, 1709–1713. [CrossRef] [PubMed]
50. Nascimento, V.M.; Silva, L.F.; Gomez, J.G.C.; Fonseca, G.G. Growth of *Burkholderia sacchari* LFM 101 cultivated in glucose, sucrose and glycerol at different temperatures. *Sci. Agric.* **2016**, *73*, 429–433. [CrossRef]
51. Alexandrino, P.M.R.; Mendonça, T.T.; Bautista, L.P.G.; Cherix, J.; Lozano-Sakalauskas, G.C.; Fujita, A.; Ramos Filho, E.; Long, P.; Padilla, G.; Taciro, M.K.; et al. Draft genome sequence of the polyhydroxyalkanoate-producing bacterium *Burkholderia sacchari* LMG 19450 isolated from Brazilian sugarcane plantation soil. *Genome Announc.* **2015**, *3*, e00313-15. [CrossRef] [PubMed]
52. Raposo, R.S.; de Almeida, M.C.M.; de Oliveira, M.D.C.M.; da Fonseca, M.M.; Cesário, M.T. A *Burkholderia sacchari* cell factory: Production of poly-3-hydroxybutyrate, xylitol and xylonic acid from xylose-rich sugar mixtures. *New Biotechnol.* **2017**, *34*, 12–22. [CrossRef] [PubMed]
53. Cesário, M.T.; Raposo, R.S.; de Almeida, M.C.M.; van Keulen, F.; Ferreira, B.S.; da Fonseca, M.M.R. Enhanced bioproduction of poly-3-hydroxybutyrate from wheat straw lignocellulosic hydrolysates. *New Biotechnol.* **2014**, *31*, 104–113. [CrossRef] [PubMed]
54. Lopes, M.S.G.; Gomez, J.G.C.; Silva, L.F. Cloning and overexpression of the xylose isomerase gene from *Burkholderia sacchari* and production of polyhydroxybutyrate from xylose. *Can. J. Microbiol.* **2009**, *55*, 1012–1015. [CrossRef] [PubMed]
55. Rocha, R.C.; da Silva, L.F.; Taciro, M.K.; Pradella, J.G. Production of poly(3-hydroxybutyrate-*co*-3-hydroxyvalerate) P(3HB-*co*-3HV) with a broad range of 3HV content at high yields by *Burkholderia sacchari* IPT 189. *World J. Microbiol. Biotechnol.* **2008**, *24*, 427–431. [CrossRef]
56. Mendonça, T.T.; Gomez, J.G.C.; Buffoni, E.; Sánchez Rodriguez, R.J.; Schripsema, J.; Lopes, M.S.G.; Silva, L.F. Exploring the potential of *Burkholderia sacchari* to produce polyhydroxyalkanoates. *J. Appl. Microbiol.* **2014**, *116*, 815–829. [CrossRef] [PubMed]
57. Da Cruz Pradella, J.G.; Taciro, M.K.; Mateus, A.Y.P. High-cell-density poly(3-hydroxybutyrate) production from sucrose using *Burkholderia sacchari* culture in airlift bioreactor. *Bioresour. Technol.* **2010**, *101*, 8355–8360. [CrossRef] [PubMed]
58. Brämer, C.O.; Silva, L.F.; Gomez, J.G.C.; Priefert, H.; Steinbüchel, A. Identification of the 2-methylcitrate pathway involved in the catabolism of propionate in the polyhydroxyalkanoate-producing strain *Burkholderia sacchari* IPT101T and analysis of a mutant accumulating a copolyester with higher 3-hydroxyvalerate content. *Appl. Environ. Microbiol.* **2002**, *68*, 271–279. [CrossRef] [PubMed]
59. Silva, L.F.; Gomez, J.G.C.; Oliveira, M.S.; Torres, B.B. Propionic acid metabolism and poly-3-hydroxybutyrate-*co*-3-hydroxyvalerate (P3HB-*co*-3HV) production by *Burkholderia* sp. *J. Biotechnol.* **2000**, *76*, 165–174. [CrossRef]
60. Mendonça, T.T.; Tavares, R.R.; Cespedes, L.G.; Sánchez-Rodriguez, R.J.; Schripsema, J.; Taciro, M.K.; Gomez, J.G.C.; Silva, L.F. Combining molecular and bioprocess techniques to produce poly(3-hydroxybutyrate-*co*-3-hydroxyhexanoate) with controlled monomer composition by *Burkholderia sacchari*. *Int. J. Biol. Macromol.* **2017**, *98*, 654–663. [CrossRef] [PubMed]
61. Silva, L.F.; Taciro, M.K.; Raicher, G.; Piccoli, R.A.M.; Mendonça, T.T.; Lopes, M.S.G.; Gomez, J.G.C. Perspectives on the production of polyhydroxyalkanoates in biorefineries associated with the production of sugar and ethanol. *Int. J. Biol. Macromol.* **2014**, *71*, 2–7. [CrossRef] [PubMed]
62. Küng, W. Wachstum und Poly-D-(-)-3-Hydroxybuttersäure-Akkumulation bei *Alcaligenes latus*. Diploma Thesis, Graz University of Technology, Graz, Austria, 1982.
63. Braunegg, G.; Sonnleitner, B.; Lafferty, R. A rapid gaschromatographic method for the determination of poly-β-hydroxybutyric acid in microbial biomass. *Eur. J. Appl. Microbiol.* **1978**, *6*, 29–37. [CrossRef]

64. Kunioka, M.; Kawaguchi, Y.; Doi, Y. Production of biodegradable copolyesters of 3-hydroxybutyrate and 4-hydroxybutyrate by *Alcaligenes eutrophus*. *Appl. Microbiol. Biotechnol.* **1989**, *30*, 569–573. [CrossRef]

65. Choi, M.H.; Yoon, S.C.; Lenz, R.W. Production of poly(3-hydroxybutyric acid-co-4-hydroxybutyric acid) and poly(4-hydroxybutyric acid) without subsequent degradation by *Hydrogenophaga pseudoflava*. *Appl. Environ. Microbiol.* **1999**, *65*, 1570–1577. [PubMed]

66. Valentin, H.E.; Zwingmann, G.; Schönebaum, A.; Steinbüchel, A. Metabolic pathway for biosynthesis of poly(3-hydroxybutyrate-co-4-hydroxybutyrate) from 4-hydroxybutyrate by *Alcaligenes eutrophus*. *Eur. J. Biochem.* **1995**, *227*, 43–60. [CrossRef] [PubMed]

67. Lee, Y.H.; Kang, M.S.; Jung, Y.M. Regulating the molar fraction of 4-hydroxybutyrate in poly(3-hydroxybutyrate-4-hydroxybutyrate) biosynthesis by *Ralstonia eutropha* using propionate as a stimulator. *J. Biosci. Bioeng.* **2000**, *89*, 380–383. [CrossRef]

68. Koller, M.; Bona, R.; Braunegg, G.; Hermann, C.; Horvat, P.; Kroutil, M.; Martinz, J.; Neto, J.; Pereira, L.; Varila, P. Production of polyhydroxyalkanoates from agricultural waste and surplus materials. *Biomacromolecules* **2005**, *6*, 561–565. [CrossRef] [PubMed]

69. Madden, L.A.; Anderson, A.J.; Shah, D.T.; Asrar, J. Chain termination in polyhydroxyalkanoate synthesis: Involvement of exogenous hydroxy-compounds as chain transfer agents. *Int. J. Biol. Macromol.* **1999**, *25*, 43–53. [CrossRef]

70. Atlić, A.; Koller, M.; Scherzer, D.; Kutschera, C.; Grillo-Fernandes, E.; Horvat, P.; Chiellini, E.; Braunegg, G. Continuous production of poly([R]-3-hydroxybutyrate) by *Cupriavidus necator* in a multistage bioreactor cascade. *Appl. Microbiol. Biotechnol.* **2011**, *91*, 295–304. [CrossRef] [PubMed]

bioengineering

MDPI

Article

Additive Manufacturing of Poly(3-hydroxybutyrate-co-3-hydroxyhexanoate)/ poly(ε-caprolactone) Blend Scaffolds for Tissue Engineering

Dario Puppi, Andrea Morelli and Federica Chiellini *

BIOLab Research Group, Department of Chemistry and Industrial Chemistry, University of Pisa,
UdR INSTM Pisa, via Moruzzi 13, 56124 Pisa, Italy; d.puppi@dcci.unipi.it (D.P.); a.morelli@dcci.unipi.it (A.M.)
* Correspondence: federica.chiellini@unipi.it; Tel.: +39-050-221-9333

Academic Editor: Martin Koller
Received: 27 April 2017; Accepted: 21 May 2017; Published: 24 May 2017

Abstract: Additive manufacturing of scaffolds made of a polyhydroxyalkanoate blended with another biocompatible polymer represents a cost-effective strategy for combining the advantages of the two blend components in order to develop tailored tissue engineering approaches. The aim of this study was the development of novel poly(3-hydroxybutyrate-*co*-3-hydroxyhexanoate)/ poly(ε-caprolactone) (PHBHHx/PCL) blend scaffolds for tissue engineering by means of computer-aided wet-spinning, a hybrid additive manufacturing technique suitable for processing polyhydroxyalkanoates dissolved in organic solvents. The experimental conditions for processing tetrahydrofuran solutions containing the two polymers at different concentrations (PHBHHx/PCL weight ratio of 3:1, 2:1 or 1:1) were optimized in order to manufacture scaffolds with predefined geometry and internal porous architecture. PHBHHx/PCL scaffolds with a 3D interconnected network of macropores and a local microporosity of the polymeric matrix, as a consequence of the phase inversion process governing material solidification, were successfully fabricated. As shown by scanning electron microscopy, thermogravimetric, differential scanning calorimetric and uniaxial compressive analyses, blend composition significantly influenced the scaffold morphological, thermal and mechanical properties. In vitro biological characterization showed that the developed scaffolds were able to sustain the adhesion and proliferation of MC3T3-E1 murine preosteoblast cells. The additive manufacturing approach developed in this study, based on a polymeric solution processing method avoiding possible material degradation related to thermal treatments, could represent a powerful tool for the development of customized PHBHHx-based blend scaffolds for tissue engineering.

Keywords: polyhydroxyalkanoates; poly(3-hydroxybutyrate-co-3-hydroxyhexanoate); poly(ε-caprolactone); polymers blend; tissue engineering; scaffolds; additive manufacturing; computer-aided wet-spinning

---

1. Introduction

Tissue engineering is a growing research area, with a few successful clinical results, aimed at developing reliable alternatives to conventional surgical strategies (e.g., auto- and allogenic tissue transplantation or artificial prosthesis implantation) for the treatment of human tissue and organ failure caused by defects, injuries or other types of damage [1]. Tissue engineering relies on the combination of cells, biomaterials and bioactive molecules to generate replacement biological tissues and organs for a wide range of medical conditions. The most common approach involves the employment of a highly porous biodegradable support, commonly referred to as the scaffold, which acts as a temporary template providing a cell adhesion substrate and mechanical support, and guiding the regeneration processes [2]. In the last two decades, a great variety of biodegradable materials and processing techniques have been investigated for the development of scaffolds with proper physico-chemical properties as well as macro-, micro- and nano-architecture features suitable for tissue growth in three dimensions [3].

Polyhydroxyalkanoates (PHAs) are microbial aliphatic polyesters widely investigated for biomedical applications due to their biodegradability and biocompatibility, as well as the wide range of mechanical and processing properties of the numerous homopolymers and copolymers belonging to this class of renewable polymers [4]. Different articles have reported on poly[(R)-3-hydroybutyrate] (PHB) and poly[(R)-3-hydroxybutyrate-*co*-(R)-3-hydroxyvalerate] (PHBV) investigations, both in vitro and in vivo, for bone tissue regeneration approaches [5–7]. Due to the relatively long alkyl side chain, poly[(R)-3-hydroxybutyrate-*co*-(R)-3-hydroxyhexanoate) (PHBHHx) exhibits lower crystallinity, a broader processing window and higher elasticity compared with PHB and PHBV [8]. Among the different investigated biomedical applications, PHBHHx has been proposed as scaffolding material for bone regeneration thanks to its piezoelectric behavior and cytocompatibility when cultured with osteoblasts and bone marrow cells [9–14]. In addition, recent articles showed that PHBHHx in the form of microgrooved membrane [15], aligned nanofibers [16] or carbon nanotubes-loaded composite materials [17] well supports the osteogenesis of human mesenchymal stem cells.

As defined by the American Society for Testing and Materials (ASTM), Additive Manufacturing (AM) refers to the process of joining materials to make objects from three-dimensional (3D) model data, usually layer upon layer, as opposed to subtractive manufacturing methodologies [18]. The introduction of a number of AM techniques, such as stereolithography and fused deposition modeling, into the tissue engineering field has allowed the enhancement of control over scaffold structure at different size scales (from macro- to micrometric scale) in terms of external shape and porous structure [19]. They involve a computer-controlled layered manufacturing process based on a sequential delivery of energy and/or materials starting from a 3D digital model to build up 3D polymeric scaffolds with a predefined geometry and internal porosity. Advanced computer-aided design and manufacturing approaches enable a high degree of automation, good accuracy and reproducibility for the fabrication of clinically-sized, anatomically-shaped scaffolds with a tailored porous structure characterized by a fully interconnected network of pores with customized size and shape. However, despite the promising results and widespread research on PHAs for tissue engineering applications, their narrow melt processing temperature window [20,21] has hindered the application of AM techniques for their processing into 3D porous scaffolds. Computer-aided wet-spinning (CAWS), a hybrid AM technique based on the computer-controlled deposition of a solidifying polymeric fiber extruded directly into a coagulation bath, was recently applied to process PHBHHx into 3D scaffolds with tailored geometry and networks of macropores as well as a homogenous microporous matrix [22,23].

The aim of this study was to investigate the suitability of the CAWS technique for the fabrication of scaffolds made of PHBHHx blended with poly(ε-caprolactone) (PCL). PCL is an aliphatic polyester that has been widely investigated for biomedical applications receiving FDA approval and CE Mark registration for a number of drug delivery and medical device applications [24]. Thanks to its good processing properties, tunable mechanical properties and slow biodegradation, PCL is seen as one of the most versatile scaffolding materials for the development of long-term biodegradable bone implants. Recent studies have investigated the blending of PHAs with PCL and other synthetic polyesters as a cost-effective strategy for combining the advantages of the two polymers and achieving additional desirable properties [21,25,26]. As an example, a research activity on PCL/PHBHHx blend membranes by solvent processing showed that by optimizing the weight ratio between the two components it was possible to enhance the resulting mechanical properties in comparison with PCL and PHBHHx alone [27]. Although the great versatility of the CAWS technique in customizing PHBHHx scaffold's shape and internal architecture, in the case of inter-fiber deposition distances larger than 200 μm (i.e., 500 and 1000 μm), a well-defined porosity along the Z axis was not achieved due to the slow solidification of the coagulating fiber. On the other hand, the optimization of PCL processing by CAWS has enabled the employment of large inter-fiber deposition distances (i.e., 500 and 1000 μm) for the fabrication of 3D scaffolds with a homogeneous porosity in the cross-section characterized by a Z axis pore size in the range of hundreds of micrometers [28–30]. Since a pore size larger than 100 μm is recommended to achieve enhanced bone tissue regeneration and vascularization [31], blending PHBHHx with a polymer showing better processing properties was investigated during the present study as an effective strategy to develop scaffolds meeting the aforementioned structural parameters requirement. For this purpose, the CAWS conditions for the fabrication of PHBHHx/PCL scaffolds with different ratios between the two blend components were investigated. Optimized PHBHHx/PCL scaffold prototypes were characterized in comparison with PHBHHx scaffolds for their morphology by means of scanning electron microscopy (SEM) under backscattered electron imaging, thermal properties by means of thermogravimetric analysis (TGA) and differential scanning calorimetry (DSC), and mechanical properties under compression using a uniaxial testing machine. The scaffold's biocompatibility was evaluated in vitro by employing the MC3T3-E1 murine preosteoblast cell line. Cell response, in terms of viability, proliferation and morphology was investigated by tetrazolium salts (WST-1) and confocal laser scanning microscopy (CLSM).

2. Materials and Methods

2.1. Materials

Poly(ε-caprolactone) (PCL, CAPA 6800, Mw = 80,000 g·mol^{-1}) was supplied by Perstorp UK Ltd (Warrington, Cheshire, UK) and used as received. Poly(3-hydroxybutyrate-*co*-3-hydroxyhexanoate) (PHBHHx, 12% mol HHx, Mw = 300,000 g·mol^{-1}) was kindly supplied by Tsinghua University (Beijing, China). PHBHHx was purified before use according to the following procedure: (i) the polymer was dissolved in 1,4 dioxane (5% w/v) under stirring at room temperature for 1 h; (ii) the solution was filtered under vacuum using filter paper; (iii) the filtrate was slowly dropped into 10-fold volume water to precipitate PHBHHx; (iv) after precipitation the polymer was collected by filtering; (v) the polymer was washed with distilled water and then ethanol, and vacuum dried and stored in a desiccator. All the solvents and chemical reagents were purchased from Sigma-Aldrich (Italy) and used as received without further purification.

2.2. Scaffolds Fabrication

PHBHHx solutions were prepared by dissolving the polymer in tetrahydrofuran (THF) at 32 °C under stirring for 2 h at a concentration of 25% w/v. For the preparation of PHBHHx/PCL solutions, PCL was dissolved in THF at 32 °C under stirring for 2 h and then the desired amount of PHBHHx was added to the polymer solution. The mixture was left under stirring for 2 h at 32 °C until a homogenous solution was obtained. Solutions with different PHBHHx/PCL weight ratios (3:1, 2:1 and 1:1) and a total concentration of the polymeric phase of 12% w/v were prepared.

Scaffolds were fabricated by means of a subtractive rapid prototyping system (MDX 40A, Roland MID EUROPE, Acquaviva Picena, Italy) modified in-house by replacing the milling head unit with a programmable syringe pump system (NE-1000; New Era Pump Systems Inc., Wantagh, NY, USA) to enable the deposition of polymeric solutions with a controlled 3D pattern (Figure 1) [29]. The 3D geometrical scaffold parameters were designed using an algorithm developed in Matlab software (The Mathworks, Inc., Natick, MA, USA). The desired polymeric solution was placed into a glass syringe fitted with a metallic needle (Gauge 23) and injected at a controlled feeding rate directly into an ethanol coagulation bath by using the syringe pump. Scaffold fabrication was carried out by employing a deposition trajectory aimed at the production of scaffolds with a 0–90° lay-down pattern, distance between fiber axis of 500 μm and layer thickness of 100 μm. The optimized initial distance between the tip of the needle and the bottom of the beaker (Z$_0$) was 1.5 mm. The effect of different processing parameters, such as the deposition velocity (V$_{dep}$) and the solution feed rate (F), on fiber collection and morphology was evaluated to produce blend scaffolds with a different PHBHHx/PCL ratio (Table 1). By employing the optimized fabrication parameters, cylindrical samples with a designed diameter of

15 mm and height of 5 mm were fabricated. The samples were removed from the coagulation bath, left under a fume hood for 24 h, placed in a vacuum chamber at about 0.5 mbar for 48 h and then stored in a desiccator for at least 72 h before characterization.

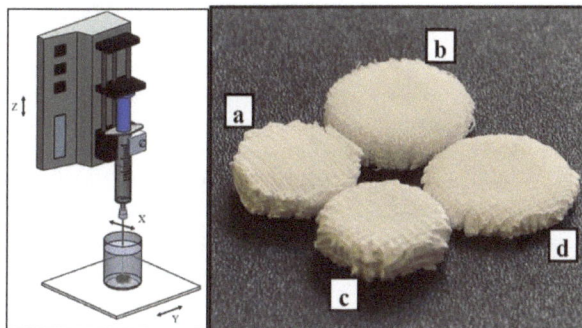

Figure 1. Schematics of the computer-aided wet-spinning (CAWS) process (left); representative image of the developed scaffolds (right): (**a**) PHBHHx; (**b**) PHBHHx/PCL 3:1; (**c**) PHBHHx/PCL 2:1; (**d**) PHBHHx/PCL 1:1.

Table 1. Optimized processing parameters, and scaffold structural parameters obtained from scanning electron microscopy (SEM) analysis.

Sample	F (mL·h^{-1})	Vdep (mm·min^{-1})	Fiber Diameter (μm)	Pore Size (μm)
PHBHHx	0.5	300	88 ± 12	485 ± 40
PHBHHx/PCL 3:1	1	560	116 ± 12	493 ± 29
PHBHHx/PCL 2:1	1	560	114 ± 15	470 ± 46
PHBHHx/PCL 1:1	1	560	120 ± 11	484 ± 18

Morphological parameters expressed as average ± standard deviation.

2.3. Morphological Characterization

The top-view and cross section (obtained by fracture in liquid nitrogen) of the scaffolds were analyzed by means of scanning electron microscopy (SEM, JEOL JSM 300, Tokyo, Japan) under backscattered electron imaging. The average fiber diameter and pore size, defined as inter-fiber distance, were measured by means of ImageJ 1.43u software on top-view micrographs with a 50X magnification. Data were calculated over 20 measurements per scaffold.

2.4. Thermal Analysis

Thermal properties of the scaffolds were evaluated by means of thermogravimetric analysis (TGA) and differential scanning calorimetry (DSC). TGA was performed using TGA Q500 instruments (TA Instruments, Milano, Italy) in the temperature range 30–600 °C, at a heating rate of 10 °C/min and under a nitrogen flow of 60 mL·min^{-1}. The scaffold's thermal decomposition were evaluated by analyzing weight and derivative weight profiles as functions of temperature. DSC analysis was performed using a Mettler DSC-822 instrument (Mettler Toledo, Novate Milanese (MI), Italy) in the range −100–200 °C, at a heating rate of 10 °C/min and a cooling rate of −20 °C/min, and under a nitrogen flow of 80 mL·min^{-1}. By considering the first and second heating cycle in the thermograms, glass transition temperature (T_g) was evaluated by analyzing the inflection point, while melting temperature (T_m) and enthalpy (ΔH) was evaluated by analyzing the endothermic peaks.

2.5. Mechanical Testing

The scaffold's mechanical properties were analyzed under compression using an Instron 5564 uniaxial testing machine (Instron Corporation, Norwood, MA, USA) equipped with a 2 kN load cell. After the treatment to remove residual solvents, as previously described, the samples were preconditioned at 25 °C and 50% of humidity for 48 h and then characterized at room temperature. The test was carried out on cylindrical samples with actual diameters of around 15 mm and actual heights of around 4 mm (50 layers). Six samples of each kind of scaffold were tested at a constant crosshead displacement of 0.4 mm·min^{-1} between two parallel steel plates up to 85% strain [32]. The stress was defined as the measured force divided by the total area of the apparent cross section of the scaffold, whilst the strain was evaluated as the ratio between the height variation and the initial height. Stress-strain curves were obtained from the software recording the data (Merlin, Series IX, Instron Corporation, Norwood, MA, USA). The compressive modulus was calculated as the slope of the initial linear region in the stress-strain curve, avoiding the toe region. Compressive yield strength and strain were considered at the yield point, and compressive strength was considered as the stress corresponding to 85% strain.

2.6. In Vitro Biological Evaluation

2.6.1. Cell Culture

Mouse calvaria-derived pre-osteoblast cell line MC3T3-E1 subclone 4 was obtained from the American Type Culture Collection (ATCC CRL-2593, Manassas, VA, USA) and cultured in Alpha Minimum Essential Medium (α—MEM, Sigma, Milan, Italy) supplemented with 2 mM·L-glutamine, 10% fetal bovine serum, 100 U/mL:100 μg/mL penicillin:streptomycin solution (GIBCO, Invitrogen Corporation, Milan, Italy) and antimycotic. Before experiments, cells were trypsinized with 0.25% trypsin-EDTA (GIBCO, Gaithersburg, MD, USA) solution and resuspended in complete α-MEM at a concentration of 3×10^4/mL. Scaffolds were seeded with 100 μL of cell suspension and the final volume was adjusted to 1 mL with complete medium. The specimens were then placed in an incubator with humidified atmosphere at 37 °C in 5% CO_2. Osteogenic differentiation was induced 24 h after seeding by culturing cells in osteogenic medium prepared with α–MEM supplemented with ascorbic acid (0.3 mM) and β—glycerolphosphate (10 mM). The culture medium was replaced every 48 h and biological characterizations were carried out weekly at days 7, 14, 21 and 28. Cells grown onto tissue culture polystyrene plates were used as control.

2.6.2. Cell Viability and Proliferation

Cell viability and proliferation were measured by using the (4-[3-(4-iodophenyl)-2-(4 nitrophenyl) -2H-5-tetrazolium]-1,3-benzene disulfonate) (WST-1) assay (Roche Molecular Biochemicals, Monza, Italy), which is based on the mitochondrial conversion of the tetrazolium salt WST-1 into soluble formazan in viable cells. WST-1 reagent diluted 1:10 was added to the culture and incubated for 4 h at 37 °C. Measurements of formazan dye absorbance were carried out with a Biorad microplate reader at 450 nm, with the reference wavelength at 655 nm. The in vitro biological test was performed on triplicate samples for each material.

2.6.3. Morphologic Characterizations by Confocal Laser Scanning Microscopy (CLSM)

The morphology of the cells grown on the prepared meshes was investigated by means of CLSM. Cells were fixed with 3.8% paraformaldehyde in PBS 0.01 M pH 7.4 (PBS 1X), permeabilized with a PBS 1X/Triton X-100 solution (0.2%) for 15 min and incubated with a solution of 4'-6-diamidino-2-phenylindole (DAPI; Invitrogen) and phalloidin-AlexaFluor488 (Invitrogen) in PBS 1X for 60 min at room temperature in the dark. After dye incubation, samples were washed with PBS 1X before being mounted on a glass slide and sealed with resin for microscopic observation. A Nikon Eclipse TE2000 inverted microscope equipped with an EZC1 confocal laser and Differential Interference Contrast

(DIC) apparatus was used to analyze the samples (Nikon, Tokyo, Japan). A 405 nm laser diode (405 nm emission) and an argon ion laser (488 nm emission) were used to excite DAPI and Alexa fluorophores, respectively. Images were captured with Nikon EZ-C1 software with identical settings for each sample. Images were further processed with GIMP (GNU Free Software Foundation) Image Manipulation Software and merged with Nikon ACT-2U software.

2.7. Statistical Analysis

The data are represented as mean ± standard deviation. Statistical differences were analyzed using one-way analysis of variance (ANOVA), and a Tukey test was used for post hoc analysis. A *p*-value < 0.05 was considered statistically significant.

3. Results and Discussion

3.1. Additive Manufacturing of Scaffolds

The fabrication process involved the deposition with a predefined pattern of an extruded polymeric solution into a coagulation bath to make a 3D scaffold using a layer-by-layer process. By optimizing the most influential manufacturing parameters, in terms of solution flow rate (F) and deposition velocity (V_{dep}) (Table 1), 3D cylindrical scaffolds were developed by processing solutions with different PHBHHx/PCL weight ratios (3:1, 2:1 or 1:1) (Figure 1).

Wet-spinning (WS) is a non-solvent-induced phase inversion technique suitable for the industrial production of continuous polymeric fibers through an immersion-precipitation process. Briefly, a polymeric solution is injected into a coagulation bath containing a non-solvent of the polymer, and the solution filament solidifies because of polymer desolvation caused by solvent/non-solvent exchange [33]. A number of studies have shown that 3D macroporous scaffolds made of synthetic or natural polymers can be obtained through physical bonding of fibers prefabricated by means of WS, using a glue and/or a thermomechanical treatment, or in a single-step fabrication process involving the continuous, randomly-oriented deposition of the solidifying fiber by means of a manually controlled motion of the coagulation bath [33–38]. Although 3D structures with high and interconnected porosity suitable for tissue regeneration processes have been developed, an accurate control over the scaffold macro- and microstructure has not been achieved by employing these methods. The CAWS technique has been proposed as a suitable AM approach to upgrade the fabrication process in terms of reproducibility, resolution, design freedom and automation degree [39]. Advanced scaffold structural features at different scale levels, in terms of external shape and internal porosity, have been developed by applying the CAWS technique to different biocompatible polymers, including PCL, three-arm star PCL, PHBHHx, chitosan/poly(γ-glutamate) polyelectrolyte complexes and a poly(ethylene oxide terephthalate)/poly(butylene terephthalate) block copolymer [22,23,28,29,40–44]. As previously discussed, the application of AM to PHAs is very limited due to the narrow thermal processing window of this class of polymers. The few exceptions of PHAs scaffolds manufactured by AM are represented by a set of nanocomposite PHA/tricalcium phosphate composite scaffolds fabricated by means of selective laser sintering [45–47] and PCL/PHBV blend scaffolds fabricated by fused deposition modeling [21]. The research activity reported in this study has led to the development of a novel AM process for the fabrication of PHA-based blend scaffolds by processing solutions containing PHBHHx and PCL in different ratios. This approach does not require thermal treatments that could cause polymer degradation as well as denaturation of bioactive agents possibly loaded into the scaffold [48]. The solvents employed are allowed by the European Medicine Agency as residues in medical products below recommended safety levels, and classified as solvents with low toxicity (i.e., ethanol) or to be limited (i.e., THF). Considering the small volume of THF and the high ethanol/THF volume ratio involved (>20), the high volatility of the two solvents, as well as the absence of thermal events in TGA and DSC curves that could be related to the evaporation of residual solvents in scaffolds, as discussed in one of the following sections, the process can be considered as meeting the basic requirements of good manufacturing practices.

3.2. Morphological Characterization

The morphology of the developed scaffolds was investigated by means of SEM analysis using backscattering electron imaging. Analysis of samples both in top view and cross section highlighted that the fabricated scaffolds were composed by a 3D layered structure of aligned fibers forming a fully interconnected network of macropores (Figure 2). Comparative analysis of SEM micrographs showed that the scaffold's fiber morphology and alignment were influenced by the composition of the wet-spinning solution. In addition, as observed in high magnification micrographs, the fibers had a microporous morphology both in the outer surface and in the cross-section due to the phase inversion process governing polymer solidification, as discussed elsewhere [28].

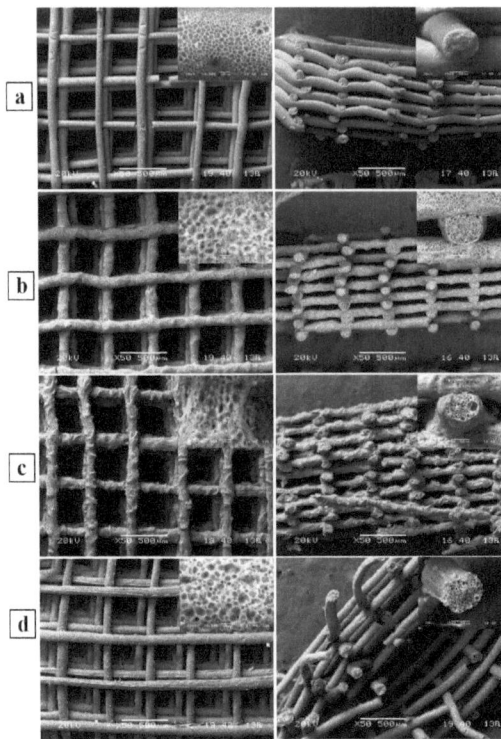

Figure 2. Representative top view (left) and cross-section (right) SEM micrographs of (**a**) PHBHHx; (**b**) PHBHHx/PCL 3:1, (**c**) PHBHHx/PCL 2:1, (**d**) PHBHHx/PCL 1:1. Inset high magnification micrographs show porosity of outer surface (left) and cross section (right) of single fibers.

PHBHHx/PCL blend scaffolds showed significantly larger fiber diameters (average value in the range of 100 to 135 μm) in comparison with plain PHBHHx scaffolds (average value of 88 μm) (Table 1). Differences in fiber diameter among the different blend scaffolds were not statistically significant. The pore size was in the range 400–500 μm, with no statistically significant differences among the different scaffolds.

Together with the possibility of easily loading a scaffold with a drug by simply adding it to the polymeric solution before processing [41], the main advantage of CAWS is represented by the obtainment of multi-scale scaffold porosities. In fact, porous structures fabricated by means of this technique are generally characterized by a fully interconnected network of macropores, with a size that can be tuned in the range of tens to hundreds of micrometers by varying the fiber lay-down pattern, and a local micro/nanosized porous morphology of the polymeric matrix that can be tailored by acting on different parameters of the phase separation process determining polymer solidification [39]. This multi-scale morphology control represents a powerful tool to tune key scaffold properties strictly related to porosity and surface roughness, such as biodegradation rate, mechanical behavior, and cell interactions.

3.3. Thermal Characterization

The thermal properties of the developed PHBHHx/PCL blend scaffolds were investigated by TGA and DSC in comparison with PHBHHx scaffolds and PCL raw polymer. TGA evaluation showed that the weight and derivative weight curves of the blend scaffolds were characterized by two main thermal decomposition events: the first one centered at around 290 °C ascribable to PHBHHx decomposition, and the other one centered at around 405 °C related to PCL decomposition (Figure 3). Thermal events relevant to evaporation of residual THF (boiling point of 66 °C) or ethanol (boiling point of 78 °C) were not detected. The thermograms of the PHBHHx scaffolds are characterized by a relatively high residue at 600 °C, that did not compromise scaffold's cytocompatibility as shown also by a previous study [22]. Nevertheless, future studies should investigate the reason and composition of this ash content and whether it may have an impact on use, particularly in extended biological testing.

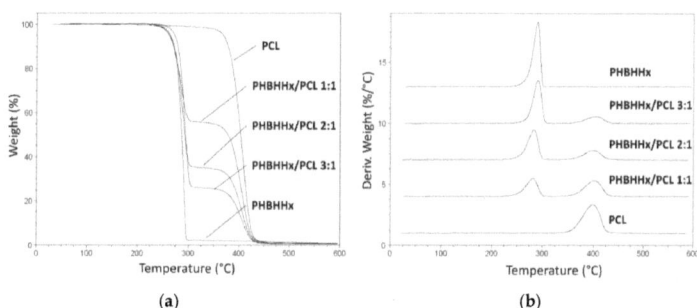

(a)   (b)

Figure 3. Thermogravimetric analysis (TGA) characterization: weight (**a**) and derivative weight (**b**) profiles vs temperature of the developed scaffolds.

The area under the first peak in derivative weight curves decreased on increasing PCL percentage, while that under the second peak increased by increasing PCL content. The resulting percentage weight losses during the first and second thermal events were close to the percentage weight in the starting polymeric solution of PHBHHx and PCL, respectively (Table 2).

Table 2. Data relevant to thermal decomposition obtained from TGA analysis.

Sample	1st Decomposition Step		2nd Decomposition STEP	
	Peak (°C)	Weight Loss (%)	Peak (°C)	Weight Loss (%)
PHBHHx	290.7 ± 1.8	97.6 ± 0.9	-	-
PHBHHx/PCL 3:1	291.3 ± 2.2	74.6 ± 1.2	405.9 ± 0.5	25.1 ± 0.4
PHBHHx/PCL 2:1	289.2 ± 1.4	65.4 ± 1.4	405.9 ± 0.7	34.2 ± 0.5
PHBHHx/PCL 1:1	285.2 ± 1.6	44.9 ± 0.8	405.3 ± 0.8	54.6 ± 0.8
PCL raw	-	-	406.6 ± 0.4	99.1 ± 0.4

Data expressed as average ± standard deviation ($n = 3$).

Representative DSC thermograms of the characterized samples are reported in Figure 4. The first heating cycle analysis was carried out to assess the thermal properties of the scaffolds in comparison to what was observed in the second heating cycle after blend melting and solidification to erase the prior thermal history. The glass transition and melting temperature of PCL (Tg_1 and Tm_1) and PHBHHx (Tg_2 and Tm_2), as well as their respective melting enthalpies (ΔH_1 and ΔH_2), were analyzed. An endothermic peak centered at around 60 °C ascribable to the melting of PCL crystalline domains is evident in both the first and second heating cycle thermograms of blend scaffolds. The endothermic peak related to the melting of PHBHHx crystalline domains is evident only in the first heating cycle thermograms, while in the second heating cycle thermograms, a pronounced glass transition of PHBHHx only is detectable. These results corroborate what was reported in previous articles about the appreciable crystalline degree of PHBHHx scaffolds by CAWS due to the relatively slow crystallization mechanism [22,23]. Polymer crystallinity influences different properties of scaffolds made of aliphatic polyesters, such as their mechanical behavior and biodegradation rate. Indeed, as widely reported in literature [3,49], crystalline and amorphous domains show different water diffusivity as well as different macromolecular deformation and rearrangement when subjected to mechanical solicitations. The quite broad endothermic peak in the first heating scan of PHBHHx scaffolds curve can be explained with the melting of different lamellar crystalline domains formed during polymer solidification, as suggested by previous articles on thermal characterization of PHBHHx films [50,51]. Endothermic peaks related to evaporation of residual solvents were not detectable in the first nor in the second heating scan.

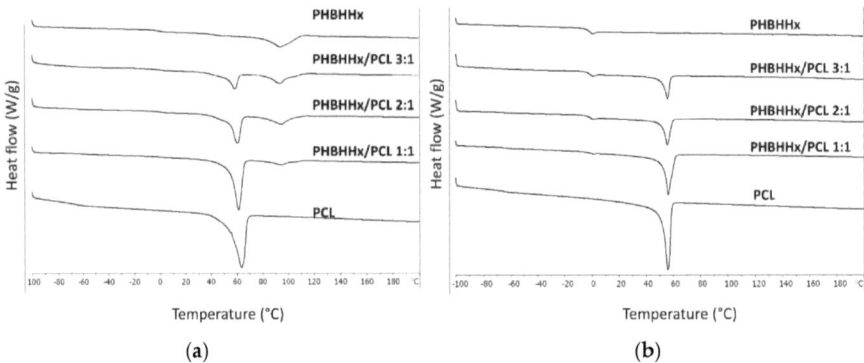

Figure 4. Representative differential scanning calorimetry (DSC) thermograms of the analyzed samples relevant to the first heating (**a**) and second heating (**b**) cycles.

By comparing data from either the first or the second DSC scan (Table 3), the effect of blend composition on endothermic peaks area was quantitatively confirmed through analysis of differences in enthalpies ΔH_1 and ΔH_2. In addition, Tg_1 and Tm_1 in the first heating cycle were significantly affected by PHBHHx/PCL ratio, in agreement with a previous article showing that by increasing PHBHHx content in the blend, the melting of the PCL component was shifted to lower temperatures [27], possibly due to a slight plasticization effect of PHBHHx on the PCL phase. Besides the previously mentioned effect on ΔH_1, differences in data from the second scan were not statistically significant [26].

Table 3. Data relevant to thermal characterization by DSC analysis.

Sample	1st Heating						2nd Heating			
	Tg$_1$ (°C)	Tm$_1$ (°C)	ΔH$_1$ (J/g)	Tg$_2$ (°C)	Tm$_2$ (°C)	ΔH$_2$ (J/g)	Tg$_1$ (°C)	Tm$_1$ (°C)	ΔH$_1$ (J/g)	Tg$_2$ (°C)
PHBHHx	—		—	−0.6 ± 0.2	93.7 ± 1.2	41.4 ± 1.8	—	—	—	−1.1 ± 0.6
PHBHHx/PCL 3:1	−70.7 ± 1.6	58.5 ± 0.9	18.6 ± 1.2	−0.2 ± 0.6	93.5 ± 1.6	18.4 ± 0.9	−66.2 ± 0.8	56.0 ± 1.2	15.4 ± 0.8	−1.1 ± 0.8
PHBHHx/PCL 2:1	−65.2 ± 1.4	60.7 ± 1.2	30.2 ± 1.8	−0.1 ± 0.3	94.3 ± 2.1	14.9 ± 1.1	−64.9 ± 1.3	55.8 ± 0.4	21.1 ± 1.4	−0.9 ± 0.4
PHBHHx/PCL 1:1	−60.2 ± 1.4	61.3 ± 0.8	53.4 ± 2.4	−0.6 ± 0.2	94.2 ± 1.9	7.5 ± 0.4	−63.4 ± 1.4	56.3 ± 1.6	36.0 ± 1.4	−1.4 ± 0.5
PCL raw	−61.4 ± 1.1	63.4 ± 0.5	96.5 ± 2.1	—	—	—	−64.7 ± 1.6	55.9 ± 1.5	83.0 ± 2.6	—

Data expressed as average ± standard deviation ($n = 3$).

3.4. Mechanical Characterization

The scaffold's mechanical properties were evaluated under compression using a uniaxial testing machine at a constant strain rate. Representative stress-strain compressive curves of PHBHHx and PHBHHx/PCL blend scaffolds are reported in Figure 5. They are characterized by three distinct regions: a roughly linear region, followed by a small plateau at fairly constant stress, and a final region of steeply rising stress. As suggested by previous papers reporting on mechanical characterization of polymeric scaffolds manufactured by CAWS [22,29,41], this three-region behavior can be explained with the sample response to the applied deformation at different structural scale levels. Indeed, the linear region is likely due to the initial response of the fiber–fiber contact points, the subsequent plateau region to the collapse of the pores network, and the final stress increase region to a further densification of the scaffold structure that behaves like a dense matrix.

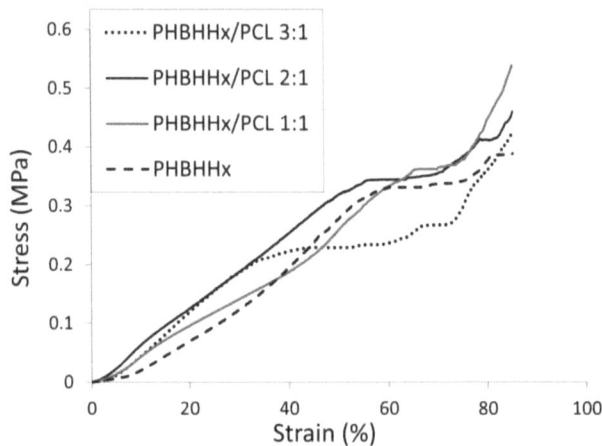

Figure 5. Representative stress-strain curve under compression (0.4 mm/min) of PHBHHx-based scaffolds.

The characterized samples showed compressive mechanical parameters (Table 4) of the same order of magnitude of PHBHHx scaffolds previously developed by CAWS [22,23] or solvent casting combined with salt-leaching techniques [52]. PHBHHx/PCL 3:1 scaffolds showed a comparable modulus but a marked drop in yield strain, yield stress and maximum stress in comparison to PHBHHx scaffolds. However, by increasing PCL content, the compressive modulus and the other mechanical parameters increased significantly. In fact, PHBHHx/PCL 2:1 and PHBHHx/PCL 1:1 scaffolds showed compressive modulus values (0.39 ± 0.14 and 0.37 ± 0.07, respectively) between that of PHBHHx scaffolds (0.16 ± 0.12 MPa) and that of PCL scaffolds with the same designed architecture (0.60 ± 0.20 MPa) [28]. In addition, PHBHHx/PCL 2:1 scaffolds showed a yield stress (0.36 ± 0.05 MPa) comparable to that of PHBHHx scaffolds and significantly higher than the other blend scaffolds. Overall, the developed scaffolds showed lower compressive strength in comparison to PHBV/PCL blend scaffolds by fused deposition modeling [21]. This difference should be mainly related to the multi-scale porous structure of scaffolds prepared using CAWS that, although the aforementioned advantages in providing a versatile tool for tuning scaffold properties and favor cells adhesion and interactions, is characterized by a higher void volume percentage in comparison to macroporous structures with a dense polymer matrix fabricated by means of melt processing.

Table 4. Compressive mechanical parameters of the developed scaffolds.

Scaffolds	Compressive Modulus (MPa)	Yield Strain (%)	Yield Stress (MPa)	Stress at 85% Strain (MPa)
PHBHHx	0.16 ± 0.12	56.8 ± 9.5	0.32 ± 0.02	0.47 ± 0.09
PHBHHx/PCL 3:1	0.17 ± 0.89	35.0 ± 9.4	0.18 ± 0.03	0.41 ± 0.09
PHBHHx/PCL 2:1	0.39 ± 0.14	57.9 ± 5.5	0.36 ± 0.05	0.48 ± 0.05
PHBHHx/PCL 1:1	0.37 ± 0.07	66.8 ± 5.5	0.36 ± 0.02	0.51 ± 0.07

Data expressed as average $\pm$ standard deviation ($n = 3$).

3.5. Biological Characterization

Investigations of viability and proliferation of cells seeded onto the developed scaffolds performed using the WST-1 assay showed an increase in the number of viable cells in all the tested samples during the 28 days of culture (Figure 6).

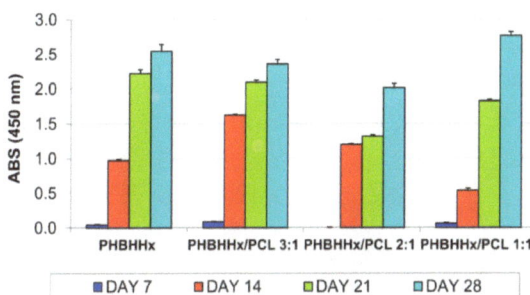

Figure 6. MC3T3-E1 cell proliferation on PHBHHx and PHBHHx/PCL based scaffolds.

The time-dependent changes in proliferation and differentiation can be divided into distinct stages of pre-osteoblast development [53]. The initial phase of osteoblast development is characterized by active replication of undifferentiated cells. At day 7, the observed limited cell proliferation was probably due to the large inter-fiber distance of the tested scaffolds that did not retain cells during the seeding procedure. During the second week, the cultures display a rapid increase in cell proliferation (Figure 6). The obtained data confirm a crucial role of cell–material interaction and cell density on cell adhesion, which influence cell proliferation during the initial stage of culturing [54].

Confocal Laser Scanning Microscopy (CLSM) characterization was employed to observe cell morphology on the scaffolds at different time points. Actin filament and nuclei were stained by Phalloidin-AlexaFluor488 and DAPI, and visualized as green and blue fluorescence respectively. Figure 7 shows the cell morphology and distribution of cells on the investigated samples after 7, 14, 21, and 28 days of culture. A good surface colonization of the scaffolds by MC3T3-E1 cells, with a variable shape and spreading can be observed, especially starting from the second week of culture. F-actin organization was consistent with early stages of cell adaptation to the material [55], exhibiting great stress fibers stretched along the cytoplasm, and a low cell number coherent with the quantitative proliferation data (Figure 6). Similar results for number and morphology of cells were detected by comparing different types of samples at early stages. After 14 days of culture the cell density appears to be increased, and at day 28 cells completely spread on the polymeric structure, with large cell cluster formations with inter-cellular connections (Figure 7).

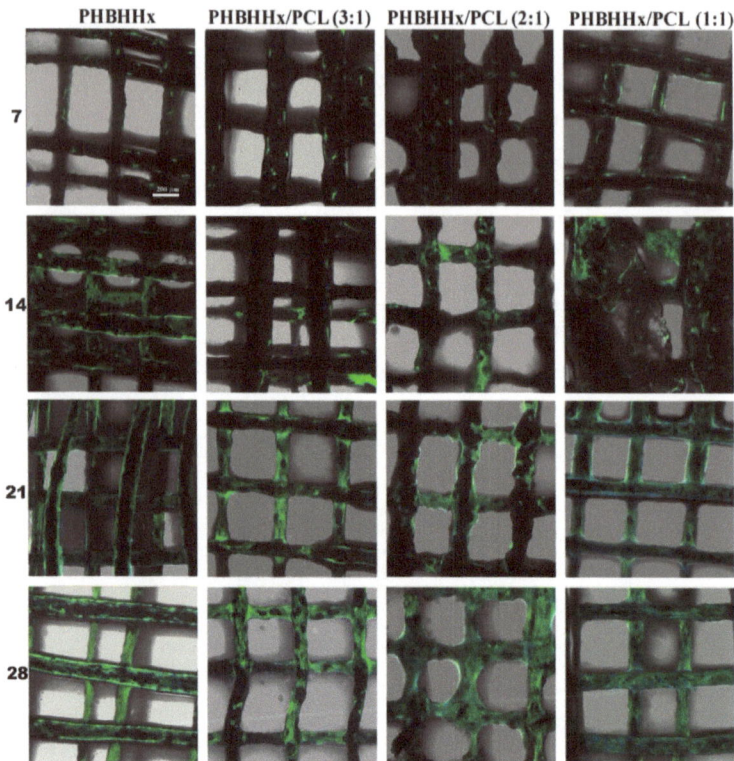

Figure 7. Confocal Laser Scanning Microscopy (CLSM) microphotographs showing MC3T3-E1 cell cultured on PHBHHx and PHBHHx/PCL based scaffolds, at different end-points.

These observations were in accordance with the differentiation pathway proposed for the preosteoblasts in vitro, where after an early growing latency, morpho-functional cellular aggregates are developed and single cell morphology is not distinguishable [56]. At the final phase of culturing, samples exhibited a nearly full cellular colonization of the available fiber surface by a wide continuous cell culture net.

4. Conclusions

The main result attained during the reported research activity is the development of an AM process based on the processing of polymeric solutions for the fabrication of PHBHHx-based blend scaffolds. This represents a novel approach to combining the advantages of PHA blending with other biocompatible polymers and the versatility of AM in supplying advanced fabrication tools for the development of scaffolds with customized macro- and microstructure. The developed manufacturing process meets both the product specification and good manufacturing practice requirements. In fact, it allows a good control of scaffold composition, external shape and internal porosity, it does not require thermal treatments that could cause material degradation, and involves the use of solvents allowed in medical device manufacturing that are completely removed from the scaffold during the fabrication and post-processing treatment.

The characterization analyses highlighted the versatility of the developed manufacturing process by demonstrating how PHBHHx/PCL blend scaffold composition, morphological features, thermal properties and mechanical parameters could be tuned in certain ranges by varying the ratio between

Bioengineering **2017**, *4*, 49

the two blend components in the starting solution. In addition, the results obtained from the performed preliminary biological evaluations indicated that the developed scaffolds are able to sustain a good cell adhesion and proliferation, and after 28 days of culture, scaffolds were fully colonized by MC3T3-E1 preosteoblast cells.

As shown by recent studies, the CAWS technique is well suited for the development of PHBHHx scaffolds with a complex shape resembling that of an anatomical part and a tailored porous structure with advanced architectural features at different scale levels (e.g., longitudinal macrochannel, local micro/nanoporosity) designed to enhance tissue regeneration processes [22,23]. The developed PHBHHx/PCL scaffolds can therefore represent advanced prototypes for the development of sophisticated PHAs-based blend constructs with tailored composition, anatomical shape, macroporosity and nanoporous morphology.

Acknowledgments: The financial support of the University of Pisa PRA-2016-50 project entitled "Functional Materials" and the Tuscany Region (Italy) funded Project "Nuovi Supporti Bioattivi a Matrice Polimerica per la Rigenerazione Ossea in Applicazioni Odontoiatriche (R.E.O.S.S.)" as part of the program POR CReO FESR 2007–2013—Le ali alle tue idee are gratefully acknowledged. PHBHHx was kindly supplied by Guo-Qiang Chen of Tsinghua University (Beijing, China) within the framework of the EC-Funded project Hyanji Scaffold in the People Program of the 7FP (2010–2013). Mairam Myrzabekova is acknowledged for her contribution during scaffolds preparation. Matteo Gazzarri and Cristina Bartoli are acknowledged for their contribution during biological characterization. Paolo Narducci is acknowledged for his support in recording SEM images.

Author Contributions: Dario Puppi and Federica Chiellini conceived and designed the experiments; Dario Puppi and Andrea Morelli performed the experiments; Dario Puppi, Andrea Morelli and Federica Chiellini analyzed the data; Dario Puppi and Federica Chiellini wrote the paper.

Conflicts of Interest: The authors declare no conflict of interest.

References

1. Puppi, D.; Chiellini, F.; Dash, M.; Chiellini, E. Biodegradable Polymers for Biomedical Applications. In *Biodegradable Polymers: Processing, Degradation and Applications*; Felton, G.P., Ed.; Nova Science Publishers, Inc.: Hauppauge, NY, USA, 2011; pp. 545–604.

2. Woodruff, M.A.; Lange, C.; Reichert, J.; Berner, A.; Chen, F.; Fratzl, P.; Schantz, J.-T.; Hutmacher, D.W. Bone tissue engineering: From bench to bedside. *Mater. Today* **2012**, *15*, 430–435. [CrossRef]

3. Puppi, D.; Chiellini, F.; Piras, A.M.; Chiellini, E. Polymeric materials for bone and cartilage repair. *Prog. Polym. Sci.* **2010**, *35*, 403–440. [CrossRef]

4. Morelli, A.; Puppi, D.; Chiellini, F. Polymers from Renewable Resources. *J. Renew. Mater.* **2013**, *1*, 83–112. [CrossRef]

5. Doyle, C.; Tanner, E.T.; Bonfield, W. In vitro and in vivo evaluation of polyhydroxybutyrate and of polyhydroxybutyrate reinforced with hydroxyapatite. *Biomaterials* **1991**, *12*, 841–847. [CrossRef]

6. Chen, G.Q.; Wu, Q. The application of polyhydroxyalkanoates as tissue engineering materials. *Biomaterials* **2005**, *26*, 6565–6578. [CrossRef] [PubMed]

7. Jack, K.S.; Velayudhan, S.; Luckman, P.; Trau, M.; Grøndahl, L.; Cooper-White, J. The fabrication and characterization of biodegradable HA/PHBV nanoparticle-polymer composite scaffolds. *Acta Biomater.* **2009**, *5*, 2657–2667. [CrossRef] [PubMed]

8. Gao, Y.; Kong, L.; Zhang, L.; Gong, Y.; Chen, G.; Zhao, N.; Zhang, X. Improvement of mechanical properties of poly(dl-lactide) films by blending of poly(3-hydroxybutyrate-co-3-hydroxyhexanoate). *Eur. Polym. J.* **2006**, *42*, 764–775. [CrossRef]

9. Wang, Y.W.; Wu, Q.; Chen, G.Q. Attachment, proliferation and differentiation of osteoblasts on random biopolyester poly(3-hydroxybutyrate-co-3-hydroxyhexanoate) scaffolds. *Biomaterials* **2004**, *25*, 669–675. [CrossRef]

10. Wang, Y.W.; Yang, F.; Wu, Q.; Cheng, Y.C.; Yu, P.H.; Chen, J.; Chen, G.Q. Effect of composition of poly(3-hydroxybutyrate-co-3-hydroxyhexanoate) on growth of fibroblast and osteoblast. *Biomaterials* **2005**, *26*, 755–761. [CrossRef] [PubMed]

11. Yang, M.; Zhu, S.; Chen, Y.; Chang, Z.; Chen, G.; Gong, Y.; Zhao, N.; Zhang, X. Studies on bone marrow stromal cells affinity of poly (3-hydroxybutyrate-co-3-hydroxyhexanoate). *Biomaterials* **2004**, *25*, 1365–1373. [CrossRef] [PubMed]

12. Jing, X.; Ling, Z.; Zhenhu An, Z.; Guoqiang, C.; Yandao, G.; Nanming, Z.; Xiufang, Z. Preparation and evaluation of porous poly(3-hydroxybutyrate-co-3-hydroxyhexanoate) hydroxyapatite composite scaffolds. *J. Biomater. Appl.* **2008**, *22*, 293–307. [CrossRef] [PubMed]

13. Garcia-Garcia, J.M.; Garrido, L.; Quijada-Garrido, I.; Kaschta, J.; Schubert, D.W.; Boccaccini, A.R. Novel poly(hydroxyalkanoates)-based composites containing Bioglass (R) and calcium sulfate for bone tissue engineering. *Biomed. Mater.* **2012**, *7*, 054105. [CrossRef] [PubMed]

14. Ke, S.; Yang, Y.; Ren, L.; Wang, Y.; Li, Y.; Huang, H. Dielectric behaviors of PHBHHx–BaTiO$_3$ multifunctional composite films. *Compos. Sci. Technol.* **2012**, *72*, 370–375. [CrossRef]

15. Wang, Y.; Jiang, X.-L.; Yang, S.-C.; Lin, X.; He, Y.; Yan, C.; Wu, L.; Chen, G.-Q.; Wang, Z.-Y.; Wu, Q. MicroRNAs in the regulation of interfacial behaviors of MSCs cultured on microgrooved surface pattern. *Biomaterials* **2011**, *32*, 9207–9217. [CrossRef] [PubMed]

16. Wang, Y.; Gao, R.; Wang, P.-P.; Jian, J.; Jiang, X.-L.; Yan, C.; Lin, X.; Wu, L.; Chen, G.-Q.; Wu, Q. The differential effects of aligned electrospun PHBHHx fibers on adipogenic and osteogenic potential of MSCs through the regulation of PPARγ signaling. *Biomaterials* **2012**, *33*, 485–493. [CrossRef] [PubMed]

17. Wu, L.-P.; You, M.; Wang, D.; Peng, G.; Wang, Z.; Chen, G.-Q. Fabrication of carbon nanotube (CNT)/poly(3-hydroxybutyrate-co-3-hydroxyhexanoate) (PHBHHx) nanocomposite films for human mesenchymal stem cell (hMSC) differentiation. *Polym. Chem.* **2013**, *4*, 4490–4498. [CrossRef]

18. ASTM. *International F2792—12a Standard Terminology for Additive Manufacturing Technologies*; ASTM: West Conshohocken, PA, USA, 2012.

19. Mota, C.; Puppi, D.; Chiellini, F.; Chiellini, E. Additive manufacturing techniques for the production of tissue engineering constructs. *J. Tissue Eng. Regen. Med.* **2015**, *9*, 174–190. [CrossRef] [PubMed]

20. Leroy, E.; Petit, I.; Audic, J.L.; Colomines, G.; Deterre, R. Rheological characterization of a thermally unstable bioplastic in injection molding conditions. *Polym. Degrad. Stab.* **2012**, *97*, 1915–1921. [CrossRef]

21. Kosorn, W.; Sakulsumbat, M.; Uppanan, P.; Kaewkong, P.; Chantaweroad, S.; Jitsaard, J.; Sitthiseripratip, K.; Janvikul, W. PCL/PHBV blended three dimensional scaffolds fabricated by fused deposition modeling and responses of chondrocytes to the scaffolds. *J. Biomed. Mater. Res. B* **2016**. [CrossRef]

22. Mota, C.; Wang, S.Y.; Puppi, D.; Gazzarri, M.; Migone, C.; Chiellini, F.; Chen, G.Q.; Chiellini, E. Additive manufacturing of poly[(R)-3-hydroxybutyrate-co-(R)-3-hydroxyhexanoate] scaffolds for engineered bone development. *J. Tissue Eng. Regen. Med.* **2017**, *11*, 175–186. [CrossRef] [PubMed]

23. Puppi, D.; Pirosa, A.; Morelli, A.; Chiellini, F. Design, fabrication and characterization of tailored poly[(R)-3-hydroxybutyrate-co-(R)-3-hydroxyexanoate] scaffolds by Computer-aided Wet-spinning. *Rapid Prototyp. J.* **2018**, *24*. unpublished.

24. Woodruff, M.A.; Hutmacher, D.W. The return of a forgotten polymer-Polycaprolactone in the 21st century. *Prog. Polym. Sci.* **2010**, *35*, 1217–1256. [CrossRef]

25. Zhao, Q.; Wang, S.; Kong, M.; Geng, W.; Li, R.K.Y.; Song, C.; Kong, D. Phase morphology, physical properties, and biodegradation behavior of novel PLA/PHBHHx blends. *J. Biomed. Mater. Res. B* **2011**, *100B*, 23–31. [CrossRef] [PubMed]

26. Chiono, V.; Ciardelli, G.; Vozzi, G.; Sotgiu, M.G.; Vinci, B.; Domenici, C.; Giusti, P. Poly(3-hydroxybutyrate-co-3-hydroxyvalerate)/poly(ε-caprolactone) blends for tissue engineering applications in the form of hollow fibers. *J. Biomed. Mater. Res. A* **2008**, *85A*, 938–953. [CrossRef] [PubMed]

27. Lim, J.; Chong, M.S.K.; Teo, E.Y.; Chen, G.-Q.; Chan, J.K.Y.; Teoh, S.-H. Biocompatibility studies and characterization of poly(3-hydroxybutyrate-co-3-hydroxyhexanoate)/polycaprolactone blends. *J. Biomed. Mater. Res. B* **2013**, *101B*, 752–761. [CrossRef] [PubMed]

28. Puppi, D.; Mota, C.; Gazzarri, M.; Dinucci, D.; Gloria, A.; Myrzabekova, M.; Ambrosio, L.; Chiellini, F. Additive manufacturing of wet-spun polymeric scaffolds for bone tissue engineering. *Biomed. Microdevices* **2012**, *14*, 1115–1127. [CrossRef] [PubMed]

29. Mota, C.; Puppi, D.; Dinucci, D.; Gazzarri, M.; Chiellini, F. Additive manufacturing of star poly(ε-caprolactone) wet-spun scaffolds for bone tissue engineering applications. *J. Bioact. Compat. Polym.* **2013**, *28*, 320–340. [CrossRef]

30. Dini, F.; Barsotti, G.; Puppi, D.; Coli, A.; Briganti, A.; Giannessi, E.; Miragliotta, V.; Mota, C.; Pirosa, A.; Stornelli, M.R.; et al. Tailored star poly (ε-caprolactone) wet-spun scaffolds for in vivo regeneration of long bone critical size defects. *J. Bioact. Compat. Polym.* **2016**, *31*, 15–30. [CrossRef]

31. Karageorgiou, V.; Kaplan, D. Porosity of 3D biomaterial scaffolds and osteogenesis. *Biomaterials* **2005**, *26*, 5474–5491. [CrossRef] [PubMed]

32. ASTM. *D1621—10 "Standard Test Method for Compressive Properties of Rigid Cellular Plastics"*; ASTM: West Conshohocken, PA, USA, 2010.

33. Puppi, D.; Piras, A.M.; Chiellini, F.; Chiellini, E.; Martins, A.; Leonor, I.B.; Neves, N.; Reis, R. Optimized electro- and wet-spinning techniques for the production of polymeric fibrous scaffolds loaded with bisphosphonate and hydroxyapatite. *J. Tissue Eng. Regen. Med.* **2011**, *5*, 253–263. [CrossRef] [PubMed]

34. Tuzlakoglu, K.; Alves, C.M.; Mano, J.F.; Reis, R.L. Production and characterization of chitosan fibers and 3-D fiber mesh scaffolds for tissue engineering applications. *Macromol. Biosci.* **2004**, *4*, 811–819. [CrossRef] [PubMed]

35. Gomes, M.E.; Holtorf, H.L.; Reis, R.L.; Mikos, A.G. Influence of the porosity of starch-based fiber mesh scaffolds on the proliferation and osteogenic differentiation of bone marrow stromal cells cultured in a flow perfusion bioreactor. *Tissue Eng.* **2006**, *12*, 801–809. [CrossRef] [PubMed]

36. Malheiro, V.N.; Caridade, S.G.; Alves, N.M.; Mano, J.F. New poly(ε-caprolactone)/chitosan blend fibers for tissue engineering applications. *Acta Biomater.* **2010**, *6*, 418–428. [CrossRef] [PubMed]

37. Neves, S.C.; Moreira Teixeira, L.S.; Moroni, L.; Reis, R.L.; Van Blitterswijk, C.A.; Alves, N.M.; Karperien, M.; Mano, J.F. Chitosan/Poly(ε-caprolactone) blend scaffolds for cartilage repair. *Biomaterials* **2011**, *32*, 1068–1079. [CrossRef] [PubMed]

38. Puppi, D.; Dinucci, D.; Bartoli, C.; Mota, C.; Migone, C.; Dini, F.; Barsotti, G.; Carlucci, F.; Chiellini, F. Development of 3D wet-spun polymeric scaffolds loaded with antimicrobial agents for bone engineering. *J. Bioact. Compat. Polym.* **2011**, *26*, 478–492. [CrossRef]

39. Puppi, D.; Chiellini, F. Wet-spinning of Biomedical Polymers: From Single Fibers Production to Additive Manufacturing of 3D Scaffolds. *Polym. Int.* **2017**. [CrossRef]

40. Puppi, D.; Migone, C.; Grassi, L.; Pirosa, A.; Maisetta, G.; Batoni, G.; Chiellini, F. Integrated three-dimensional fiber/hydrogel biphasic scaffolds for periodontal bone tissue engineering. *Polym. Int.* **2016**, *65*, 631–640. [CrossRef]

41. Puppi, D.; Piras, A.M.; Pirosa, A.; Sandreschi, S.; Chiellini, F. Levofloxacin-loaded star poly(ε-caprolactone) scaffolds by additive manufacturing. *J. Mater. Sci. Mater. Med.* **2016**, *27*, 1–11. [CrossRef] [PubMed]

42. Puppi, D.; Migone, C.; Morelli, A.; Bartoli, C.; Gazzarri, M.; Pasini, D.; Chiellini, F. Microstructured chitosan/poly(γ-glutamic acid) polyelectrolyte complex hydrogels by computer-aided wet-spinning for biomedical three-dimensional scaffolds. *J. Bioact. Compat. Polym.* **2016**, *31*, 531–549. [CrossRef]

43. Chiellini, F.; Puppi, D.; Piras, A.M.; Morelli, A.; Bartoli, C.; Migone, C. Modelling of pancreatic ductal adenocarcinoma in vitro with three-dimensional microstructured hydrogels. *RSC Adv.* **2016**, *6*, 54226–54235. [CrossRef]

44. Neves, S.C.; Mota, C.; Longoni, A.; Barrias, C.C.; Granja, P.L.; Moroni, L. Additive manufactured polymeric 3D scaffolds with tailored surface topography influence mesenchymal stromal cells activity. *Biofabrication* **2016**, *8*, 025012. [CrossRef] [PubMed]

45. Duan, B.; Wang, M.; Zhou, W.Y.; Cheung, W.L.; Li, Z.Y.; Lu, W.W. Three-dimensional nanocomposite scaffolds fabricated via selective laser sintering for bone tissue engineering. *Acta Biomater.* **2010**, *6*, 4495–4505. [CrossRef] [PubMed]

46. Duan, B.; Wang, M. Customized Ca–P/PHBV nanocomposite scaffolds for bone tissue engineering: Design, fabrication, surface modification and sustained release of growth factor. *J. R. Soc. Interface* **2010**, *7*, S615–S629. [CrossRef] [PubMed]

47. Bin, D.; Wai Lam, C.; Min, W. Optimized fabrication of Ca–P/PHBV nanocomposite scaffolds via selective laser sintering for bone tissue engineering. *Biofabrication* **2011**, *3*, 015001.

48. Puppi, D.; Zhang, X.; Yang, L.; Chiellini, F.; Sun, X.; Chiellini, E. Nano/microfibrous polymeric constructs loaded with bioactive agents and designed for tissue engineering applications: A review. *J. Biomed. Mater. Res. B* **2014**, *102*, 1562–1579. [CrossRef] [PubMed]

49. Middleton, J.C.; Tipton, A.J. Synthetic biodegradable polymers as orthopedic devices. *Biomaterials* **2000**, *21*, 2335–2346. [CrossRef]

50. Yang, H.-X.; Sun, M.; Zhou, P. Investigation of water diffusion in poly(3-hydroxybutyrate-co-3-hydroxyhexanoate) by generalized two-dimensional correlation ATR–FTIR spectroscopy. *Polymer* **2009**, *50*, 1533–1540. [CrossRef]

51. Ding, C.; Cheng, B.; Wu, Q. DSC analysis of isothermally melt-crystallized bacterial poly(3-hydroxybutyrate-co-3-hydroxyhexanoate) films. *J. Therm. Anal. Calorim.* **2011**, *103*, 1001–1006. [CrossRef]

52. Wang, Y.W.; Wu, Q.; Chen, J.; Chen, G.Q. Evaluation of three-dimensional scaffolds made of blends of hydroxyapatite and poly(3-hydroxybutyrate-co-3-hydroxyhexanoate) for bone reconstruction. *Biomaterials* **2005**, *26*, 899–904. [CrossRef] [PubMed]

53. Wutticharoenmongkol, P.; Pavasant, P.; Supaphol, P. Osteoblastic Phenotype Expression of MC3T3-E1 Cultured on Electrospun Polycaprolactone Fiber Mats Filled with Hydroxyapatite Nanoparticles. *Biomacromolecules* **2007**, *8*, 2602–2610. [CrossRef] [PubMed]

54. Kommareddy, K.P.; Lange, C.; Rumpler, M.; Dunlop, J.W.C.; Manjubala, I.; Cui, J.; Kratz, K.; Lendlein, A.; Fratzl, P. Two stages in three-dimensional in vitro growth of tissue generated by osteoblastlike cells. *Biointerphases* **2010**, *5*, 45–52. [CrossRef] [PubMed]

55. Hutmacher, D.W.; Schantz, T.; Zein, I.; Ng, K.W.; Teoh, S.H.; Tan, K.C. Mechanical properties and cell cultural response of polycaprolactone scaffolds designed and fabricated via fused deposition modeling. *J. Biomed. Mater. Res.* **2001**, *55*, 203–216. [CrossRef]

56. Quarles, L.D.; Yohay, D.A.; Lever, L.W.; Caton, R.; Wenstrup, R.J. Distinct proliferative and differentiated stages of murine MC3T3-E1 cells in culture: An in vitro model of osteoblast development. *J. Bone Min. Res.* **1992**, *7*, 683–692. [CrossRef] [PubMed]

MDPI AG

St. Alban-Anlage 66

4052 Basel, Switzerland

Tel. +41 61 683 77 34

Fax +41 61 302 89 18

http://www.mdpi.com

Bioengineering Editorial Office

E-mail: bioengineering@mdpi.com

http://www.mdpi.com/journal/bioengineering

www.ingramcontent.com/pod-product-compliance
Lightning Source LLC
Chambersburg PA
CBHW050916210326
41597CB00003B/122